Forensic Science Today

Second Edition

Dr. Henry C. Lee
George M. Taft
Kimberly A. Taylor
Jeanette Hencken

Tucson, Arizona

This publication is designed to provide accurate and authoritative information in regard to the subject matter covered. It is sold with the understanding that the publisher is not engaged in rendering legal, accounting, or other professional service. If legal advice or other expert assistance is required, the services of a competent professional person should be sought.

—From a *Declaration of Principles* jointly adopted by
a Committee of the American Bar Association
and a Committee of Publishers and Associations.

The publisher, editors and authors must disclaim any liability, in whole or in part, arising from the information in this volume. The reader is urged to verify the reference material prior to any detrimental reliance thereupon. Since this material deals with legal, medical and engineering information, the reader is urged to consult with an appropriate licensed professional prior to taking any action that might involve any interpretation or application of information within the realm of a licensed professional practice.

Copyright ©2009 by Lawyers & Judges Publishing Co., Inc. All rights reserved. All chapters are the product of the Authors and do not reflect the opinions of the Publisher, or of any other person, entity, or company. No part of this book may be reproduced in any form or by any means, including photocopying, without permission from the Publisher.

Lawyers & Judges Publishing Company, Inc.

P.O. Box 30040 • Tucson, AZ 85751-0040
(800) 209-7109 • FAX (800) 330-8795
website: www.lawyersandjudges.com
email: sales@lawyersandjudges.com

Library of Congress Cataloging-in-Publication Data

Forensic science today / Henry C. Lee ... [et al.]. -- 2nd ed.
   p. cm.
  Includes bibliographical references and index.
  ISBN-13: 978-1-933264-87-5 (student : alk. paper)
  ISBN-10: 1-933264-87-X (student : alk. paper)
  ISBN-13: 978-1-933264-73-8 (instructor w/ cd-rom : alk. paper)
  ISBN-10: 1-933264-73-X (instructor w/ cd-rom : alk. paper)
  1. Forensic sciences. 2. Criminal investigation. 3. Evidence, Criminal. I. Lee, Henry C.
HV8073.L3695 2009
363.25--dc22
                          2009006614

ISBN: 978-1-933264-87-5 (student)
ISBN: 978-1-933264-73-8 (instructor w/ CD-ROM)
Printed in the United States of America
10 9 8 7 6 5 4 3 2 1

Table of Contents

Acknowledgments v
Foreword.. ix
Introduction.. xi

Chapter 1   Introduction to the World of Forensic Science
                 and Criminalistics.. 1
Chapter 2   Crime Scene Investigation I: The Basics............................. 13
Chapter 3   Crime Scene Investigation II: The Physical Evidence 29
Chapter 4   Fingerprints 45
Chapter 5   Trace Evidence... 53
Chapter 6   Drug Evidence 67
Chapter 7   Forensic Serology 77
Chapter 8   DNA Evidence... 85
Chapter 9   Bloodstain Patterns 101
Chapter 10   Questioned Documents 111
Chapter 11   Imprints and Impressions.. 123
Chapter 12   Toolmark and Firearms Evidence 139
Chapter 13   Fire and Arson Evidence... 155
Chapter 14   Chemical Evidence... 163
Chapter 15   Digital Evidence... 169
Chapter 16   Forensic Art 183
Chapter 17   Forensic Pathology 191
Chapter 18   Forensic Toxicology.. 199
Chapter 19   Forensic Anthropology... 211
Chapter 20   Forensic Entomology 217
Chapter 21   Forensic Dentistry.. 225
Chapter 22   Forensic Engineering.. 231
Chapter 23   Forensic Psychology 237
Chapter 24   Crime Scene Reconstruction 241
Chapter 25   Jurisprudence: Law and Forensics................................... 251

Glossary 257
Resources 267
About the Authors.. 275
Index.. 277

Acknowledgments

The authors wish to thank the many forensic professionals, teachers, and others who have given their time and encouragement to help us turn the dream of a forensic science textbook into a reality.

We are extremely grateful to Jeff Salyards and Michael Buratowski for contributing the new chapter on Digital Evidence and for their insight and assistance with classroom activities to accompany that chapter.

Our deepest thanks to forensic artist Karen T. Taylor for reviewing the chapter on Forensic Art and contributing all the photographs and images found in that chapter.

For their thoughtful responses to our teacher survey, we are indebted to Jalinda Hogan, Sandy Peterson, Patt Mical, Joan Twillman, and any others whose names we may have missed.

A special thanks to Judge Haskell M. Pitluck for contributing an especially meaningful and thought-provoking foreword to this book.

We thank Wayne D. Lord of the Federal Bureau of Investigation for permission to adapt an entomological case profile from his work, *Case Histories of the Use of Insects in Investigation*.

We are also indebted to Valerie Shook and Joseph Sudol, assistants to Dr. Lee and message coordinators extraordinaire.

A special thanks to the following who have spent endless hours in their search for ways to make the learning experience positive and exciting: Debbie Fancher, Lesley Hammer, Kim Bunselmeyer, Rhonda Kropf, Sheri Roach, Susan Fisk, Sondra Dexter, Nanch Whitmore, Greg Fullmer, Terry Jorgensen, Elaine Lawson, Scott Shumway, Gail Chagnon, Robyn Hill, Barbara Running, Bill Booth, Sandy Davis, B. Jarvi, Connie Wurtz, Donna Dart, Chuck Heath, Jill Hayes, Jody Solomonson, Martha Lehe, Liz Burck, Kim McCoy, Jill Crystal, Rick Farrell, Pam Lloyd, Dennis Strait, Patricia Gersich, Sharon Clawson, Susan Hoyer, Kimberly Cheek, Mary Margaret Stephens, Marilee Miner, Sue Liebner, Pam Collins, Shirley Mills, Richard Burmeister, Lois Nance, Cheri Stephens, Judy Campbell, Terri Cowart, M. Klopfer, Kathy Baker, Jill Hayes, Martha Lehe, Kerry Davis, Mark Benedetti, Deborah Martens, Linda Churchill, Lynda Adams, Sandy Stephens, Marva Stevens, Janice Ott, Karen

Tabios, Marilita Sanders, Laurie Martin, Michelle Bifelt, Susan Magestro, Sherry Annee, Kay Pitluck, Anthony Rossilli, Marty Peters, Liane Ryan, M. Elizabeth Beks, Jane Sandstrom, Laura Brunner, Dionne Levine, Wendy Ehnert, Christine Hanson, Beth Miller, Judge Niesje Steinkruger, Jeanne Swartz, Karen Taylor, Barbara Weekley, Joan Twillman, Cindy Harrison, Jill Brunner, Alice Sonju, Jack Hurd, Bob Shem, Marie Schneider, Leesa Wingo, Lisa Mitchell, and Barbara Behnke.

We would also like to thank the following people for providing valuable information in their fields of expertise: Abirami Chidambaram, Criminalist, DNA Section, Scientific Crime Detection Laboratory, Anchorage, Alaska provided a paper entitled, "DNA Analysis in Forensic Investigations," presented at the Forensic Science Education Conference in Anchorage, AK June 2004. Jim Wolfe, State of Alaska Scientific Crime Detection Laboratory provided a paper entitled, "Shoeprint and Tiretrack Evidence," presented at the Alaska Summer Justice Institute at the University of Alaska, Anchorage, June 2004. Donald Chinn, Laboratory Director, DEA Western Laboratory and Ashraf Mozayani, Ph.D., Pharm.D., Houston, Texas both provided information on toxicology and drug evidence which formed the basis of the section on drug-facilitated rape. Dr. Elizabeth Miller, Department of Anthropology, California State University provided information on forensic anthropology. Dr. Christine Hanson, Department of Anthropology, University of Alaska, Anchorage provided a paper entitled, "Age, Sex and Stature From Human Skeletons," presented in June 2004.

A big warm thanks to the team at Lawyers and Judges Publishing Company, Inc. We are especially grateful to Steve Weintraub not only for bringing us together to create a book, but also for giving us complete creative license to take the idea and run with it. We also thank Carla Schuder for her creative layout of the first edition, and Eric Salo for pulling everything together for the second edition.

Finally, the authors wish to give thanks to those who have made a difference both in the writing of this book and in their lives:

Dr. Henry C. Lee dedicates this book to Mrs. An Fu Lee—mother, educator, and pioneer. Dr. Lee also expresses thanks to his colleagues of the Connecticut State Police Forensic Laboratory, and the faculty of the forensic science program at the University of New Haven, for their contributions. Also, he wants to thank Margaret Lee, Sherry Lee, Ted Hersey, Dr. Stanley Lee, Romie Lee, David, Rachel, and Alex for their continued encouragement and support.

George Taft would like to thank the members of the Alaska Scientific Crime Detection Laboratory with special thanks to Dale and Turner Pippin for their innovative ideas to improve Forensic Techniques. Also many thanks to Karen Taylor, a nationally recognized Forensic Artist, for her help with the Forensic Art chapter. Science teachers who have been an inspiration to George are Deb Fancher, Debbie Brewer, Leesa Wingo, Lesley Hammer and Jeanne Swartz. Thanks to Bailey Jackson, a chemistry student, for her encouragement with this Edition. Also, thanks to Connie White and Bob Woodrich for their continued support and help with CSI Montana. Many thanks to the Co-Authors, and to my daughter Carey Hook and son Scott Taft who have allowed me to pursue a very interesting Forensic Career.

Acknowledgments

Kimberly Taylor dedicates this book with love to her husband and daughter. Kimberly is especially grateful to her husband Martin Taylor, Ph.D. not only for providing entomological and DNA expertise as well as the maggot lab exercise for the Forensic Entomology chapter, but also for his unending moral support and picking up the slack during deadline crunch time. Kimberly thanks her co-authors for the opportunity to work with them on this special project. She also thanks Ron Ellis, Ph.D. for reviewing the manuscript and providing insight and vocabulary definitions for the Chemical Evidence chapter; Gregory Ellis, Ph.D., Esq. for his help with DNA and PCR analysis and updates on the Frye and Daubert cases; Sharlene Ellis for her insight on chain of custody issues; Sue West for her encouragement; Sue Behan for help with childcare; and Julie and Naomi Taylor for their teaching perspective.

Jeanette Hencken would like to thank her family, especially her husband Tom and daughters, Emily and Julia, for all the many things they have done to support her in her quest to learn about forensic science and share what she has learned with others. Jeanette is very grateful to Harold Messler of the St. Louis Metropolitan Police Department Crime Lab for the many opportunities and hours of mentoring in forensic science and for time spent as an Intern. Many thanks to Major Mike Copeland of the Franklin County Missouri Sheriff's Department for the many learning opportunities and insight into crime scene work provided by him. She would also like to thank Mary Fran Ernst and Julie Howe for the experience of working with them on forensic education conferences and the Masters Conference for Advanced Death Investigation and the many learning opportunities this has provided.

▲ National Law Enforcement Memorial in Washington, DC, with names of those law enforcement officers killed in the line of duty.

Foreword

Before there was a TV show known to all as *C.S.I.* to interest viewers in forensics, Dr. Henry Lee began his career in Forensic Science. As a criminalist whose career has spanned decades, Dr. Lee's willingness to introduce forensics to those who know little about the subject and to help practicing forensic science professionals become more proficient has enhanced the profession.

Dr. Lee is sought after to speak to groups worldwide on many forensic subjects. He has authored many books, but this is one of the first books written specifically for students anxious to learn more about forensics from one of the masters of the highly complex science of forensics.

Studies have shown that between grade school, high school and college, American students often fall behind many similarly aged students outside the United States in science and math scores. No one is sure why this occurs. However, it is no secret that many American students, for whatever reason, have little interest in pursuing study and careers in science and mathematics.

It is also known that forensic science has become a successful tool, not only to interest young people in science and math, but also to facilitate and make it easier to teach science concepts. The real life news accounts of forensic science aiding in the solving of actual crimes, coupled with the multitude of television programs extolling the success of forensic science, have stimulated the interest of many young students, as well as adults, in forensic science. While many instructors have started forensic programs in various learning environments across the country, other than a few workshops spearheaded by the American Academy of Forensic Sciences to assist teachers in preparing such courses, little has been done to solidify the process. While most of these instructors have prepared their course material from various sources they have gathered, there has been no definitive introductory textbook for this level.

That has now changed. Dr. Henry Lee, one of the foremost forensic scientists in the world, George Taft, an eminent forensic scientist in his own right, Kimberly Taylor, a writer and attorney, and Jeanette Hencken, an award-winning high school forensics teacher, have written this text – now in its second edition, which gives an introductory explanation of a cross section of forensic science.

Forensic Science Today

The authors state: "Science can be fun". As a student progresses through this text, he or she will learn that science can be fun, and that they can also learn by being involved with the subject. Interspersed with the concepts are examples of actual cases to show the applications of the concepts.

Forensic science is more than memorizing terms and procedures. It is also learning how to apply the knowledge to the overall picture. Practitioners need to learn to think outside the box to come up with solutions to the problem. For example, twenty years ago, DNA was an unknown concept as applied to forensic science and the legal profession. Today, DNA is a major part of scientific investigation and detection and the legal system. There are many other emerging and advancing scientific areas that will assist in solving future problems within the forensic and legal professions. With such advances in science, forensic science needs bright young minds to go into the field to pass the torch to future generations.

This text is an excellent starting point to open the door to the fascinating world of forensic science. The companion material and the resources with references to excellent web sites add to the opportunities to prepare students for the steps to the next challenging level in forensic science.

Haskell M. Pitluck
Retired Circuit Court Judge
Past President American Academy of Forensic Sciences

Introduction

Welcome to the exciting world of forensic science! You are about to embark upon a journey of discovery that will take you behind the scenes of criminal investigation and prepare you—should you so choose—for a future career in forensic science.

During this journey, you will join world-renowned forensic scientist Dr. Henry Lee as he explains the science behind solving famous murder cases that most people only see on T.V. Importantly, you will learn new skills of science and logic, and sharpen other skills you already have, like your intuition. Our philosophy is that you should be truly engaged when learning about forensic science. *Forensic Science Today* reflects this philosophy and teaches forensic science in an informative and interest-sustaining manner. Because forensic science is a practical science using a variety of different skills, you will learn about it in a hands-on manner. And because forensic scientists work as part of a team, the focus in this curriculum is on teamwork conducted in the classroom.

The first edition of *Forensic Science Today* was met with such enthusiasm by students and teachers alike that we have revised and updated it to make it even better. We have been joined by a leading high school forensic science teacher to add more science and engaging laboratory and classroom exercises. We have added more real-life case profiles and have introduced Dr. Lee's "Logic Trees" to help students visualize the steps in solving crimes involving different types of physical evidence. We have added an important new chapter on Digital Evidence written specially for *Forensic Science Today* by two of the country's top computer forensic experts. We have supplemented the materials with eye-catching Powerpoint presentations that will make the class even more interesting and fun. Finally, we have reorganized the chapters in an order that teachers will find helpful as they plan their syllabus.

This book takes students through an in-depth exploration of solving crime through the application of forensic techniques. We begin with the basics of Crime Scene Investigation followed by a study of Physical Evidence and the many roles it plays in criminal investigation. Then, we will explore a myriad of forms of physical evidence. These include Fingerprints, Trace Evidence, Drugs, Serology, DNA, Blood Stain Patterns, Questioned Documents, Imprints and Impressions, Toolmarks and Firearms, Fire and Arson, Chemical Evidence, and Digital Evidence. In each of these chapters, students learn how to accurately document, preserve, collect and analyze the different types of physical evidence.

Forensic Science Today

Forensic Science Today is one of the first textbooks to introduce students to the wide array of career choices in forensics. So in addition to learning about physical evidence, we will explore many of the different specialties in forensic science that are in practice today. These include Forensic Art, Forensic Pathology, Forensic Toxicology, Forensic Anthropology, Forensic Entomology, Forensic Dentistry, Forensic Engineering, Forensic Psychology, Crime Scene Reconstruction, and Jurisprudence. You will learn not only what a professional in each of these specialties does, but also which college courses and other types of training you need to complete if you would like to pursue a career in that field.

Each chapter in this book begins with a set of objectives indicating what you should be able to achieve after working through the chapter. There is also a set of key terms to pay special attention to as you read through the chapter. At the end of each chapter you will find definitions of scientific and technical words and phrases introduced in that chapter. These words and phrases are also defined in the Glossary at the back of the book.

At the end of this book you will find a comprehensive set of Resources, which lists books and websites that provide further information for more in-depth research or project work. This includes websites pertaining to crime prevention and survival. It is our hope that if your life has been touched by crime in any way, you can turn to some of these websites for help and further guidance.

In the real world, forensic scientists succeed when they apply a combination of knowledge, experience, skills, and intuition to the situation at hand. An open mind, creativity, and curiosity are essential tools in the field of forensic science. During our exploration of the exciting world of forensic science, we will teach you how to use these tools and apply your knowledge in new ways. We hope that someday this exploration will enable you to be the one who solves a crime using these tools.

Let the journey begin!

Chapter 1

Introduction to the World of Forensic Science and Criminalistics

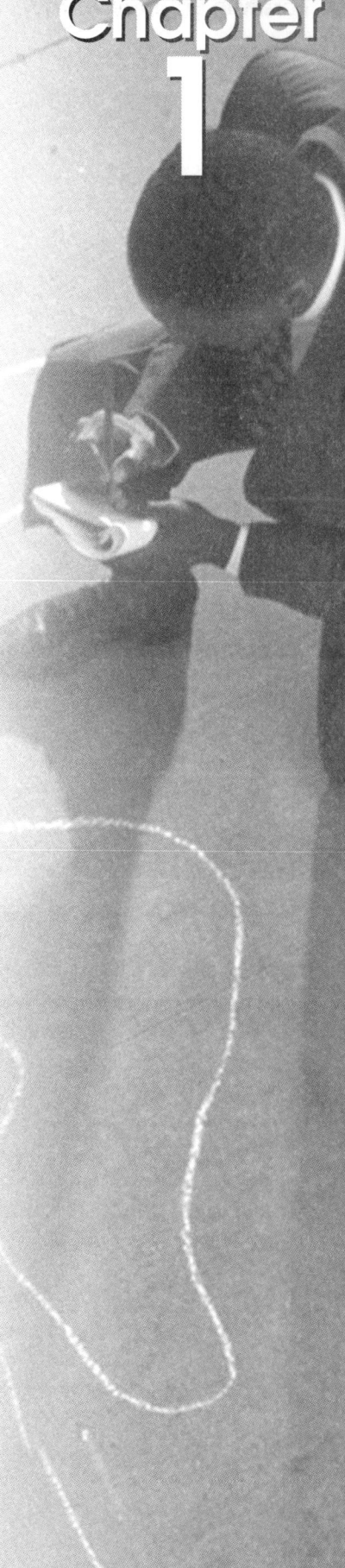

CHAPTER OBJECTIVES

- ✔ Explain the basic concepts of forensic science.
- ✔ Identify the characteristics that are helpful for a career in forensic science.
- ✔ Discuss the day-to-day life of forensic scientists, giving examples of different places a forensic scientist might work.
- ✔ Understand the role of forensic scientists as experts who testify in court.
- ✔ Identify the educational and training requirements for careers in forensic science.
- ✔ Give a brief definition of ten different specialties in the field of forensic science.
- ✔ Define the following vocabulary words and phrases: *evidence*, *expert witness*, and *forensic science*.

"Wherever he steps, whatever he touches, whatever he leaves, even unconsciously, will serve as a silent witness against him. Not only his fingerprints or his footprints, but his hair, the fibers from his clothes, the glass he breaks, the tool mark he leaves, the paint he scratches, the blood or semen he deposits or collects. All of these and more, bear mute witness against him. This is evidence that does not forget. It is not confused by the excitement of the moment. It is not absent because human witnesses are. It is factual evidence. Physical evidence cannot be wrong, it cannot perjure itself, it cannot be wholly absent. Only human failure to find it, study and understand it, can diminish its value."

—Professor Edmond Locard

WHAT IS FORENSIC SCIENCE?

You may be familiar with popular crime-solving television shows hitting the airwaves with increasing frequency each season, such as the fictionalized programs *CSI*, *Cold Case*, *Law & Order*, *Numb3rs*, *Without A Trace*, *Criminal Minds*, *NCIS*, *Bones* and *Fringe*; and reality TV shows such as *Trace Evidence*, *Forensic Files*, and *TruTV* programs. In these programs, we watch a variety of people in different roles working together as a team to solve crimes and restore justice. These people are using forensic science to discover the facts of what really happened when a crime was committed.

Forensic Science Today

KEY TERMS

Evidence
Expert Witness
Forensic Science
Civil Matter
Criminal Matter
Criminalistics

When a criminal case goes to court, prosecution and defense lawyers may literally spend hours presenting a single piece of evidence, such as a drop of blood or a strand of hair. What the jury does not see is the diligence and hard work of the forensic scientists who work behind the scenes to recover and analyze this evidence. Forensic professionals are not limited to crime scene investigators or the pathologists who perform autopsies; they also include chemists, anthropologists, engineers, and even insect specialists who help find the clues that provide the missing link to a crime. There are also many other specialties and subspecialties that we will learn about as we work through this book.

Together these professionals form the backbone of the criminal investigation, and without them, many crimes would go unsolved. Additionally, forensic scientists are involved in civil litigation, war crime investigation, human rights issues, national security, consumer protection, product safety, environmental protection, and historical identification.

What is forensic science? Although a dictionary definition of forensic science might be "science in the service of law," in common usage forensic science usually refers to the scientific study of evidence to discover the facts in a criminal or civil matter. With the advent of new technology in recent years, the field of forensic science has expanded rapidly, and the size and scope of forensic laboratories across the nation have grown tremendously. The law enforcement community also requires a great deal more training and expertise in order to deal with the growing variety of useful physical evidence.

CASE PROFILE
Cold Case: The Murder of Concetta "Penney" Serra

New Haven, CT: On Friday, July 16, 1973, twenty-one year old Concetta "Penney" Serra took the day off from her job as a dental assistant to go shopping in downtown New Haven. After Penney stopped by her father's garage to do some bookkeeping, Penney borrowed her father's blue Buick and headed out to do some shopping on Chapel Street. Penney pulled into a parking garage on Temple Street at 12:42 P.M. and parked the car on the ninth level. At 12:43, another car entered the parking garage from the George Street entrance. As Penney made her way through the garage to get down to Chapel Street, witnesses later said that Penney was chased on foot by a tall skinny man with long dark hair. Penney screamed for help as she ran through the garage. At 1:00 P.M., an employee from the New Haven parking authority returned from a lunch break. On the tenth level, the employee spotted the body of someone lying in the fetal position at the base of stairwell. Realizing that the person was dead, the employee called the police.

When the police arrived, they found Penney Serra dead. Her blue dress was covered in blood. Cuts and scrapes were found on Penney's wrist, finger, knee, and face. An autopsy later revealed that Penney died of a small deep wound through her 5th and 6th ribs that penetrated her right ventricle of the heart. Penney's chest was full of blood. The medical examiner told police that it was impossible to determine the weapon, only that it was three inches in length. The medical examiner said that it took only a minute for Penney Serra to bleed to death.

Police cornered off the crime scene and found her unlocked Buick parked at an erratic angle. Penney's car seat had blood on the outside door handle and door surfaces, on the steering shaft and driver's side floor. Police also found Penney's purse, her wallet with $14.75, her shoes, a parking stamp with an entry time of 12:42 P.M., and unopened envelopes containing invoices for

continued on next page...

a dental patient. Behind the driver's seat, police found more bloodstains on the floor and on a Rite Aid tissue box. Police found a trail of blood drops and splatters leading away from the Buick toward the stairwell up to the ninth and tenth levels. On the seventh level, police found a set of keys. Near the keys was a man's handkerchief covered in blood. Bloodstains were found on levels seven, six, and five. Police were able to determine that the assailant's blood type was type-O. (This was before DNA testing became commonly available).

From the car's license plates, police traced the car to Penney's father John Serra and contacted him about his daughter. Police learned that Penney had an on-off relationship with a man named Phil DiLieto. After his alibi checked out, he was eliminated as a suspect. Police were then presented with another suspect, Anthony Golino. Golino's wife stated that during one of their vicious arguments he wanted to "do her like Penney Serra." Golino's blood type was type-O; however, after some speculation about whether or not Golino and Serra had a relationship, Golino was eliminated as a suspect. Police also interviewed Martin Cooratal. His dental bill was found on the dashboard of the Buick belonging to Serra. Cooratal was spotted in the parking garage and matched the description of the suspect. However, like Phil DiLieto, Cooratal's alibi was solid.

For nearly two decades Penney's murder had gone unsolved. Penney's father John Serra was very critical about how the investigation was handled and put a lot of pressure on the state's attorney to find Penney's killer, so prosecutors looked to Dr. Henry Lee to investigate.

Dr. Lee and his staff began to reexamine all of the physical evidence left at the crime scene. On September 10, 1989, Dr. Lee reconstructed the crime scene at the Temple Street garage. With the help of his staff, Dr. Lee reviewed documents, photographs, and diagrams. According to a witness, the attacker ran back to the ninth floor and started Penney's Buick and drove down to the eighth floor and parked it at an extreme angle. The attacker then went down to the 7th floor, got into his own vehicle and drove it out of the garage leaving a blood smeared ticket with the garage attendant at 1:01 P.M.

Dr. Lee also examined latent fingerprints – prints not visible to the naked eye. Dr. Lee used chemicals to make these latent fingerprints visible. Still, it would take another five years to get a break in the case.

Then, in another case, Edward Grant of Waterbury was arrested, booked, and fingerprinted for beating his fiancée so badly that she required hospitalization. Grant, who had been badly wounded in Vietnam and had a metal plate in his head, had been subject to violent mood swings. Grant's fingerprints were matched against latent fingerprints left on the bloody tissue box in Penney Serra's car. Grant, who denied his involvement, offered no explanation as to how his prints were found at the crime scene. So detectives offered him a chance to exonerate himself by asking for a sample of his blood. The sample was sent to Dr. Lee's lab. Not only was Grant's blood type-O, but based on newly available DNA testing it was also a 300-million-to-one chance the blood belonged to Edward Grant. Sadly, by the time Edward Grant was arrested, Penney's father John Serra had passed away.

At the trial, Dr. Lee testified about the crime scene reconstruction, latent fingerprints, and DNA blood evidence. Dr. Lee explained to the jury that the killer chased Penney Serra, who was running barefoot, up several levels of the garage into the stairwell where she ran into a dead end at the 10th floor. The killer stabbed Penney in the heart and she died on the steps. Dr. Lee explained the concept of primary versus secondary crime scenes. The murderer then ran down several levels, got into Penney's car, and left blood all over the car. The killer tried to stop some bleeding on his hand, a defensive wound probably inflicted by Penney, and grabbed the tissues in the back seat to soak up the blood. The killer then parked Penney's car on a sharp angle, got out, ran down another level and got into his own car and fled the scene leaving another bloody print on a parking ticket as he left. The fingerprints and DNA left at the scene matched Edward Grant. After a long trial, the jury found Edward Grant guilty of first-degree murder. Thanks to the advances in forensic science and Dr. Henry Lee, a pioneer in the field, Penney Serra's killer was brought to justice at last.

Forensic Science Today

WHAT ARE THE ORIGINS OF FORENSIC SCIENCE?

The term "forensic" is derived from Latin, meaning "of or before the forum." This refers to the ancient Roman practice of resolving disputes through speech in a public forum. Forensic science has taken a long and varied path toward development into the field it is today. In China in 1247, early accounts of using medicine to solve crime were written in the book *Xi Yuan Ji Lu (Collected Cases of Injustice Rectified)*. This was the birth of forensic pathology as we know it today. In the field of toxicology, Mathieu Orfila was a Spanish physician who was the first to systematically use autopsy material and chemical analysis as legal proof of poisoning. In 1813, Orfila wrote *Toxicologie générale (General Toxicology)*, the first major publication to document symptoms of poisoning, how poisons work in the body, and how to detect them.

In the 1830's, Alphonse Bertillon devised a method of identifying criminals by their physical characteristics. Anthropometry, as it was known, used measurements of the head and body, facial characteristics, and mug shots of offenders to uniquely identify each individual criminal who passed through the criminal courts. This was before the widespread use of fingerprinting to identify offenders. Can you think of problems that could arise with the use of body characteristics to identify criminals?

Fingerprinting became commonly used as a means of identifying individuals after William Herschel, a magistrate in colonial India, began requiring their use for various purposes in 1877. These purposes included registering land deeds, collecting pensions, and ensuring that convicts did not hire substitutes to serve their prison time. In 1888 Francis Galton designed a form for recording inked fingerprint impressions and defined the three main pattern types. In the early 1900's, fingerprinting was adopted by Scotland Yard. Scotland Yard then taught the St. Louis Police Department how to use fingerprints at the 1904 World's Fair in St. Louis, Missouri, thus spreading the practice to the U.S.

▼ Forensic scientists Dr. Baden and Dr. Lee examine human remains.

Around the same time that fingerprinting came into vogue, Dr. Edmond Locard pioneered the use of trace evidence to solve crimes. Dr. Locard set up the first known crime laboratory and is famous for postulating the Locard Exchange Principle. This principle states that when two items come into contact, there will be an exchange of substances. Dr. Locard was also known as the "Sherlock Holmes of France." Meanwhile, the original Sherlock Holmes came alive as a fictional detective in readers' imaginations with the publication of Arthur Conan Doyle's crime stories beginning in 1877. Based in London, Sherlock Holmes was famous for using deductive reasoning, logic, and careful observation in order to solve crime. Even today, these skills are very important in the field of forensic science.

In 1893, Dr. Hans Gross of Austria wrote *System Der Kriminalistik (Criminal Investigation)*, the first work to comprehensively describe the use of physical evidence in solving crime. Now known as the "Father of Criminalistics," Dr.

Chapter 1
Introduction to Forensic Science and Criminalistics

Gross is the first person to coin the term "criminalistics" to refer to scientific methods of identifying, apprehending, and prosecuting criminals. Several decades later, Paul Kirk helped establish criminology as an academic discipline through his work at the University of California, Berkeley. Specializing in microscopy, Kirk wrote the important textbook, *Criminal Investigation* and provided key blood spatter evidence resulting in the acquittal of Sam Sheppard in his wife's famous murder in 1955. (This case was the inspiration for a television series and subsequent movie, both called *The Fugitive*).

▲ Forensic science display in shopping mall in Wasilla, Alaska.

One of the most important and recent discoveries in the history of forensic science is the use of DNA to identify individuals with nearly 100% accuracy. In 1986, DNA was used to solve a crime for the first time when DNA profiling identified Colin Pitchfork as the murderer of two young girls in England. Interestingly, the same case was also the first time that DNA was used to exonerate an innocent suspect before Pitchfork was identified as the culprit.

With the march of time and emergence of new technologies, forensic science is an ever-evolving field and we can be sure that more exciting discoveries are just around the corner.

WHY STUDY FORENSIC SCIENCE?

The field of forensic science is exciting. It's challenging. It offers great opportunity for career advancement. Most importantly, it can be immensely rewarding on a personal and professional level. Imagine being the person who helps crack the case, who helps bring a sense of justice to families and communities rocked by a violent criminal act. Or imagine being the person who helps set an innocent individual free from prison with post-conviction forensic testing. Technology is becoming more and more exciting and allows us to discover what would previously have been impossible-to-find clues. With the huge variety of exciting jobs in the field of forensic science, you are guaranteed to discover the career of a lifetime.

WHERE DO FORENSIC SCIENTISTS WORK?

Since there is such a wide variety of specialties in forensic science, there is also a wide variety of places where you might someday work. Some forensic scientists work in laboratories on an eight-to-five, weekday basis. Some work out in the field and may work different hours, and still others are "on call" and work after their regular shift and receive overtime or "comp" time pay. Forensic scientists may work at crime scenes, in morgues and medical examiner's offices, at universities, in police departments, in offices, at mass disasters, or as independent consultants.

WHAT DO FORENSIC SCIENTISTS DO?

As you may have guessed by now, forensic scientists do a wide variety of work, depending on which discipline—type of work in the field—they choose. However, all forensic scientists share a common goal: to use all available information in a non-biased manner in order to determine the facts and, therefore, the truth. Some forensic scientists work in the civil arena of the courts—on issues such as the validity of a signature on a will or the authenticity of a document. The focus of this book, however, is on criminal applications of forensic science, also known as "criminalistics" (see *Different Specialties in Forensic Science*, below).

What's the difference between a *criminal matter* and a *civil matter*? Both are involved in the court system and legal process. The difference is that criminal matters refer to acts that are punishable under the *criminal code* (i.e., "crimes"), whereas civil matters involve things like contracts, real estate, and business matters.

Forensic Science Today

In the criminal arena, forensic scientists work on all aspects of cases, and their work may serve either the prosecution or defense. Forensic scientists provide information and work with police officers, prosecutors, defense attorneys, detectives, and federal agents from agencies such as the FBI (Federal Bureau of Investigation), DEA (Drug Enforcement Administration), CIA (Central Intelligence Agency), USPIS (United States Postal Inspection Service) and BATFE (bureau of Alcohol, Tobacco, Firearms and Explosives).

Forensic scientists work with **physical evidence** to help solve crimes. In Chapter 3 (Crime Scene Investigation Part II: The Physical Evidence), we will learn in detail about the stages of analysis of physical evidence. In brief, forensic scientists must *recognize* the potential physical evidence at a crime scene, *document, collect,* and *preserve* physical evidence, *identify* what the evidence is, *classify* the evidence by comparing it to known standards, *individualize* the evidence (demonstrate that the evidence is absolutely related to a particular source), *evaluate* the evidence by conducting laboratory tests and other analysis, and *reconstruct* the chain of events in a crime by analyzing the physical evidence and interpreting patterns at the crime scene. Some forensic scientists are involved in all of these stages of analysis; others may focus on one particular stage.

Finally, ethics are an extremely important part of working in forensic science. No matter which side they work for, forensic professionals must remain neutral at all times and prevent their personal feelings and opinions from getting in the way of discovering the hard facts. They must also make sure their work is the very best they can do: tests must be conducted correctly, written reports must be accurate, and testimony must be truthful, complete, and objective. After all, with long prison terms or even a death sentence at stake in violent crime trials, lives may depend upon the work that forensic scientists do.

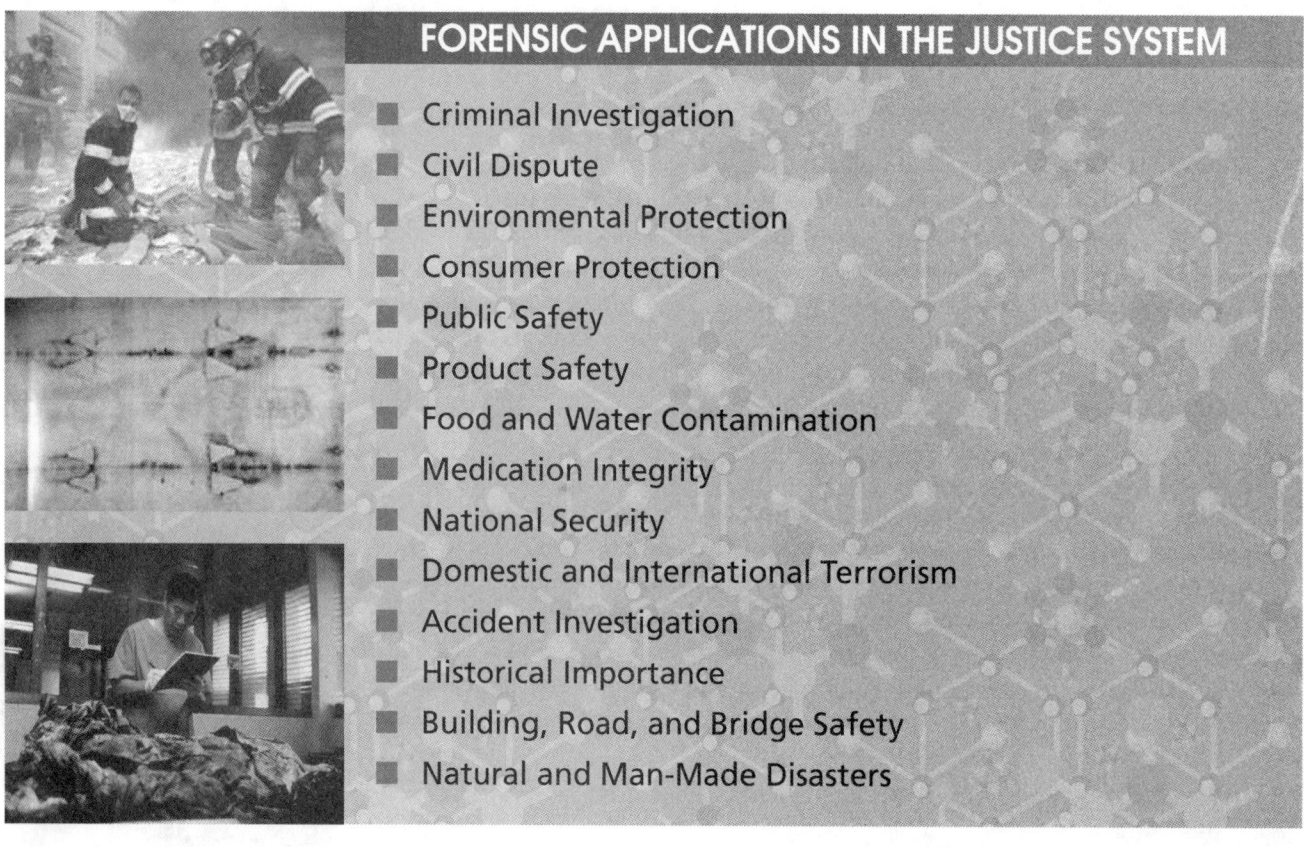

FORENSIC APPLICATIONS IN THE JUSTICE SYSTEM

- Criminal Investigation
- Civil Dispute
- Environmental Protection
- Consumer Protection
- Public Safety
- Product Safety
- Food and Water Contamination
- Medication Integrity
- National Security
- Domestic and International Terrorism
- Accident Investigation
- Historical Importance
- Building, Road, and Bridge Safety
- Natural and Man-Made Disasters

Chapter 1
Introduction to Forensic Science and Criminalistics

DIFFERENT SPECIALTIES IN FORENSIC SCIENCE

Some of the many exciting disciplines and job opportunities in the field of forensic science include:

Criminalistics. Criminalistics include many different subspecialty areas involved in solving crimes. A criminalist has traditionally been regarded as a jack-of-all-trades, drawing upon his or her expertise in many different areas of forensic science to discover the facts about what really happened in a crime. A criminalist collects, identifies, and analyzes physical evidence found at crime scenes. Subspecialties within the field of criminalistics include crime scene reconstruction, forensic drug analysis, forensic chemistry, trace evidence analysis, firearms examination, latent fingerprint examination, voice analysis, forensic serology, DNA analysis, impressions analysis, and questioned document examination.

Crime Scene Investigation. Crime scene investigators are criminalists who gather information at crime scenes, search for evidence, collect the evidence, and preserve the evidence so that it remains exactly as it was when found at the crime scene. They also conduct field tests and reconstruct the sequence of events in order to help solve the crime.

Forensic Medicine: Pathology and the Medical Examiner. A pathologist is a doctor who studies injuries and disease, to determine the cause and manner of sickness or death. A medical examiner is often referred to as a forensic pathologist, who determines the cause, manner, and mode of death by examining the body of a corpse during autopsy.

Forensic Toxicology. A forensic toxicologist identifies and analyzes poisons, drugs, and chemicals, and then studies the effect of such substances on the human body or on animals.

Forensic Anthropology. A forensic anthropologist studies and examines bones in order to identify human remains as well as to determine the age, race, and other information about the remains.

Forensic Entomology. A forensic entomologist identifies unknown insects and bugs found on a decomposing body, and studies their life cycles in order to determine how much time has passed since death occurred.

Forensic Engineering. Forensic engineers use their engineering skills to determine material or structural failures, and to reconstruct events such as building collapses, train wrecks, and car accidents.

Forensic Dentistry. Forensic odontologists are dentists who identify human remains by matching teeth with dental records, and who examine and identify bite marks on both perpetrators and victims.

Forensic Art. Forensic artists couple their knowledge of human anatomy with special techniques in age progression, composite imagery, facial reconstruction, and digital technology to identify missing children, criminal suspects, or unidentified homicide victims.

Forensic Psychology and Psychiatry. A forensic psychologist or psychiatrist specializes in studying human or criminal psychology and behavior, and profiles criminals as well as victims.

Jurisprudence. Jurisprudence is the philosophy of law. A lawyer may specialize in forensic issues working as a prosecutor, criminal defense attorney, or judge on the criminal or civil bench.

Additionally, people working in many other fields are called upon to lend their expertise in solving particular aspects of a crime. Some of these people might include:

- Accountant (to detect fraudulent financial activities)
- Aviation accident investigator (to determine the cause of airplane accidents)
- Computer specialist (to investigate online stalkers and the use of computers in crimes)
- Environmental scientist (to determine the facts in crimes against the environment, such as illegal dumping of hazardous waste or illegal killing of and trade in endangered species)
- Medical investigator (to examine medical documents and suspicious deaths)
- Image enhancement specialist (to work with new imaging technology to enhance photos, video, and other secondary materials)
- Forensic nurse (to conduct forensic examination on assault and abuse victims)
- Photographer (to document crime scenes and other evidence using photo enhancement, nonvisible light photography, alternate light sources, video conversion and digital processing and clarification)
- Polygraph examiner (to conduct "lie-detector" tests on suspects and witnesses)
- Radiologist (to examine the x-rays of crime victims, suspects, or unknown material)
- Speech scientist (to recognize speech patterns and identify voices in recordings)

NOTHING BUT THE TRUTH: TESTIFYING IN COURT AS AN EXPERT

Forensic scientists are often called upon as experts to testify in court. Because they are expert witnesses, they can give their professional opinions based not only on testing of the evidence, but also on their own knowledge and expertise. Expert witnesses are the only witnesses allowed to state their opinion in court. All others must only tell what they personally heard or saw. The more experienced a forensic scientist is, the more persuasive his or her testimony will be to the court and jury.

Are there limits to what a forensic scientist can say in court? There are limits, but these are determined on a state-by-state basis. For instance, the California Evidence Code states that reliability of scientific opinions is determined by one factor: the opinion must be based on principles generally accepted by experts in the field.

However, federal courts (and some state courts) use a different standard. The U.S. Supreme Court has ruled that in order for a forensic scientist's expert testimony to be admissible (allowed) in federal court, the testimony must be both "relevant" and "reliable." This is a two-pronged test, and the trial judge is considered the "gatekeeper" of scientific evidence. Therefore, the judge will first determine whether the forensic evidence is relevant to the case. Next, the judge will use the following questions as a guide to determine if the evidence is also reliable:

- Is the scientific theory or technique falsifiable, refutable and testable using empirical testing methods?
- Has it been subjected to peer review and publication?
- What is the known or potential error rate?
- What are the standards concerning its operation?
- Is it generally accepted by a relevant scientific community?

The Federal Rules of Evidence also state that an expert witness may only testify if:

1. The testimony is based upon sufficient facts or data,
2. The testimony is the product of reliable principles and methods, and
3. The witness has applied the principles and methods reliably to the facts of the case.

Thus, a forensic scientist must be prepared not only to present evidence to the court, but also to explain how the evidence was tested, what techniques were used, the error rates and standards of operation, and anything else the judge might need to know. As always, ethics and truthfulness are of utmost importance when testifying under oath in a court of law. Forensic scientists must present all the facts to the court in an objective and fair manner.

When is a case heard in federal court as opposed to a state court?
If state lines are crossed during the commission of a crime (such as interstate kidnapping), or if a federal statute (such as the prohibition against bringing illegal drugs into the country) is violated, then the crime falls under the jurisdiction of the federal courts. If a crime is committed in one state and violates a state law, then it will be tried in a state court at the county level.

YOU'LL LOVE A CAREER IN FORENSIC SCIENCE IF YOU:

- ✔ Are curious and persistent
- ✔ Enjoy solving mysteries and puzzles
- ✔ Work well with details
- ✔ Enjoy being challenged
- ✔ Enjoy working hard and doing your best
- ✔ Enjoy using your skills of logic to solve problems

Forensic Science Today

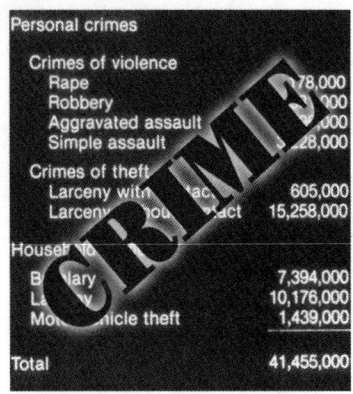

▲ Annual Crime Report by the Justice Department.

THE "*CSI* EFFECT"

Since 2000, the television series, *CSI: Crime Scene Investigation* and its spin-offs have grown in the public consciousness as its popularity has continued to rise. Now seen in over 200 countries with an audience of two billion viewers, it has introduced members of the public to forensic techniques previously only known to specialists. But is this necessarily a good thing? Researchers are investigating whether some jury members who watch *CSI* have unreasonable expectations of the forensic techniques used in criminal cases. In real life, a crime is never solved in an hour. Investigating a crime scene can take days; and analysis of physical evidence such as DNA or trace can take weeks. Most people also do not realize that real-life forensic analysis will sometimes yield equivocal results; that is, an item of evidence can lead in more than one direction.

Additionally, *CSI* has changed the way some criminals commit crimes. Now, some criminals attempt to destroy evidence of their crime such as blood or DNA. However, any attempt to clean up after a crime is readily apparent to forensic investigators, and this will also be considered evidence of that crime. On a positive note, *CSI* as a cultural phenomenon has also led to an increase in applications to forensic science research programs at universities. The more well-trained forensic scientists we have working in the labs, the faster we can work on the backlog of cases awaiting analysis and bring justice to those affected by (and those who commit) crimes.

HOW CAN I BECOME A FORENSIC SCIENTIST?

As you will soon discover, there are many different types of forensic science, and just as many exciting careers. The most important thing is to keep an open mind with your career options and remember that people who are happy with their work do it much better. Think about the things you truly love to do. Do you like to spend most of your time outdoors, or do you prefer working with computers? Are you good with small, intricate details, or are you particularly talented at seeing the big picture? Whatever your talents are, and wherever your interests may lay, chances are you will find your niche—your special place where you belong—in the world of forensics and the dozens of exciting, enriching, and rewarding opportunities it has to offer.

With this in mind, begin laying a strong foundation for future studies, right now. It is a good idea to take as many science courses as you can. Don't worry if you think you're not the "science type." As you will quickly learn, *science can be fun*. Some people may avoid careers using science because they didn't particularly like math or science as a student. Later, they realize that the rich rewards of an exhilarating career are well worth the extra effort of learning skills that may come as more of a challenge.

As you proceed through this book, you will also learn that there are other skills you can acquire to prepare yourself for a career in forensic science. Not only is science important, any course that sharpens your skills of analysis will be useful. Analysis means taking a set of data or information and interpreting it to come to a conclusion. Thus, interpreting 19th Century poetry in your literature class is just as important as learning the principles of chemistry or physics. What other skills do you think are important in forensic science?

Chapter 1
Introduction to Forensic Science and Criminalistics

There are many paths to a career in forensic science. In 2003, the American Academy of Forensic Sciences promulgated accreditation standards for forensic science educational programs. This was based on recommendations by the U.S. National Institute of Justice. When planning your coursework, be sure to check with the department head that your program meets these standards. Whether you attend classes at your local community college or go to an out-of-state university; whether you go to school full-time or part-time; or whether you take time off to work, raise a family or travel the world, with a bit of planning and hard work you could soon be working in a forensic science career you will love.

For more helpful hints and ideas, go to the website of the American Academy of Fo-

CHECKLIST FOR A CAREER IN FORENSIC SCIENCE

You will need:
- A bachelor's (four-year) degree from a college or university (some forensic sciences require advanced degrees such as a Masters or Doctorate). Choose a major in science, unless your field of interest is in anthropology, art, jurisprudence, or psychology. No matter which major you choose, be sure to include courses in chemistry, biology, math, and writing.
- Good note-taking skills
- The ability to write an understandable scientific report
- Good communication skills for testifying in court—if you are shy, rest assured! Your knowledge and expertise at your job will bring confidence. Plus, you can take classes in public speaking or join the drama club or debate team.
- Good ethics and personal integrity

TECHNICAL DEFINITIONS

Evidence: Any tangible material that can be used to prove the facts of a matter in a court of law. Evidence can be something like a piece of clothing, the testimony of an eye-witness, tire tracks in the dirt, microscopic human cells, or documented data such as a record of phone calls or mileage of a car.

Expert Witness: Someone who testifies in court as a qualified expert in a particular subject area.

Forensic Science: The direct application of scientific knowledge and techniques to matters of law. Forensic scientists use logic and follow careful scientific steps in order to solve issues in criminal and civil litigation.

Chapter 2

Crime Scene Investigation Part I: The Basics

CHAPTER OBJECTIVES

- ✔ Visualize the different stages in crime scene investigation as a flow chart.
- ✔ Understand and discuss what happens at a crime scene, beginning with the call to the first responding officer.
- ✔ Differentiate between various types of crime scenes.
- ✔ Describe the steps in surveying a crime scene.
- ✔ Identify the four parts of crime scene documentation.
- ✔ Identify different methods used in searches for physical evidence.
- ✔ Understand the importance of developing a system to recognize what is and what is not physical evidence.
- ✔ Identify the goals of the crime scene search.
- ✔ Explain the two theories (linkage and transfer) behind crime scene searches.
- ✔ Recognize and explain the six pattern methods for searching crime scenes.
- ✔ Define the following vocabulary words and phrases: *chain of custody, crime scene survey, linkage theory, Locard's Theory of Exchange, primary crime scene, primary transfer, secondary crime scene, secondary transfer,* and *transfer theory.*

WHAT HAPPENS AT A CRIME SCENE?

Once a crime has been committed, there are many stages in which forensic scientists, police detectives, and prosecutors each play a unique role before the case goes to court. When someone calls 911 to report a crime, the police dispatcher notifies police officers over the radio. It is the duty of the first responding officer to the scene of a crime to first check the status of the victim. If the victim requires medical assistance, the first responding officer will call for an ambulance and give first aid. He or she will also secure the scene, apprehend any suspects, interview the witnesses, and report to superior officers about the incident.

One of the most important duties of the first responding officer is to *secure and protect the integrity of the crime scene.* This cannot be emphasized enough. Too many crimes have gone unsolved, or criminals set free, because the evidence at the crime scene was disturbed or treated improperly. There are many opportunities for mistakes, so it is absolutely essential that police officers *do not disturb the evidence.* There have been cases where police officers have stepped in blood and mixed up their own footprints with those of the suspect, improperly handled evidence, or done other things at the

Forensic Science Today

KEY TERMS

Chain of Custody
Primary Scene
Secondary Scene
Scene Security
Crime Scene Search
Documentation
Linkage Theory
Transfer Theory
Locard's Theory of Exchange
Primary Transfer
Secondary Transfer

scene of the crime that have ruined entire criminal investigations. If a window is open or a telephone is off the hook, the responding officer must leave these things exactly as they were found, unless it is necessary to prevent harm to the evidence (such as closing a window to stop rain from coming in and damaging a ransom note or blood stains).

Next to arrive on the scene are the criminal investigators, such as police detectives and crime scene investigators. The first responding police officers on duty report their findings to the detectives, who then take over the case investigation. The focus of this chapter, however, is on the crime scene investigators or the CSI Team who work the crime scene. When working a crime scene, investigators gather information, search for evidence, document the evidence, collect the evidence, and preserve the evidence so that it remains exactly as it was when found at the scene. They also conduct field tests and reconstruct the sequence of events in order to solve the crime.

Crime scene investigators use a logical, systematic approach to solving a crime. Rather than coming up with an instant theory, they must use logical thinking to form a hypothesis, which can be tested and either proved or disproved. Forensic investigators and scientists must always safeguard the evidence through every single step of collection, labeling, testing, and presenting at trial. These steps are referred to as keeping the evidence in the "chain of custody." (Chain of custody is explained in more detail in Chapter 3.) If the chain of custody becomes broken at any point due to improper handling or labeling, the evidence may no longer be admissible in court, which could result in the acquittal of a guilty suspect.

In this chapter, we will learn the basic concepts of crime scene investigation and advanced techniques of searching for physical evidence. In Chapter 24 we will learn about reconstructing crimes. For an in-depth examination of investigating crime scenes that goes beyond the scope of this introductory course, we recommend that you read *Henry Lee's Crime Scene Handbook*. (See *"Resources"* at the end of this book for more information).

CASE PROFILE
Murder in the Park

Hartford, CT: Bank vice president, Diane Gellenbeck, left her office to go to a lunch meeting with another bank manager. She never made the meeting. A few hours later, her body, with multiple gunshot wounds, was found by a jogger in Keney Park around 1:00 P.M. Hartford Police were called. Detectives searched for witnesses in the park; however, nobody initially seemed to have noticed anything. Crime scene investigators secured the scene and looked for evidence. Detectives also contacted the Medical Examiner as well as Chief Forensic Scientist Dr. Henry Lee to assist in the investigation of the case.

At the park, Dr. Lee found Ms. Gellenbeck's body face down in the grass on the edge of a public golf course. Her business suit and beeper were still in place. Dr. Lee noticed that her suit jacket had several areas of rips and damages. Her skirt was caked with soil and leaves, which told him that the victim must have been moved from another location in the park. In other words, the location of the victim's body is a secondary crime scene. The Medical Examiner identified at least eight, possibly nine bullet wounds, some made from a distance, others obviously at very close

continued on next page...

range. It seemed Ms. Gellenbeck's killer had shot her in the back as she ran away, then chased her, moved in closer and shot her in the head at a very close distance execution-style.

Police searched the park and golf course and found Ms. Gellenbeck's purse and panties in the nearby woods. Because Dr. Lee saw no signs of rape, he suspected an attempted rape had turned to a murder when the victim resisted. Near the purse and underwear, police found nine .38 caliber shell casings. The caliber meant the weapon was probably a five or six-shot revolver. To Dr. Lee, it meant a particularly vicious person must have emptied the weapon, reloaded, and fired again, determined to kill the woman who tried to get away from him.

After looking for witnesses, police found Anthony Bibbens, who told police that he heard gunshots in the park, and then a few minutes later noticed a red car leave the park and stop at the corner light. Bibbens gave police what he could remember of the license plate number. The car, a red Mercury Zephyr, was then traced back to Rosa Billington, who told police she had given the car to her boyfriend for the day. Police seized the car but found no immediate evidence that tied the car to the murder. Police interviewed the boyfriend who said he had nothing to do with the killing, a dead end for police.

At the same time, Dr. Lee was able to pinpoint when Ms. Gellenbeck had left her office. Tape from a security camera showed her entering the building's parking garage at 12:15 P.M. Yet her car had never been moved. Dr. Lee deduced that Ms. Gellenbeck had left the garage in someone else's vehicle. Dr. Lee ordered that every parking receipt collected between 12:00 noon and 1:00 P.M. be checked for fingerprints. More than 100 ticket stubs yielded prints that could be traced. Police checked prints that Dr. Lee uncovered against their database. They got a match. A thumb print on the ticket stamped out at 12:20 P.M. matched exactly the thumb print in the police file on Daniel Webb, a 28-year-old with a record of attempted rapes and other assaults dating back to his teenage years. Webb was Rosa Billington's boyfriend and the person she loaned the car to. At the time police had interviewed Webb they were unable to tie him to the crime.

The matching thumbprint gave probable cause for a warrant to search the car Webb had been driving on the day of the murder. In the car, Dr. Lee found Diane Gellenbeck's fingerprints on the inside of the car windshield and strands of her hair on the car seat. Now police could place Diane in the car and Rosa could place her boyfriend in the car. When arrested for murder, Daniel Webb said he was innocent and fought the charges.

Then, police put out another call for witnesses to come forward. William Woodard, it turned out, had been hitting golf balls in the park that day with his 7-year-old daughter, but was supposed to be at the office and he was frightened by what he saw. So he had kept quiet, hoping police would not need him. Now he decided to talk. Woodard told police the he saw a woman run from a parked car and a man got out and shot her several times, killing her.

At the trial, Dr. Lee took the jury step-by-step through the physical evidence including the brilliant and efficient fingerprint work on the parking stubs that had led police to Daniel Webb.

The jury convicted Webb and sentenced him to death.

Forensic Science Today

▲ Police and forensic scientists respond to a crime scene. (top)

▲ Crime scene investigators measure and sketch the crime scene. (bottom)

TYPES OF CRIME SCENES

There are many different ways to classify crime scenes. The first is based on where the crime was committed and where the physical evidence has been found. In many situations, crimes result in both "primary" and "secondary" scenes. The primary scene is generally where the criminal act occurred, such as a homicide. The secondary scene can then be, for example, in the field where the victim's body was dumped. A secondary crime scene can also be in the suspect's garage where fibers from the victim's clothing were found, or around the victim's home where the suspect left footprints outside the victim's window. Thus, a crime can result in several secondary crime scenes, and each one must be logically and systematically searched for evidence.

Crime scenes can also be classified according to the *type* of crime that was committed (i.e., robbery, assault, and so on), the *condition* of the crime scene (organized or disorganized), the *location* of the crime (indoor versus outdoor), or the *activity* of the crime scene (active versus passive).

Moreover, a crime scene can be as large as a whole building or an airplane, or as small as a note or a broken fingernail. Finally, remember that the *human body* itself can also be a crime scene. Here, we can find hairs and fibers, DNA, body fluids, gunshot residue, pollen, soil, debris, drugs, and even insects—each of these items tells its own story about what happened during the commission of the crime.

GENERAL CRIME SCENE PROCEDURES

Investigating a crime scene involves teamwork, concentration, and an open mind. It is the first step in discovering the facts of what really happened when a crime was committed. A crime scene tells a story; and investigating the scene is meant to begin the process of answering the questions: *Who did it? How did they do it? Why did they do it? When and where was the crime committed?* While there are a great many facts to determine in the scope of crime scene investigation, CSI Team investigators first seek to discover the nature and patterns of the crime, and the different types of physical evidence left behind by the suspect and victim. Crime scene investigation must follow a precise sequence of steps in order to locate and identify all evidence, and ensure that all legal and scientific requirements have been met and documented.

SURVEYING THE CRIME SCENE

After the first responding officer has secured the crime scene, the lead investigator and the case officers will conduct a preliminary survey of the crime scene, also known as the "walk-through." According to *Henry Lee's Crime Scene Handbook*, these are the tasks that should be performed during the crime scene walk-through:

- ■ Determine what type of crime scene it is (primary/secondary, indoor/outdoor, organized/disorganized) as well as the physical boundaries of where the crime scene begins and ends. Also decide what type of additional personnel and equipment will be needed to fully investigate the crime scene.

- Note the different types of evidence present at the crime scene. Make a mental note of the weather conditions, whether light switches are on or off, whether doors are open, closed, or locked, whether the heating or air conditioning is turned on, whether any odors are present, and so on.
- Make a mental note of the points of entry and exit, the target area of the crime scene, and any damage that may have been done.
- Begin to mentally reconstruct—or put together—the events that might have led up to the crime. This is only a preliminary mental reconstruction; and it is important not to form a rigid or fixed theory of how the crime occurred. It is also important to remember that *scientific* crime scene investigation is *objective and systematic*. Otherwise, you might risk getting "tunnel vision" which could lead the investigation in the wrong direction.

During the walk-through, investigators should also try to answer the questions: *Who? What? Where? When? How?* and *Why?* After surveying the crime scene, the crime scene investigator will work with other members of the team and share the findings with them. Other agencies that the investigator may communicate the findings to include the prosecutor's office, forensic laboratory, and in the case of a questionable (or untimely) death, the Medical Examiner or Coroner's Office.

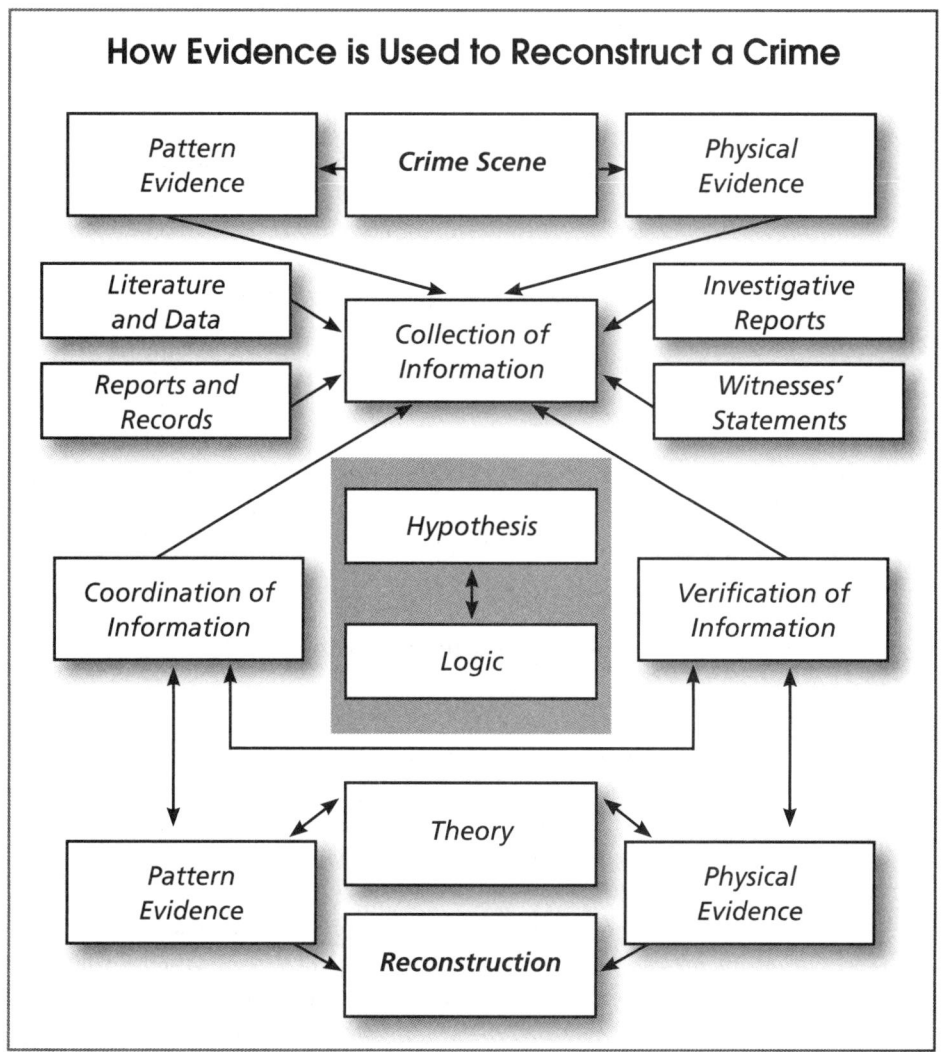

Forensic Science Today

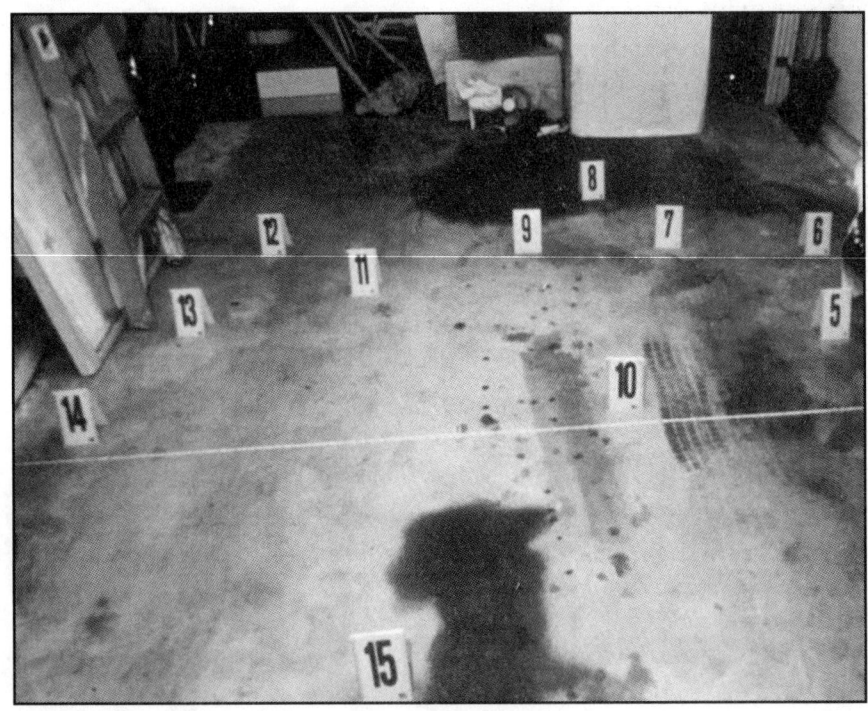

▶ Evidence markers used to identify the physical evidence found at a double-homicide scene. (Courtesy of Deborah Messina, Director of CSP Forensic Lab.)

GOALS OF THE CRIME SCENE SEARCH

When searching a crime scene, investigators should keep in mind the eight key roles that physical evidence plays in the world of forensic investigation, as discussed later in this chapter. The goals of the crime scene search are to recognize, document, collect, and preserve physical evidence that does any combination of the following:

1. Provides information on the "Corpus Delicti" ("body of the crime")
2. Provides information on the "Modus Operandi" (MO or "mode of operation")
3. Links a suspect with a victim
4. Links a person with a crime scene
5. Disproves or supports an alibi or the testimony of a witness
6. Identifies a suspect
7. Provides investigative leads
8. Identifies unknown substances

THE TWO THEORIES BEHIND CRIME SCENE SEARCHES

Before we learn about the practical aspects of where and how to look for physical evidence, it is important to understand the two underlying theories of crime scene searches. These two theories, the "Linkage" theory and the "Transfer" theory, guide crime scene investigators as they adopt a scientific and systematic approach to searching for evidence. These theories are explained below.

The Linkage Theory

The linkage theory is based on the idea that if investigators can establish positive links between the physical evidence and/or victim, and/or crime scene, and/or suspect, then the case can be resolved.

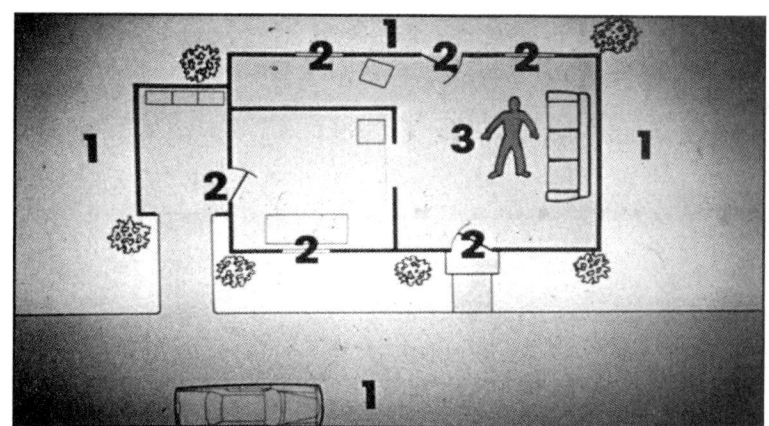

◀ The zone search is a common method for searching indoor crime scenes.

The more connections that can be made between two or more of these components, the greater the likelihood that the case can be solved. For instance, bloody fingerprints may be found both on a homicide victim's wrists and on the doorknob of the crime scene. If the fingerprints belong to the suspect, then he is linked to both the victim and the crime scene, as well as to the blood that had been on his fingertips at the time the prints were made. This builds a stronger case than if the fingerprints were only found on the doorknob, and not on the victim, especially if the blood was the suspect's own blood and not that of the victim.

The Transfer Theory—Locard's Theory of Exchange

The second theory behind crime scene searches is that of *Locard's Theory of Exchange*, which holds that transfer evidence is created whenever two surfaces come into contact with one another. The science underlying this idea is that matter is mutually exchanged across the contact barrier when two objects make contact. In other words, whenever you leave a room you leave something behind—whether it is soil from your shoes or a hair from your head—and you also take something with you, such as fibers from the carpet.

There are two types of transfer evidence: *primary transfer* and *secondary transfer*. Primary transfer is what happens when a homicide victim bleeds onto the floor. Secondary transfer—through an intermediate object or person—is what happens when the suspect (or a negligent police officer) steps in that blood and leaves bloody footprints around the crime scene.

OTHER THINGS TO KEEP IN MIND

In addition to the linkage theory and transfer theory, crime scene investigators should also keep the following concepts in mind before and during crime scene searches:

- **Viewing the crime scene as a whole.** Investigators should remember to look at the crime scene as a whole and not just as a series of individual evidence collections. They must keep an open mind; yet refrain from wandering aimlessly around the scene. They must look in places they normally wouldn't look upon entering a scene. They should also look at a crime scene from different vantage points, such as from floor level as well as looking down from above. It is also just as important to look for what is present (such as fingerprints) as it is to look for what is missing (such as the victim's wallet).

- **Is it a primary or secondary crime scene?** Earlier in this chapter we learned about primary crime scenes—where the crime occurred—and secondary crime scenes—where other valuable evidence may be found. For example, in a burglary case the primary crime scene would be the house that had been burglarized, and the secondary crime scene would be the dumpster where the victim's wallet was disposed of after the cash was removed. Or, in a homicide case the primary crime scene is where the murder occurred and the secondary crime scene would be where the body was later dumped. Crime scene investigators need to determine whether the crime scene they have arrived at is primary or secondary. If it turns out to be the secondary crime scene, the top priority becomes to find the primary crime scene before the perpetrator has had a chance to clean it up or make any other alterations to it.

- **You only get one chance to search the scene.** Crime scene investigators usually only get one opportunity to search the crime scene before "releasing" it. Therefore, each search must be done thoroughly and systematically. If investigators think they may have missed something during the initial crime scene search, then they should search the area again, perhaps using a different method. Until the crime scene is released, it must be guarded 24 hours a day by the police to maintain the integrity of the scene. Sometimes this is necessary until further testing is done on evidence in the forensic laboratory or until an autopsy has been conducted.

- **Do not alter the crime scene in any way.** Although it may seem obvious, you may be surprised to learn how often crime scenes are accidentally altered by emergency response personnel or crime scene investigators, thereby destroying valuable evidence or rendering it legally inadmissible in court. Remember that, in your search for evidence, you must always follow proper protocol and procedures for documenting, preserving, and collecting evidence. Do not change anything about the crime scene unless absolutely necessary to maintain the integrity of the scene (such as closing a window to keep the rain from washing away blood or trace evidence).

METHODS OF SEARCHING CRIME SCENES

Experienced crime scene investigators understand that there is no single ideal way to search a crime scene. Rather, a combination of skills, training, experience, and intuition will guide the investigator in choosing which of several different types of crime scene search methods that are available. A skilled investigator will take into account what is unique about a particular crime scene, as well as physical limitations that the crime scene might have. Below are the six major types of crime scene search patterns that investigators may choose from. These search patterns are illustrated in Figure 1 on page 22.

1. **Line method.** You may have seen television footage of large-scale searches for a missing person, where hundreds of people walk slowly, side by side across a field or other large area. This is known as the line method, which is one of the easiest methods to employ and works especially well in large outdoor areas. In the modified line (strip) method, the area is divided into a rectangle, which is then divided into strips. Members of the search team search along straight lines in these strips, usually very close to one another so that nothing is missed. In large-scale searches, volunteers from the community can even participate so long as they are trained in how to conduct the search properly without destroying evidence.

Chapter 2
Crime Scene Investigation Part I: The Basics

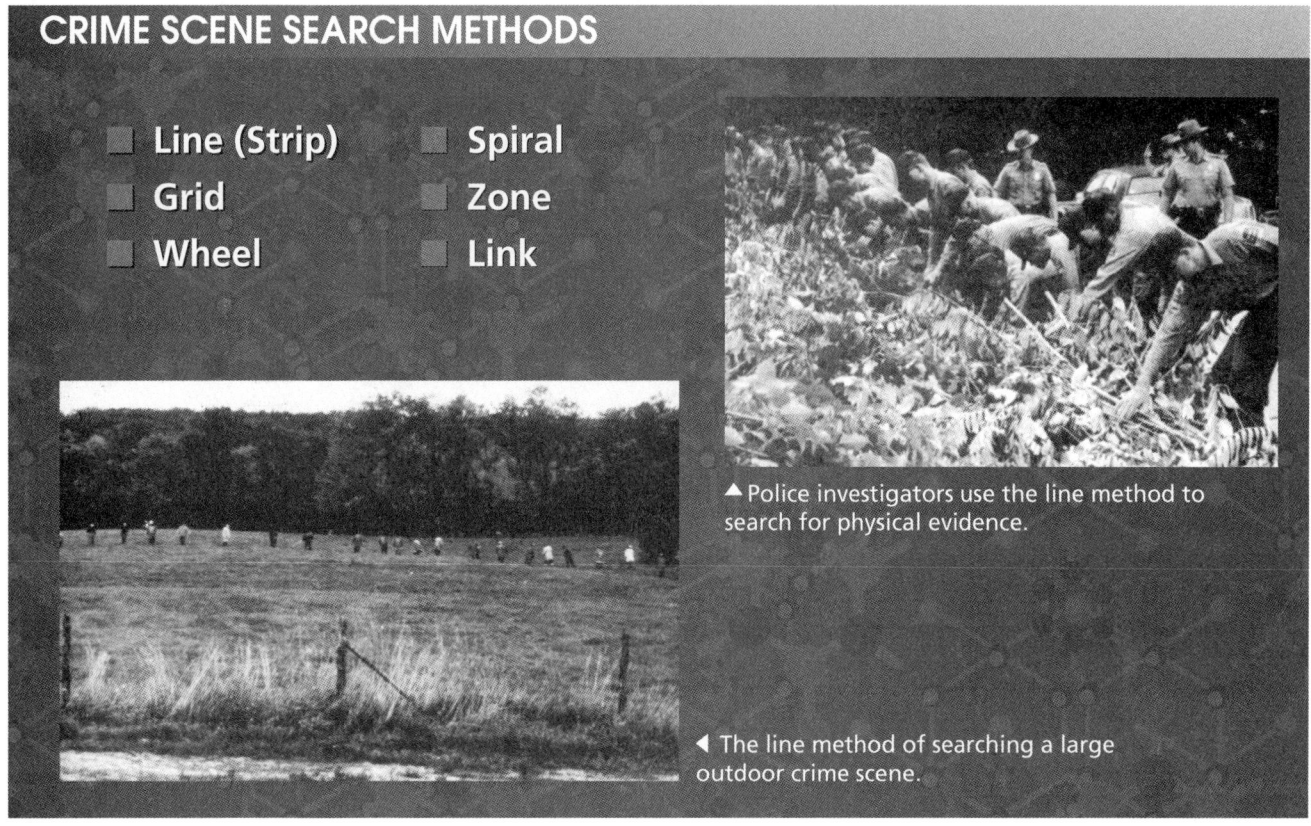

CRIME SCENE SEARCH METHODS

- Line (Strip)
- Grid
- Wheel
- Spiral
- Zone
- Link

▲ Police investigators use the line method to search for physical evidence.

◄ The line method of searching a large outdoor crime scene.

2. **Grid method.** The grid method is another modified version of the line (strip) method: once an area has been searched by the line (strip) method, the area is searched perpendicular to the original lines of search, so that both sets of lines, if they were overlaid, would form a grid. While this is obviously more time-consuming than the line method, it is more comprehensive and has the additional benefit of each section being searched by more than one person. What one person might miss, another might catch.

3. **Wheel method.** With the wheel method, the crime scene is viewed as a wagon wheel with spokes going out from the middle. Investigators begin at the center of the wheel and slowly search outwards, or move from the outside parameters in toward the center. Because the distance between searchers becomes greater as they move out further from the center of the wheel, this type of search method is not as useful for large areas, and is used in special situations where warranted.

4. **Spiral method.** The spiral method of crime scene searches requires investigators to view the scene as a circle, similar to the wheel method. However, in this case they either work inward in a spiraling direction toward the center, or outward spiraling toward the outside. This is also limited to certain circumstances since one danger in this method is that investigators may step on or otherwise destroy evidence as they walk to the center or outer area to begin the search.

5. **Zone method.** The zone method divides the crime scene area up into logical zones that can be individually searched by team members. Indoor crime scenes are good places to employ the zone method, since different people can search different rooms. Once a room is searched by one person, it can then be searched again by a different investigator in order to maximize the chances of getting every piece of evidence.

Forensic Science Today

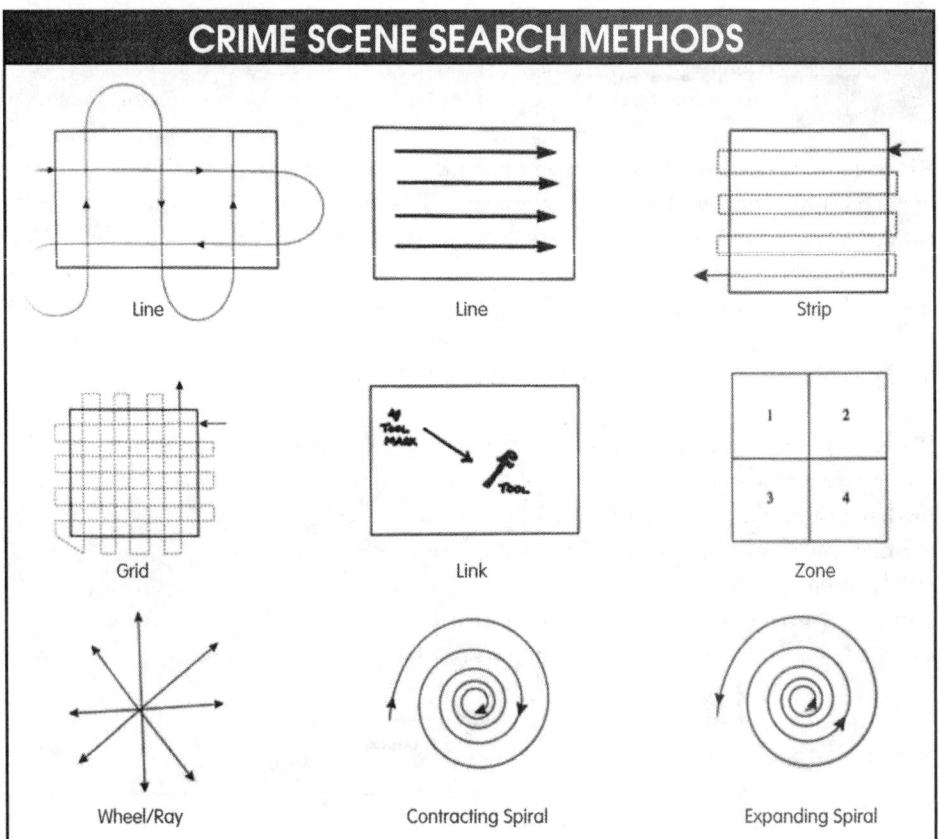

▶ Figure 1. Diagrams for different methods of searching crime scenes.

6. **Link method.** The link method correlates to the linkage theory discussed earlier in this chapter. This method of searching does not follow a geometric pattern like the first five patterns; rather, it is a systematic logical way of looking at each piece of physical evidence and determining where to look from there. For instance, a tool mark on a door would lead the investigator to search for the tool that made it. Once the tool is found, it would be tested for fingerprints so that the identity of the person using the tool may be discovered. This method can also be used in laboratory examination of large items of evidence. Additionally, the link method can be done in conjunction with any of the five geometric search patterns.

DOCUMENTING THE CRIME SCENE

Next, crime scene investigators need to document the crime scene. This serves three purposes: (1) to record and document the *location* of physical evidence; (2) to record and memorialize the overall *condition* of the crime scene; and (3) to allow investigators, forensic scientists, and attorneys to go back in time to the crime scene many months (sometimes even years) down the road during the courtroom trial and appeals process.

Crime scene investigators use several different methods to document the crime scene, in order to ensure that as much information has been collected as possible. Below are some of the methods commonly used to document crime scenes.

Taking meticulous written notes. The crime scene investigator should write down the date and time of the investigation, when he or she arrived at the crime scene, who was already present at the crime scene upon arrival, a description of the crime scene (including weather, smells, and location), a description of the victim (clothing, position, and

Chapter 2
Crime Scene Investigation Part I: The Basics

wounds), list of individuals on the CSI Team and what each member is doing, and other observations (such as doors ajar, windows broken, stains on the carpet, and so forth). Essentially this is a record of all that happened at the scene while the investigator was there.

Video taping the crime scene. A good video camera can work wonders in documenting the crime scene, and it allows investigators to review the video evidence with advanced technology to see what investigators may have missed at first glance. The video tape should be done without any verbal commentary (except certain factual information such as date, time, address, and direction of the camera) by the videographer, and it should include both general and specific areas. The crime scene video should also include a view of the crime scene from the victim's point of view.

▲ Dr. Lee reviewing case evidence for Anchorage Police while on layover at Anchorage International Airport.

Photographing the crime scene. Still photography (taking pictures) is also very important in providing a visual record of the crime scene and evidence, and is done immediately after videotaping the scene. It is always better to err on the side of taking too many photographs rather than not enough, since you never know what may become vitally important down the road at trial (such as a dish of melting ice-cream at the O.J. Simpson crime scene, which would have helped reconstruct the correct time of the murder). Minimum photographs should include:

- Photos of each piece of evidence, in a close-up showing only that piece of evidence,
- Photograph of the evidence taken with other items from the scene to show the perspective of the location of the evidence,
- An overview photo of the entire scene, and
- Photos of possible entries and exits and of rooms adjoining the crime scene area.

TECHNIQUES FOR CRIME SCENE DOCUMENTATION

- **Notes**
- **Videotape**
- **Photograph**
- **Diagram**
- **Audio Recording**

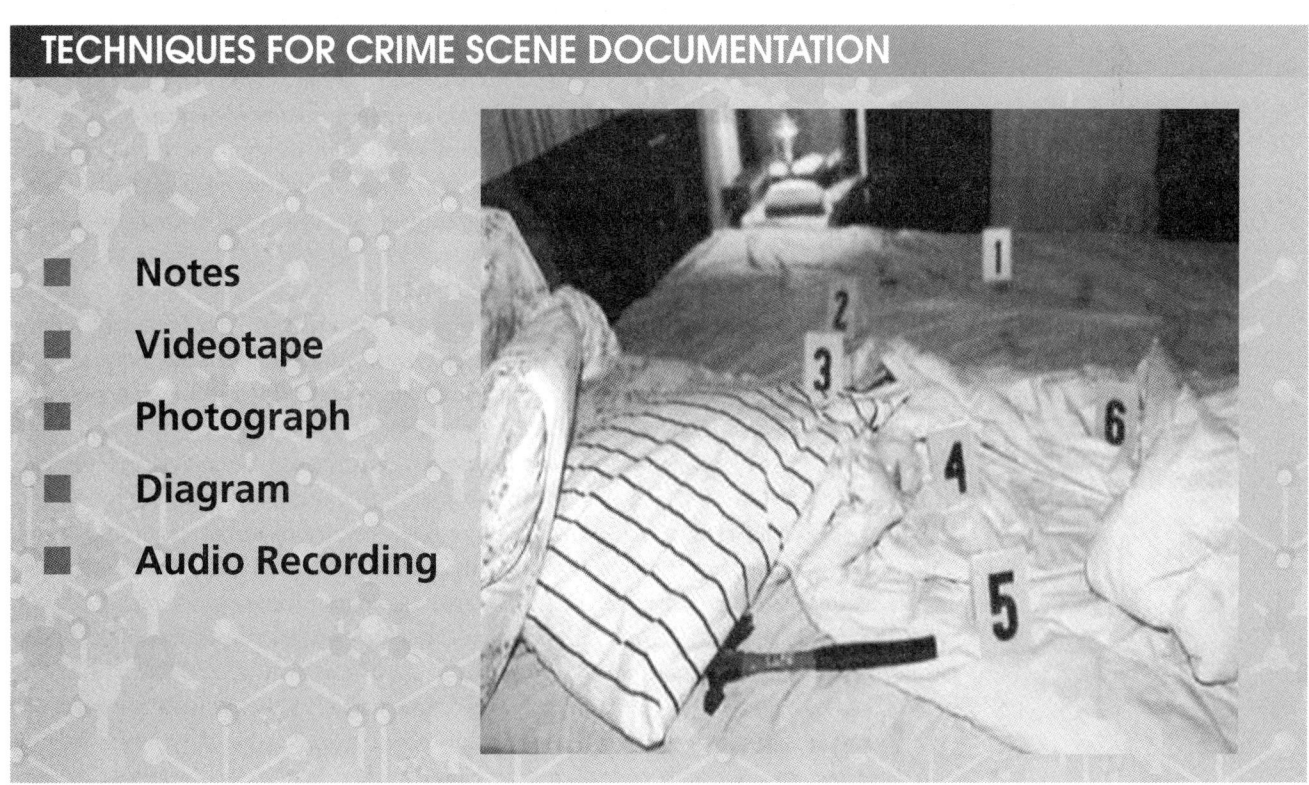

Making a diagram of the crime scene. The next step in documenting a crime scene is to make a diagram or sketch of the crime scene, showing the spatial relationship of objects that photography sometimes fails to capture. For instance, it is possible to take photographs of individual rooms in a house, but without a floor plan-like diagram or sketch, it is impossible to see how each room relates to the other rooms in terms of distance, size, and other considerations. Spatial distortions are important to recognize since they can get in the way of reconstructing the trajectory (flight path) of a bullet or where a drop of blood may have come from. After the search for physical evidence has concluded, the diagram of the scene should include a minimum of the following information:

- All pieces of evidence, drawn in or shown with symbols,
- All relevant items in the crime scene (table, chairs, sofa)
- A north directional marker, and
- Two measurements from fixed points to each piece of evidence.

IN SUMMARY: CRIME SCENE BASICS

Take a few moments now to review what crime scene investigators do at the scene of the crime. This will be of use to you as we take a more in-depth look at physical evidence in the next chapter.

We have also learned about many of the different types of physical evidence; and we learned that almost anything can be considered evidence of a crime. Physical evidence can be a solid, liquid, or a gas; it can occur in the form of patterns or in the form of an actual physical object. In other words, as long as something exists in the physical world and is relevant to a crime, anything can be useful to prove the facts in a court of law.

But if almost anything can be considered evidence, does it follow that every single item in a room in which a crime occurred should be collected and handed over to the nearest forensic laboratory for testing? Absolutely not! If this were to happen, the laboratory would be overwhelmed and overburdened with the processing of many items that would end up having no probative (proving something) value. On the other hand, if crime scene investigators err on the side of collecting too little evidence, then valuable clues to the crime could be lost forever.

Therefore, it is very important that crime scene investigators develop a system to recognize what is and what is not relevant physical evidence. It is also important for investigative team members to understand the goals of a crime scene search, and to regularly practice and update their systematic searching skills.

After careful documentation of the crime scene, crime scene investigators will begin their detailed and methodical search for physical evidence of the crime. Physical evidence can tell us many things about a crime: it shows us how a crime was committed, it demonstrates the personal "signature" or "MO" (mode of operation) of a particular criminal, it tells us how people, objects, and places are linked to one another. It helps prove whether a witness or suspect is telling the truth, it identifies suspects, and it helps to identify unknown substances found at a crime scene. Physical evidence also provides

Chapter 2
Crime Scene Investigation Part I: The Basics

general investigative leads, and can help investigators reconstruct what actually happened during the crime. In Chapter 3 (Crime Scene Investigation Part II: The Physical Evidence) this book focuses specifically on recognition, collection and testing of physical evidence.

Experienced crime scene investigators know to keep an open mind and always "expect the unexpected." Physical evidence can be used to link the suspect and victim to one another and to the crime scenes. It is just as important to look for what is missing from a crime scene (such as the murder weapon), as it is to look for unusual things that are present (like a bloody footprint).

Crime scene investigators use several different patterns when searching for evidence. These include the line, grid, spiral, ray, zone, and link methods. How do crime scene investigators know which type of search method to use? This is where the CSI Team's instinct and experience come into play, based on the type of crime scene, whether the crime scene is indoors or outdoors, available personnel, and other factors. Remember, just about anything can be a clue to a crime; and it is important to "leave no stone unturned."

REAL CASE SCENARIOS

- While searching a crime scene on a remote river the suspect was nowhere to be found. Later, however, the suspect was seen coming straight down the same river toward the investigators in a boat loaded with guns and weapons. This shows that investigators can never be too careful; and personal safety is always a top concern.

- Police and investigators were called to respond to a domestic violence investigation. While searching the house for the husband, they found that he had committed suicide with a shotgun. Blood and tissue were all over the bed and throughout the bedroom. The husband was later confirmed to be an AIDS patient. Investigators must always take precautions when handling biological evidence for this reason.

Forensic Science Today

SAFETY IS PARAMOUNT

Crime scene investigators must always remember that their physical safety and that of others is of primary importance. Especially in homicides, bombings, arson, and other crimes involving harm to the body, investigators should take all reasonable care to avoid contact with biohazardous material, such as blood and body fluids that could contain the AIDS virus or other communicable diseases. The following are some basic safety guidelines that all crime scene investigators should follow when searching a crime scene for physical evidence:

- ✔ Wear latex gloves and coverall gowns when conducting crime scene searches.

- ✔ Wear surgical masks if airborne particles such as blood droplets, chemicals, drugs, or other harmful substances may be present.

- ✔ Use double latex gloves, protective eye wear, and surgical masks when collecting or handling liquid blood, body fluids, dried blood particles, blood contaminated evidence, or bodies of the deceased.

- ✔ Wear latex gloves, a surgical gown, eye-coverings, and a surgical mask when attending an autopsy or dealing with decomposed materials at the crime scene.

- ✔ When processing the crime scene, always be on the alert for sharp objects and broken objects or surfaces.

- ✔ When conducting a crime scene search, do not put your hands in areas where you are unable to see.

- ✔ Under no circumstances should anyone at the crime scene be allowed to smoke, drink, or eat.

- ✔ When liquid blood and body fluids are collected in bottles or vials, these containers must be prominently labeled "Blood Precautions."

- ✔ Blood and body fluid-stained clothing and objects should be dried and packaged in double bags and labeled properly. If evidence is collected from a possibly infected person or scene, label the package "Caution—Potential AIDS (or Hepatitis) Case."

- ✔ When explosives, bombs, toxic chemicals, or biological agents are involved, contact the bomb squad or forensic specialists for advice.

- ✔ If possible, use only disposable items at a crime scene where infectious blood or harmful chemicals are present. Decontaminate all non-disposable items after each use.

- ✔ Destroy any reports, labels, or evidence tags that may have been splashed with blood, chemicals, or biological agents, first copying the information onto clean forms.

- ✔ After completing the search of a crime scene, investigators should clean their hands with a diluted household bleach solution and with a soap and water solution. Any contaminated clothing and footwear should be properly disposed of.

Chapter 2
Crime Scene Investigation Part I: The Basics

TECHNICAL DEFINITIONS

Linkage Theory: A theory of crime scene investigation based on the idea that if investigators can establish positive links between the physical evidence and/or victim and/or crime scene and/or suspect, then the case can be resolved.

Transfer Theory: Pattern evidence and trace evidence can be transferred between the victim and suspect, with other objects, and the crime scene.

Locard's Theory of Exchange: The principle that transfer evidence is created whenever two surfaces come into contact with one another.

Primary Transfer: The transfer of evidence directly from one surface to another.

Secondary Transfer: The transfer of evidence from one surface to another through an intermediate object or person.

Chain of Custody: The correct procedure for ensuring that evidence is properly tracked, labeled, and identified at all times so that it may be admissible as evidence in a court of law.

Primary Crime Scene: The original place where the crime occurred.

Secondary Crime Scene: An additional crime scene other than the primary crime scene where evidence may also be found.

Crime Scene Survey: The initial "walk-through" conducted by investigators upon arriving at a crime scene.

Chapter 3

Crime Scene Investigation Part II: The Physical Evidence

☞ CHAPTER OBJECTIVES

- ✔ Understand the nature and value of physical evidence.
- ✔ Identify and explain the different roles that physical evidence plays in criminal investigation.
- ✔ Identify and explain the different types of physical evidence.
- ✔ Understand the different ways of classifying physical evidence.
- ✔ Understand how to properly collect evidence.
- ✔ Understand the importance of keeping evidence in the proper chain of custody.
- ✔ Identify and explain the different stages in the analysis of physical evidence.
- ✔ Define the following vocabulary words and phrases: *crime scene investigator, known standard, physical evidence, transient evidence, pattern evidence, conditional evidence, transfer/trace evidence, classification of evidence, recognition, identification, comparison, individualization,* and *reconstruction*.

WHAT IS PHYSICAL EVIDENCE?

Virtually any material may be considered physical evidence. It may be something as small as a particle of pollen, or as large as a train. It can be solid, liquid, or gas. It can occur in the form of patterns or in the form of an actual physical object. Anything that exists in the physical world and is relevant to a crime can be useful to prove the facts in a court of law.

The value of physical evidence has been demonstrated in all aspects of criminal investigation; and law enforcement officials have become increasingly dependent on forensic laboratory results for evidence not obtainable by other means. As science and technology continue to advance, the importance and value of physical evidence in criminal investigation will also continue to grow.

▶ Physical evidence submitted by police to a forensic laboratory for examination.

Forensic Science Today

KEY TERMS

- Chain of Custody
- Physical Evidence
- Known Standard
- Transient Evidence
- Pattern Evidence
- Conditional Evidence
- Transfer/Trace Evidence
- Classification
- Recognition
- Identification
- Comparison
- Individualization
- Reconstruction

However, it is critical to remember that evidence is also subject to the human fallibility of the police officers, technicians, and others that handle it. For instance, while investigating the murder of O.J. Simpson's former wife, Nicole Brown, police officers tracked their own footprints in the victim's blood and turned the victim body-up when they should have left her face-down. In the U.S. and Australia, numerous convicted prisoners are being set free because their DNA evidence was mishandled in forensic testing laboratories. Whenever evidence is lost, mislabeled, or damaged through the mistakes of people who handled it, the value of such evidence in determining the facts of what really happened is diminished.

Ironically, the decisions about the usage of physical evidence in criminal investigations are not usually made by forensic scientists, but—in crime scene searches and initial investigation stages—by police officers, criminal investigators, or evidence technicians. And during the trial process, the use of physical evidence is usually determined by prosecution and defense attorneys, and its ultimate admissibility is determined by the judge. There is no guarantee that either of these groups will sufficiently understand the potential of physical evidence and make proper decisions. As more police officers and attorneys acquire updated information and receive special training in forensic sciences, this situation will improve, enabling better use to be made of physical evidence.

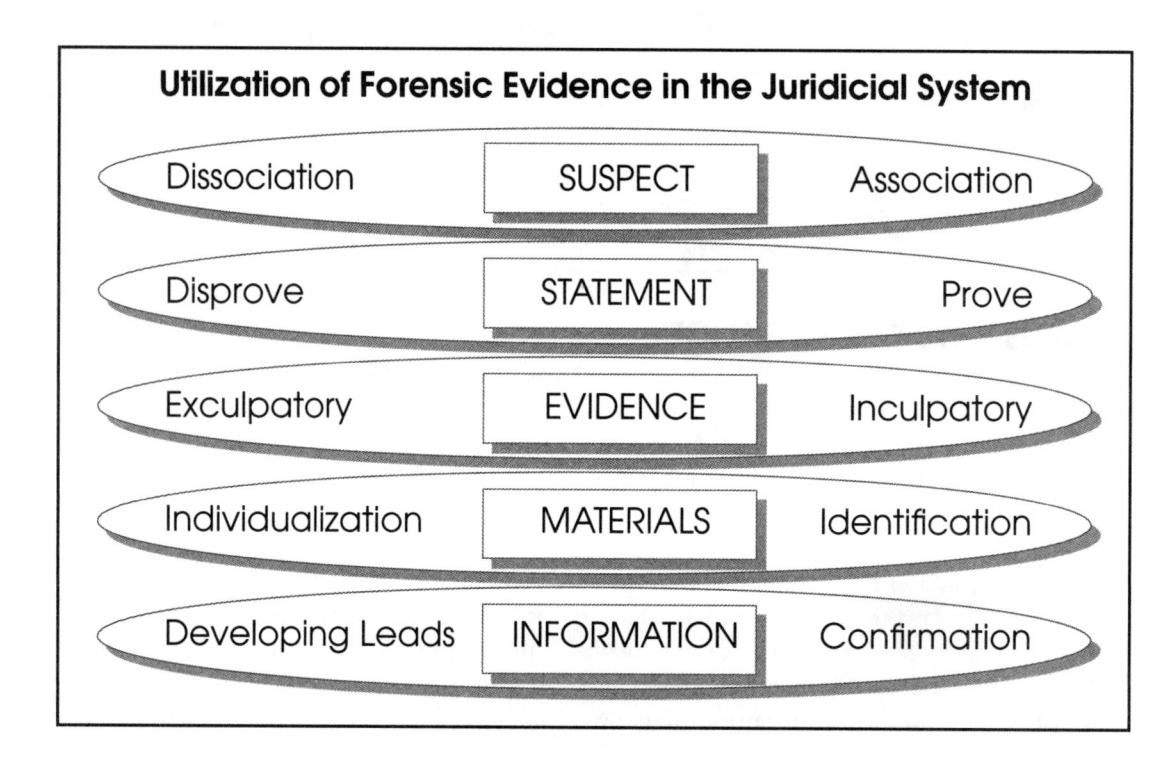

Utilization of Forensic Evidence in the Juridical System

Dissociation	SUSPECT	Association
Disprove	STATEMENT	Prove
Exculpatory	EVIDENCE	Inculpatory
Individualization	MATERIALS	Identification
Developing Leads	INFORMATION	Confirmation

Chapter 3
Crime Scene Investigation Part II: The Physical Evidence

WHAT ROLE DOES PHYSICAL EVIDENCE PLAY IN CRIMINAL INVESTIGATION AND THE WORLD OF JUSTICE?

The general role of physical evidence is to provide information about the facts of what actually happened during the commission of a crime. Physical evidence helps criminal investigators to solve crimes; and it helps courts come to the best decision in criminal cases. In particular, physical evidence plays eight key roles in the world of criminal investigation and justice:

1. **Providing information on the "Corpus Delicti" (Latin term for "body of the crime").** You will find that in the world of law, Latin terms are used to signify important legal concepts or phrases. This dates back to a time when, centuries ago, the practice of law was in its infancy and relied upon Latin as the language of the educated. Thus, corpus delicti refers to those essential facts that show that a crime has taken place. For example, a house that looks like it has been ransacked could simply be very messy. The corpus delicti—facts showing a burglary has in fact taken place—would include a broken window, toolmarks on the window from a forcible entry, and the fact that valuable items are now missing. Similarly, in an assault case, the victim's blood, a weapon, or torn clothing could be important pieces of physical evidence.

2. **Providing information on the "Modus Operandi" ("mode of operation").** Modus operandi is another Latin term, which refers to the characteristic way in which a particular criminal will commit a crime. This term is often shortened to "MO," and can be thought of as a criminal's "signature" left behind at the scene of the crime. Establishing a perpetrator's MO is a vital link in solving crimes; and physical evidence can help investigators determine this critical piece of information. For instance, in the above burglary case, perhaps the burglar always enters through the bathroom window, and only takes female underwear from the bedroom, leaving behind things of value in the rest of the house. Perhaps other houses in the same neighborhood have been broken into in the same way. This pattern of physical evidence establishes the burglar's MO. Cases like this that have been treated separately can sometimes be connected by careful documentation of similar MO.

3. **Linking a suspect with a victim.** Physical evidence found on a suspect and/or a victim can show that there is a link between these two individuals. This is one of the most common and important linkages that physical evidence can help to establish, and is particularly important in violent crimes. For instance, a victim's blood, hairs, particles of skin, fibers from clothing, cosmetics, and other items from the victim can be transferred to the suspect or his clothing. Likewise, evidence may be transferred from the perpetrator to the victim, such as the suspect's DNA in the form of skin found under the victim's fingernails from scratching the perpetrator in self-defense, or bullets from a suspect's gun that may have lodged in a victim's body. While physical evidence can prove a *link* between the two individuals, it does not necessarily *prove* that the suspect committed the crime. It is up to the investigators to use this information, together with all other available evidence, to reconstruct the crime; and it is then up to the prosecution to prove to a jury beyond a reasonable doubt that the suspect committed the crime in a particular manner.

4. **Linking a person with a crime scene.** Similarly, physical evidence can show that a person is linked with a crime scene. This is another one of the most important links that needs to be made in solving a crime, and can be shown in different ways. For example, a perpetrator may leave evidence behind at the crime scene in the form of bullets from his gun, tire marks, footprints, shoeprints, or fingerprints; or substances like blood, hairs, and fibers. Or, material from the crime scene can be transferred to the perpetrator's body, clothing, or belongings, which can also link him to the crime. This type of evidence includes soil, carpet fibers, pet hairs, or even grasshopper legs in the cuff of his pants! (This actually happened in a real case: a suspect was convicted of a crime based on the presence in the suspect's leg cuff of a species of grasshopper that only lived in the area where the crime was committed.)

5. **Disproving or supporting an alibi or the testimony of a witness.** Physical evidence can often indicate conclusively whether a person's version of events is credible or whether that person might be lying. For instance, if a suspect claims that he was at home in bed at the time of a bar-room stabbing, but a time-stamped security camera shows him in the bar at the time of the crime, then the suspect's alibi has been disproved. Likewise, if a suspect in a hit-and-run case says he hit an animal with his car but the blood on the fender turns out to be of human origin, then his statement may be proven false.

6. **Identifying a suspect.** People can leave behind individual markers that can only belong to them, such as fingerprints or DNA, which can identify them as suspects in a crime. Fingerprints are the traditional form of identifying suspects. But DNA, as we'll learn later in this book, can be found not only in blood and semen, but also in chewing gum, on cigarette butts, on the back of licked stamps and envelopes, and in other places. Investigators are increasingly using DNA evidence to discover the identity of a suspect by matching DNA found at a crime scene with either a person in custody or with DNA identities contained in national DNA databases. At present, fingerprint matching is much more common since fingerprint databases contain a larger database of people who have their prints on record for various reasons such as applying for a job in the public sector. DNA databases, on the other hand, are currently restricted in many states to convicted felons; however, there is a movement among the states to include in the DNA database all persons who have been *arrested* for a crime regardless of whether they were also convicted. This is an emerging field and there are sure to be many new developments as courts grapple with the issues of the rights of people not to have their private DNA information available to the public, versus the need of the criminal justice system to protect society from violent repeat offenders.

7. **Providing investigative leads.** Physical evidence, at its most fundamental level, provides the critical starting point for all criminal investigation. It can be helpful in directing an investigation along a productive path. Physical evidence can point in any number of directions of investigation; and each route the investigator explores can reveal more evidence that further reveals the facts of what happened. For example, not only can a bloody fingerprint be matched with a suspect in a fingerprint database; but it can also help link an unidentified body with the crime through a DNA match of the blood contained in the print.

Accreditation of Crime Labs
In order to maintain the highest standards of practice, some states now require crime laboratory accreditation by the American Society of Crime Laboratory Directors Laboratory Accreditation Board (ASCLD-LAB). There are currently 387 laboratories accredited in the ISO and Legacy programs.
For more information, please see the ASCLD-LAB website listed in the "Resources" section at the end of this book.

Chapter 3
Crime Scene Investigation Part II: The Physical Evidence

8. **Identifying unknown substances.** Finally, examining physical evidence can provide information about the identity of a particular substance, which itself can tell the investigator a great deal about what happened in a crime. For instance, forensic testing of a liquid found at a fire scene can tell investigators whether the liquid was a fire accelerant, something commonly used in arson cases to speed up the course of a fire. Testing a white powdery substance can reveal that cocaine was present in a suspect's car; or testing unknown fibers found on the victim's body can reveal carpet fibers that are similar to those found in the suspect's home. In this way, physical evidence plays yet another important role in solving crime.

▲ A detective and crime scene technician documenting evidence found at a crime scene.

CASE PROFILE

On a dark and rainy night a young male on a bicycle was left dead in the road—the victim of a hit and run. A small portion of the grille belonging to the car that hit the victim was found, with a number printed on it that identified the make and model of the vehicle. A review of Motor Vehicle Department records indicated that there were less than fifty vehicles in the county that met this description. Investigators started checking out all these vehicles and one was found that had damage on the left front fender. The case investigators spoke to the owner's father, who had been told by his son that he had recently hit a deer with the car. Unfortunately, the car had been completely washed and cleaned out. The investigators were given permission to inspect the vehicle; and under the hood they found a human tooth lodged in the frame. The tooth was removed and passed on to a forensic odontologist, who was able to identify it as having come from the victim. The various pieces of evidence in this case—the damaged grille with identifying numbers on it, the damage to the left front fender of the car matching the grille, and the human tooth —were all links in the chain showing that the victim was hit by that particular car.

TYPES OF PHYSICAL EVIDENCE FOUND AT CRIME SCENES

Now that we know what physical evidence can be used for, we must learn how criminal investigators and forensic scientists treat the different types of physical evidence that they might come across in the course of investigation. Not only are there different *types* of physical evidence, there are also different ways to *classify* it. In this section, we will look at the many types of physical evidence. In the next section, we will look at how evidence is classified.

There are five types of physical evidence: transient, pattern, conditional, transfer, and associative. These categories are not set in stone; rather, they provide guidance to investigators in the course of their investigation. As you will see, sometimes evidence can fall into more than one category. These five types of evidence are explained in detail below.

1. **Transient Evidence.** Transient evidence is any type of evidence that is temporary in nature, such as smoke from a fire or the smell of perfume lingering in a room. This form of evidence can be easily changed or lost; therefore it is extremely important that crime scene investigators document this evidence as soon as possible. Most transient evidence can be recorded by notes and verified by other officers at the crime scene. Sometimes transient evidence can be recorded by photography or videotaping, or even collected and preserved to prevent further damage. Examples of transient evidence include:

   - Odor (such as odor from food, smoke, a decomposing body, perfume, cyanide, urine, cigarettes, or cigars).
   - Temperature (in the room, of a car, of a cup of coffee or a bowl of ice-cream left on the kitchen table, body temperature of a cadaver, or temperature of a fire).
   - Imprints and indentations (such as a footprint in moisture, a fingerprint in butter or ice-cream, tooth marks on meltable material, or tire marks on wet beach sand).
   - Markings (such as post-mortem lividity before its fixes, wet blood spatters on movable objects, burn mark of a smoldering cigarette butt, marks on blood-stained clothing found in the water).

2. **Pattern Evidence.** While there are a variety of physical patterns that can be found at crime scenes, most such patterns are in the forms of imprints, indentations, striations, markings, fractures, or deposited material. Some examples of pattern evidence include blood spatter, glass fractures, fire burns, furniture positions, projectile trajectories, track-trails, tire or skid marks, clothing or articles, modus operandi, and powder residue patterns. Pattern evidence has not attracted as much attention as it should, nor has its potential been fully realized. Often, forensic scientists do not take this type of evidence seriously enough and think of it as being "not very scientific." However, pattern evidence at a crime scene is extremely valuable in the reconstruction of events. It can very often be used to prove or disprove a suspect's alibi or a witness's version of what took place. It can associate or disassociate the involvement of persons or objects in particular events, and it can provide investigators with new leads.

3. **Conditional Evidence.** The third type of physical evidence commonly encountered at crime scenes is "conditional" evidence, which is generally produced by an event or action. One way to remember this category is to think of the evidence as relying upon a certain condition to be present before the evidence can occur. For instance, the lighting conditions of a crime scene are considered conditional evidence. Smoke conditions such as the color, direction, and density of the smoke are considered conditional evidence. Note that other smoke conditions fall into the category of transient evidence. Other forms of conditional evidence include fire (color, direction, and temperature of the flame), location (of a bullet casing in relation to a victim's body, of injuries, of clothing), condition of a vehicle (whether the door is locked or opened, radio is on or off, how much mileage is on the odometer), and the condition of a deceased body (condition of rigor mortis, degree of decomposition, body temperature). If conditional evidence is not documented correctly, it may become changed or lost. Conditional evidence is extremely important for the reconstruction of events and verification of an alibi.

Chapter 3
Crime Scene Investigation Part II: The Physical Evidence

◀ Forensic scientists examine physical evidence at a crime laboratory. *(Courtesy of Sharon Lee Plotkin, North Miami Police Crime Scene Unit.)*

4. **Transfer Evidence.** Transfer evidence is the classical type of physical evidence. It is generally produced by physical contact between persons and/or objects, and results in some type of material being transferred from one person/object to another person/object. Transfer evidence is a large category that encompasses a wide range of materials. Fingerprints are transfer evidence because they are transferred from the finger of a person to the surface upon which the print is formed. Hairs, blood, dust, fibers, paint chips, glass, soil, and other particles can be transferred from a suspect to a victim, from a victim to a suspect, from any person to a crime scene, or from a crime scene to any person or object.

5. **Associative Evidence.** This type of evidence is often large and visible in nature, and is often used to associate a person, vehicle, or object with a scene or events. Examples of associative evidence would include a vehicle, television set, camera, wallet, jacket, ID card, credit cards, and so on. Associative evidence can be traced to the owner by a serial number, photographs, or barcode checking.

CLASSIFICATION OF PHYSICAL EVIDENCE

Not only are there five basic *types* of physical evidence, there are also many different methods of *classification*. These are classifications according to different perspectives, or different ways of looking at the evidence and what it means. For example, some evidence only has class characteristics such as a drug or a single fiber. Some of the resulting classifications are more practical than others. However, no single classification is completely satisfactory, because no single classification can take all the different perspectives into account. Again, these classifications are not set in stone and evidence may fall into more than one category, just as it can be considered more than one type (such as smoke being considered transient, as well as gas, arson, or conditional evidence). The seven different ways of classifying evidence are listed below.

1. Classification by Type of Crime—homicide evidence, rape evidence, arson evidence, and so forth.
2. Classification by Type of Material—biological, chemical, physical, pattern.
3. Classification by General Nature of the Evidence—evidence with class characteristics or evidence with individual characteristics.
4. Classification by the Physical State of the Evidence—gas, liquid, and solid evidence.
5. Classification by the Type of Question to be Answered.
6. Classification According to the way the Evidence was Produced.
7. Classification According to the Specific Type of Evidence—hair, fiber, blood, and so on.

COLLECTION OF PHYSICAL EVIDENCE

Each piece of evidence must be collected in a separate, appropriate container while the collector is wearing gloves. The collection container should be chosen to prevent the evidence from falling out, being broken or otherwise damaged. Some examples are as follows:

- Cartridge cases, bullets and glass in a rigid container,
- Hair, fiber and similar trace evidence in a manila envelope or zip-closure container appropriate for its size,
- Arson evidence in a mason jar or new paint can to prevent loss of possible accelerants, and
- Blood samples or evidence with body fluids on them in a paper bag after air-drying.

Gloves should be changed between pieces of evidence to prevent the transfer of substances from one piece of evidence to another. After evidence is placed in an appropriate container, the container must be labeled and sealed. The minimum amount of information to be placed on the label is a brief description of the evidence, the collector's name, the date and time of collection. The container is then sealed with evidence seal tape. The seal is placed across the opening to prevent leakage and opening of the container from this point. On the seal is placed the initials of the collector and the date of collection. Many police agencies also record information on the evidence container about the victim, possible suspects, type of crime, and identification number of the person collecting the evidence.

THE IMPORTANCE OF KEEPING EVIDENCE IN THE PROPER CHAIN OF CUSTODY

As we learned in Chapter 2, first responding officers and criminal investigators at the scene of a crime must take certain steps and precautions in handling evidence so that its value is not lost. Evidence must be documented and recorded in notes, sketches, photographs, and videotape. It must then be very carefully collected and preserved according to strict procedures that vary with the type of evidence collected. Field tests are sometimes conducted on the evidence to determine its nature and how it must then be handled. Finally, evidence is transported to the forensic laboratory for examination, testing, and analysis.

Thus, as physical evidence changes hands during each stage of investigation and analysis, it must be carefully monitored and kept within the **chain of custody**. This is a legal term used in court to denote the strict adherence to rules regarding transfer of collected evidence from one person or agency to another. In other words, chain of custody is the record of everything that happened to the evidence from the moment it is discovered until it arrives in court or is destroyed after the final dispensation of the case.

Keeping the evidence in the chain of custody begins with proper identification and labeling of the evidence in its proper container, including the initials of the person who collected and labeled the evidence. When the evidence is handed over from the crime scene technician to the forensic laboratory, forms must be filled out and signed by both the crime scene investigator and forensic laboratory personnel indicating that they transferred the article of evidence from one person to another at a specific time and location. Modern laboratory databases, or bar code systems, can also be used to keep a record of the chain of custody of evidence.

Why is it so important to keep track of the chain of custody of evidence? One reason is to prevent evidence from being lost. The main reason, however, is because in order to introduce evidence into the record in a court proceeding, the chain of custody must first be established to prove that the evidence presented in court is indeed the exact same evidence as found at the crime scene, and has not been tampered with, altered, misinterpreted, or mishandled in any way. If the chain of custody becomes broken at any point due to improper handling or labeling, the evidence may no longer be admissible in court, which could result in the acquittal of a guilty suspect.

Logic Tree: Death Scenes

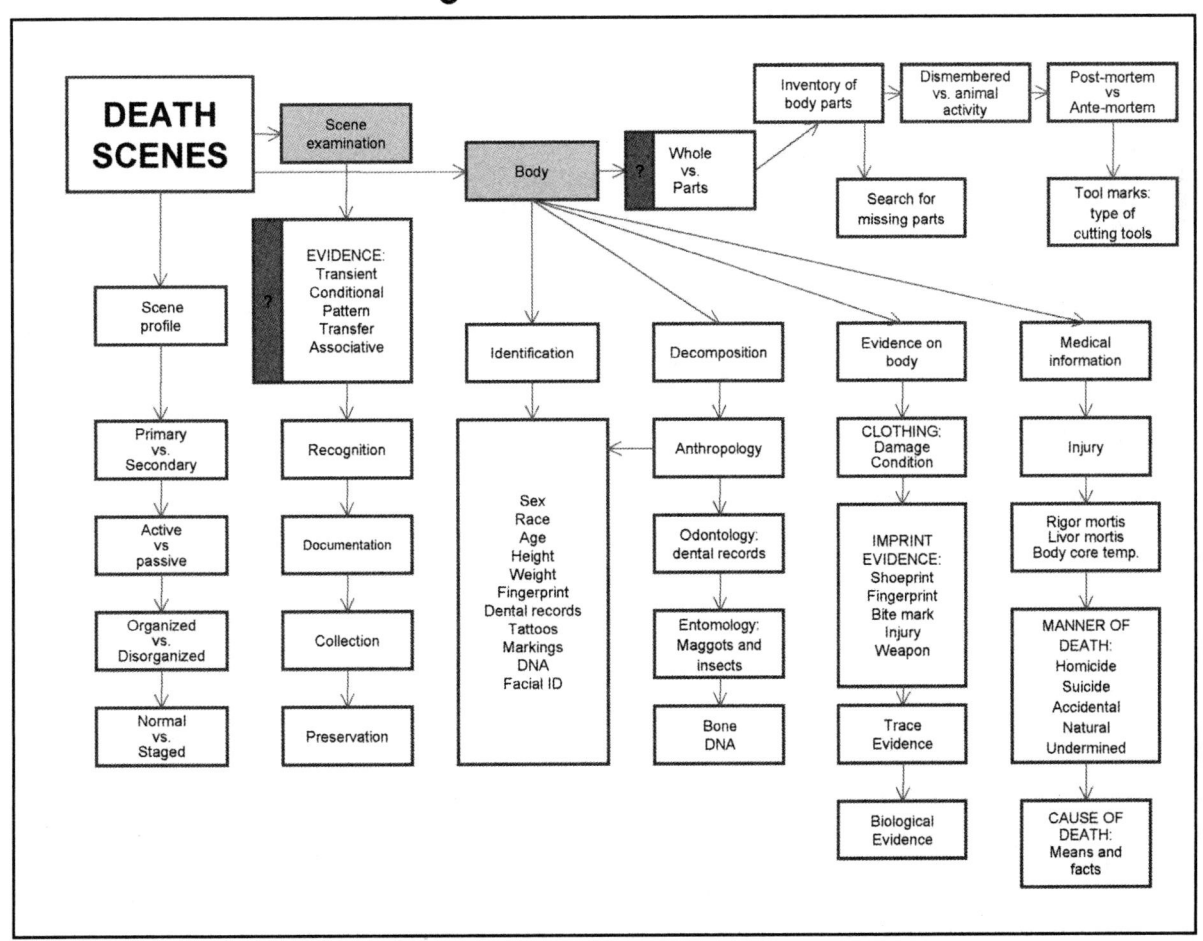

CASE PROFILE
Guilty by Lack of Evidence

Hopkinton, RI: On January 6, 2002, the body of a woman was found stuffed into a garbage bag on the side of Grassy Pond Road in the Hopkinton village of Rockville, Rhode Island. The woman's body was so decomposed that investigators were not sure if they could determine who she was or how long her body had been there. The body was taken to the local morgue. The medical examiner estimated that the woman had been dead three to four months, but noted that the body had been wrapped so tightly in plastic that 60% of her BMI (body mass index) was still intact.

Upon further examination, a tattoo of a dove with the word "Tarzan" underneath it was discovered on the woman's ankle. To help identify the woman, Hopkinton Police Chief John S. Scuncio sent a drawing of the tattoo to several newspapers and TV stations asking the public to help identify the woman. Gloria Rodriguez recognized the tattoo. Her sister-in-law, Carmen Rodriguez, a Hartford woman who had been missing for 3 ½ months, had a tattoo just like it. Soon, a positive identification was made: Carmen Rodriguez, the 32-year old mother of four had been murdered.

Hartford detectives needed to determine when Carmen was last seen alive. Witnesses said she had been seen dancing with several men at Kenny's Bar on Capital Avenue. She was last seen leaving the club around 1:30 A.M. on September 22, 2001. Around the same time Carmen Rodriguez disappeared, another Hartford woman - Latina Florence - was attacked by a man who attempted to kidnap her, but she managed to fight him off. Florence never reported the incident to police until she heard about the murder of Carmen Rodriguez.

Detectives questioned Florence about her attacker. She later identified Edwin "Ned" Snelgrove, Jr. – a door-to-door meat salesman from Berlin, CT as the perpetrator.

Investigators learned that Snelgrove had been convicted of killing his former girlfriend, Karen Osmun on Christmas Day in 1983 in New Brunswick, N.J. He had also been convicted of stabbing and strangling Mary Ellen Renard of Bergen County, NJ in 1987; although Renard had survived. Snelgrove wasn't charged with killing Karen Osmun until Mary Ellen Renard identified him as her attacker.

Detective Michael Sheldon of the Hartford police department went to Snelgrove's home to question him about Carmen Rodriguez. Snelgrove then admitted to dancing with her at Kenny's Bar. Detective Sheldon pushed further by saying that the bartender saw them leaving together. Snelgrove said that although he gave Rodriguez a ride, he soon lost interest and dropped her off at a gas station a few blocks away from the club. Under suspicion, Edwin Snelgrove, Jr. was charged with the murder of Carmen Rodriguez. With Snelgrove sitting in jail, the clock started ticking, so investigators went to work on the case.

The Hartford Chief State Attorney's office made the decision to take over the murder investigation of Carmen Rodriguez. The Connecticut State Police Major Crime Squad was called in to investigate. They needed an expert to link Snelgrove with Rodriguez's murder, so they turned to Dr. Henry Lee, Chief Emeritus of the Connecticut State Forensic Lab, to assist them with their investigation.

Because Dr. Lee was about to testify at the trial of Kennedy nephew Michael Skakel, he knew he had to organize a team of forensic scientists to search for evidence where Carmen Rodriguez's body was found. Dr. Lee assembled a group from the University of New Haven's Henry C. Lee Institute to search the woods off of Grassy Pond Road. The forensic scientists searched along Route 138. Despite an exhaustive search, no evidence was found.

continued on next page...

Chapter 3
Crime Scene Investigation Part II: The Physical Evidence

At the same time, Inspector Jim Rovella of the Chief State's Attorney's Office along with the Connecticut State Police Major Crimes Squad got a warrant to search Snelgrove's Berlin, CT home. Police seized several items, including clothing, two mannequin heads, a wire garrote (strangulation cord), and Snelgrove's tan colored Ford Escort. All items were sent to Dr. Henry Lee for examination.

At the forensic lab, Dr. Lee and staff examined Snelgrove's car for any trace of evidence. Snelgrove had admitted to giving Carmen Rodriguez a ride; could Dr. Lee uncover signs of a violent struggle? Dr. Lee searched the car for hair and fibers to compare with the clothing Rodriguez had been wearing the night she was murdered. After a thorough search, Dr. Lee discovered that the car was beyond clean. Dr. Lee believed the car had been vacuumed numerous times.

Next, Dr. Lee filled the car with superglue fumes to reveal any latent fingerprints. To everyone's astonishment, there were no fingerprints left by Rodriguez. No fingerprints on the passenger side handle, seat belt, or window. This made Dr. Lee highly suspicious – Snelgrove himself had admitted to giving Carmen Rodriguez a ride, so why weren't her fingerprints in the car? Snelgrove's Ford Escort was then sprayed with blood enhancing reagents. It tested negative for the presence of blood.

Then, Dr. Lee examined items seized from Snelgrove's home. The garrote that investigators believed Snelgrove had used to strangle his victims was examined for the presence of epithelial skin cells that could possibly link the garrote to the victims. None of the victim's DNA was found on the garrote. When Dr. Lee examined the mannequin heads seized from Snelgrove's home, he discovered various markings drawn on the face and neck. Dr. Lee determined that the markings were pressure points. Prosecutors believed that Snelgrove used the garrote on the mannequin heads to practice strangling women. Had Snelgrove regained control of his life and changed his murderous tendencies, or had he developed into an efficient killer?

Prosecutor David Zagaja obtained letters that Snelgrove had sent to a high school friend, George Recck, while he was incarcerated in New Jersey. The letters discussed Snelgrove's fantasies of strangling women. In one letter Snelgrove wrote, *"There was something I couldn't tell you or even my parents last winter when I was out on bail. This 'incident' last August was not the first time I lost control of myself with a girl. You'll never believe what I am about to tell you – I had actually got away with murder…I could not stop my hands from squeezing her throat as hard as I could."* In another letter, Snelgrove compared himself to Ted Bundy. *"Bundy was stupid after the act. He kept maps, schedules and pamphlets of the hotels, beaches and ski resorts he visited. He even purchased gas with credit cards (stupid!) This left a documented trail for the police."* Regarding Mary Ellen Renard, Snelgrove wrote *"I botched it up. She didn't die. If she had died, my name wouldn't have even made it on the suspect list."*

The trial moved forward and Prosecutor Zagaja called Dr. Henry Lee forward as a key witness. Dr. Lee testified that even by Snelgrove's own admission Carmen Rodriguez was inside his car when he gave her a ride the night of Sept. 22, 2001. Therefore, there should have been some trace evidence linking her to the interior of Snelgrove's car. Dr. Lee addressed the lack of evidence in open court. The jury was impressed. After deliberating, the jury returned with a verdict of guilty of first-degree murder. Superior Court Justice Carmen Espinosa later sentenced Edwin Snelgrove to life in prison without possibility of parole. A loud applause was audible from the families of Snelgrove's victims. Judge Espinosa said, "The court is convinced that if he gets out on the streets again, he will kill again." The circumstantial evidence may not have been enough to convict Edwin Snelgrove, Jr., but thanks to the testimony and investigative work done by the investigator, he'll spend the rest of his life behind bars.

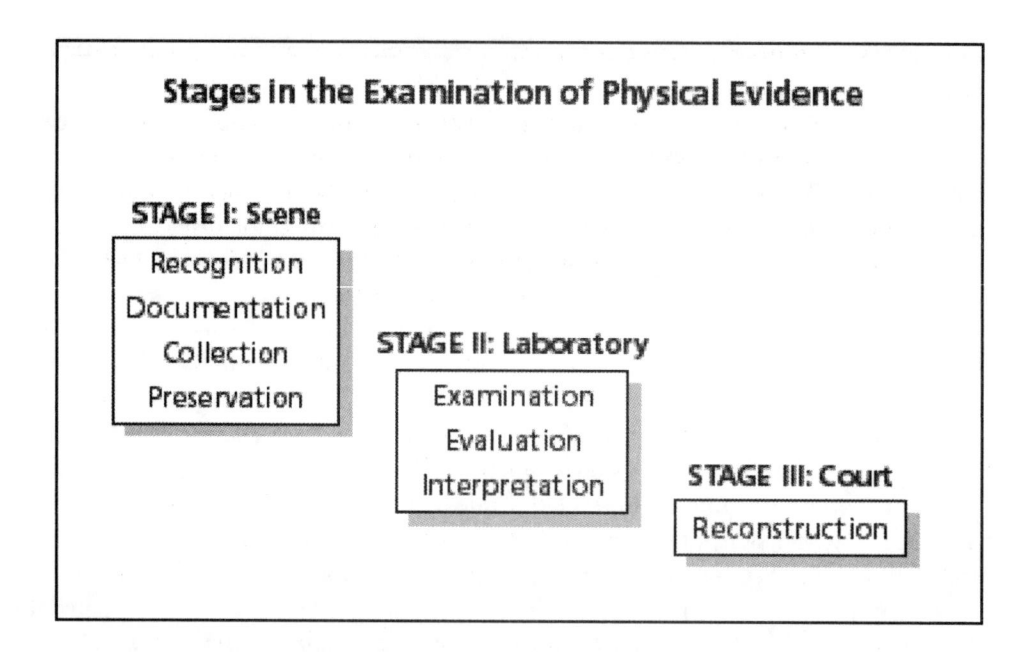

Stages in the Examination of Physical Evidence

STAGE I: Scene
- Recognition
- Documentation
- Collection
- Preservation

STAGE II: Laboratory
- Examination
- Evaluation
- Interpretation

STAGE III: Court
- Reconstruction

CLASS VERSUS INDIVIDUAL EVIDENCE

Physical evidence can also be classified as either belonging to a "class" of evidence, or as being "individual" evidence. Class evidence shares the same characteristics with other members of the same class, such as drugs with similar chemical properties or white cotton fiber that has the same properties as many other white cotton fibers. Individual evidence, on the other hand, consists of objects or items that are unique and one-of-a-kind, such as an individual's DNA or fingerprint. This is explained in further detail below.

STAGES IN THE ANALYSIS OF PHYSICAL EVIDENCE

From the crime scene to the forensic laboratory, physical evidence undergoes several stages of analysis before its full value is realized in solving crime. These stages are explained below.

Recognition. First, crime scene investigators must *recognize* that an item is evidence of a crime. Both investigator and forensic laboratory analyst must be able to separate items with important evidentiary value from unrelated background materials. If evidence is not recognized as evidence, then it will not be collected or analyzed and its value will be lost forever.

Identification. Next, investigators and lab analysts must *identify* the class characteristics of an object by taking measurements and analyzing its physical, chemical, and biological properties. In some cases, simple identification is enough, such as of a weapon or drug. However, identification is usually the beginning of a more extensive forensic analysis. In this step a hair could be determined to be light brown, human hair.

Chapter 3
Crime Scene Investigation Part II: The Physical Evidence

Comparison. If further analysis is required after identification, then the characteristics of the evidence must be compared to those of a *known standard*. A known standard is something from a known source to which evidence from an unknown source (the crime scene) is compared. For example, a strand of hair may be found at a crime scene. Police have arrested a suspect, and they take samples of his hair in order to compare it to the hair found at the scene of the crime. The hair from the arrested suspect is the known standard to which the hair from the crime scene is compared. If the hairs are similar, then the suspect can possibly be linked to the crime scene.

Individualization. Next, certain types of evidence may be *individualized*, that is, shown to be unequivocally derived from a unique source. For instance, laboratory analysis of fiber found on a victim's clothing can show not only that the fiber came from the suspect's carpet, but sometimes also reveal the manufacturer of the carpet itself! Or, a muddy footprint can be individualized to show that it came from a suspect's shoe when the patterns and grooves, including unique marks of wear and tear, match exactly those found at the crime scene.

Reconstruction. Finally, physical evidence can be used to help *reconstruct* a crime or to determine the sequence of events. This is based on the results of crime scene examinations, laboratory analyses, deductive and inductive logic, statistical data, pattern analysis, data mining, and other types of information. When reconstructing a crime, forensic professionals bring together all these aspects to reconstruct the sequence of events. We will learn more about reconstruction in greater detail in Chapter 24.

QUESTIONED EVIDENCE VERSUS KNOWN STANDARDS

Any type of evidence that comes from a questionable or unknown source is considered "Questioned Evidence." Evidence coming from a known source such as hairs taken from a suspect's head or fingerprints taken from a victim at the hospital is considered a "Known Standard." These classifications are reflected in the labeling of evidence by crime scene investigators and other forensic professionals. For example, three pieces of evidence from unknown sources at a crime scene might include a drop of blood, a hair, and a bullet. Some labs individually label these items beginning with the letter "Q" to indicate they come from a questioned source. Thus, they would have the labels Q-1, Q-2, and Q-3 respectively. Likewise, three items of evidence taken from the suspect and victim might include hair from the suspect, an oral DNA swab from the suspect, and a blood sample taken from the victim. They would be labeled beginning with the letter "K" to indicate they come from a known standard. They would be labeled K-1, K-2, and K-3 respectively.

◀ Crime scene investigation in Alaska near the Kuskokwim River.

CASE PROFILE
The Woodchipper Murder Case

In 1989, Richard Craft was convicted of murder and sentenced to 99 years in prison, even though a body had never been found. How can this be? What follows are clues to one of the most bizarre crimes ever seen by the police who investigated the case. Can you guess what happened? (Hint: Ever seen a movie called *Fargo*?)

Richard Craft was a commercial pilot and part-time police officer who lived in Connecticut with his wife, Helle, a flight attendant. Both Richard and Helle traveled worldwide for their jobs. They lived what outwardly appeared to be a happy, quiet life in the suburbs with their three children. However, all was not well beneath the surface. Helle was unhappy with Richard's increasingly abusive behavior toward her and the children. When she discovered that Richard had been carrying on an affair, she began divorce proceedings against him. Neither of them had yet moved out of the family home.

On a snowy night in November, Helle was dropped off at home by a friend after returning from an overseas flight. That was the last time any of Helle's friends saw her alive. Weeks went by and Helle's close circle of friends became extremely concerned about her whereabouts. Richard responded casually to their desperate inquiries, saying that she had gone to Denmark to visit her ailing mother. But they knew Helle would never go this long without communicating. The last straw came when Christmas came and went with no word from Helle, and it was confirmed that she had never gone to Denmark.

Meanwhile, Helle's blue Toyota remained in the airport parking lot where it had been since the night after she had arrived. In the days prior to Helle's disappearance, Richard had purchased a new freezer and a large, new pickup truck; and had arranged for the rental of a woodchipping machine. Richard and Helle had recently purchased a large wooded lot near their home and Richard liked to spend time out there chopping wood.

Eventually, Helle's circle of friends banded together and hired a private investigator to work with Helle's divorce attorney and uncover the truth. The local police chief treated it as a missing person case rather than a murder case, and adopted a "wait and see" approach. The fact that they knew Richard as one of their own did not help speed things along, either. Finally, the private investigator began digging up information. The state attorney's office then requested that the Connecticut state police take over the case investigation.

At last, police and forensic experts were led to the banks of Lake Zoar where, exactly six weeks after Helle was last seen alive, they found enough evidence to get Richard Craft convicted for the murder of his wife, even though no body had ever been found.

Police recovered evidence from the Crafts' house, Richard Craft's vehicle, and in and around Lake Zoar. Some of this evidence is listed below. Can you guess what happened? How do you think Richard killed his wife? How do you think he disposed of her body? What forensic techniques were used to recognize and analyze the evidence?

Evidence from the House
- Brown stains on a clean, folded towel in the laundry room.
- One smear on the side of the mattress of Richard and Helle's bed.
- Several quarter-sized and smaller brown spots on one side of the bed.
- Ashes in the fireplace appearing to be the remains of paper and a fibrous material.
- Two deep tire tracks in the frozen mud outside the house.

Evidence from Richard's Vehicle
This evidence was excluded from the trial since police search of the vehicle was not covered by a warrant.

continued on next page...

- Strands of blonde hair.
- Fragments of bone.
- Fragments of flesh.
- Bits of light blue cloth.
- Woodchips.

Evidence from Lake Zoar
- Bits of light blue cloth.
- Envelopes addressed to Helle Crafts.
- 2,660 strands of human, Caucasian head hairs, blond in color.
- One metal dental crown with a portion of jawbone attached to it.
- A portion of a human tooth.
- Wood chips.
- One piece of calvarium skull bone, one-inch long, cut in a traumatic manner.
- Sixty small bone fragments.
- 3 ounces of human tissue.
- Blue-green fibers.
- Chain saw.
- One human toe joint.
- One portion of a human finger with fingernail polished in color matching that found at Helle's home.
- One portion of a toe nail.

TECHNICAL DEFINITIONS

Crime Scene Investigator: Someone trained in the field of forensic crime scene investigation who goes to the scene of a crime and documents, collects, and preserves evidence for forensic testing in a laboratory.

Known Standard: Something from a known source to which evidence from an unknown source is compared.

Physical Evidence: Anything that may be recognized, documented, collected, and analyzed in order to determine the facts of what happened during a crime.

Transient Evidence: Any type of physical evidence that is temporary in nature.

Pattern Evidence: Evidence resulting when an object or a person come into contact with a surface, leaving behind a pattern, or resulting from a physical change in shape or pattern.

Conditional Evidence: Evidence that demonstrates conditions of a crime scene, body, or physical stages of evidence.

Transfer/Trace Evidence: Evidence that is produced by physical contact between two objects or two surfaces, when material is transferred from one object to another.

Classification of Evidence: Different ways of classifying and placing the evidence in an organized fashion.

Recognition: The process of recognizing the potential value of items of evidence related to a crime.

Identification: The process of identifying the class characteristics of an item of evidence by taking measurements and analyzing its physical, chemical, and biological properties.

Comparison: The process of comparing an item of evidence with known standards or alibi standards to determine where the evidence came from.

Individualization: The process of showing that an item of evidence is unequivocally derived from a unique source.

Reconstruction: The process of putting together the facts of a crime and sequence of events that occurred at a crime, based on a thorough analysis of all physical evidence, witness statements, and other facts.

Chapter 4

Fingerprint Evidence

☞ CHAPTER OBJECTIVES

- ✔ Identify different case scenarios in which fingerprint evidence is helpful for solving a crime.
- ✔ Describe the different types of ridge characteristics (minutiae) found in fingerprints.
- ✔ Identify the eight different fingerprint patterns created by general ridge patterns in the skin.
- ✔ Identify the three different types of fingerprints found at crime scenes.
- ✔ Explain how to find and examine latent fingerprints at the crime scene.
- ✔ Explain how to examine latent fingerprints at the forensic laboratory.
- ✔ Define the following vocabulary words and phrases: *AFIS, latent fingerprints, patent fingerprints, plastic fingerprints,* and *ridge impressions.*

Since the early 1900's, police detectives have used fingerprints to identify suspects and solve crimes. Even today, fingerprints remain one of the most valuable types of physical evidence in criminal investigations. This is because each person has a unique set of fingerprints unlike any other person on the planet. Thus, when a person's prints are found at a crime scene, you can be pretty sure that he or she was there at some point in time. Fingerprints are used not only to link a suspect to a scene, but also to determine whether a suspect has been arrested on a prior occasion. Fingerprints are also useful in identifying unknown bodies, victims of mass disasters, and amnesia victims who cannot remember their own identity.

Fingerprints may be visible to the naked eye (patent), or they may be invisible (latent). No two people have identical fingerprints, not even identical twins! Fingerprints are formed while the human fetus is still inside its mother's womb, at around 100 to 120 days of development and are not genetically predetermined. These basic print patterns remain the same for each individual throughout the lifetime. Moreover, fingerprints can be stored on a database accessible throughout the world. This database is called AFIS—which stands for "Automated Fingerprint Identification System." This is why fingerprints make such good evidence of identity and have been readily accepted as such by courts around the world for over 100 years.

Forensic Science Today

KEY TERMS

Automated Fingerprint Identification System (AFIS)
Latent Fingerprint
Patent Fingerprint
Plastic Fingerprint
Ridge Characteristics (Minutiae)

AFIS: AUTOMATED FINGERPRINT IDENTIFICATION SYSTEMS

Computer-based identification systems using the storage and retrieval of fingerprint files are used in many jurisdictions around the world in order to compare fingerprints. When prints are found at a crime scene, they can then be compared to prints kept on file in AFIS to see if they match up with a potential suspect, victim, or witness. AFIS systems can also digitally enhance the computerized images of fingerprints through background removal, fill-in, elimination of dust particles, or elimination of other interferences in the print.

There are three different types of AFIS concept designs: (1) Index systems that are based on a variety of information including prints, mode of operation, and ridge characteristics; (2) ridge-angle systems that are based on the angle of ridge flow on certain points of the fingerprint; and (3) encoded minutiae systems that map out the location and angle of bifurcations and ridge endings in fingerprints, which can then be compared against known samples.

Like many areas of forensic science, fingerprinting is undergoing exciting new developments with the advent of new technology. A recent advance is the Live-Scan System, which allows police officers to compare a suspect's fingerprints taken at the point of arrest with millions of fingerprints in the AFIS files. This can now be done in a matter of minutes and dramatically reduces the amount of time spent manually searching for fingerprint matches.

CAN YOU CHANGE YOUR FINGERPRINTS?

Many people have had their fingerprints altered in order to continue a life of "undetected" crime. From having acid placed on their fingertips to surgically removing the skin on fingers deeply enough to prevent re-growth of the original ridges, people have tried just about everything.

The most famous case of fingerprint alteration is John Dillinger. He paid a doctor to change his prints. The doctor placed acid on the tips of his fingers, causing them to be scarred deeply enough (below the top two layers of skin) that the ridges did not regrow. His original rolled prints were still used to identify him at his death since he also had some surgical changes to his face.

Roscoe Pitts and Donald Roquierre are just two of those who chose the surgical route to changing their prints. Pitts had the inner surface of his fingers removed in order to graft abdominal skin onto them so they would have no ridges. Despite this extensive surgery and recovery, he was caught. He still left palm prints at the scene.

Roquierre tried a slightly more creative technique. His surgeon cut squares of skin from his fingers, turned them upside down and sewed them back into his fingertips. His prints no longer were identical to the prints law enforcement had on file for him. However, all that was necessary to find the same points on the lifted prints was to turn the original tenprint card upside down.

Still others have had squares of skin removed from their feet and used to replace the skin on fingertips. These examples all have one thing in common. They were all caught. The question here is why anyone would have such a need to commit crimes that they would go through all the pain and risk of surgery and recovery? It has to be easier to get a job!

Chapter 4
Fingerprint Evidence

DIFFERENT FINGERPRINT PATTERNS
There are eight different types of fingerprints created by **general ridge patterns** in the skin:

1. Radial loop
2. Ulnar loop
3. Plain arch
4. Tented arch
5. Plain whorl
6. Central pocket loop whorl
7. Double loop whorl
8. Accidental whorl

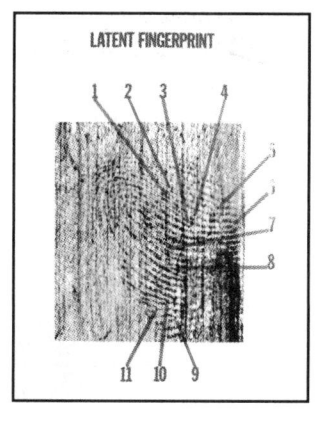

▲ A latent fingerprint was developed from a crime scene. Eleven unique ridge minutiae were found.

Additionally, there are eight categories of **minutiae** patterns, also called ridge characteristics:

1. *Ridge ending*—a ridge that abruptly ends.
2. *Bifurcation*—a single ridge dividing into two ridges.
3. *Lake or enclosure*—a single ridge that bifurcates and then comes back together to continue as a single ridge.
4. *Short ridge*—a ridge that ends after a short distance.
5. *Dot or island*—An independent ridge that has the same approximate length and width.
6. *Spur*—A long ridge that bifurcates with a shorter ridge branching off it.
7. *Bridge*—A short ridge that runs between two longer parallel ridges.
8. *Ridge crossing*—the place where two ridges cross over one another.

When comparing lifted prints to scanned or rolled prints, fingerprint examiners use these patterns. The place a pattern is located on a finger is called a point.

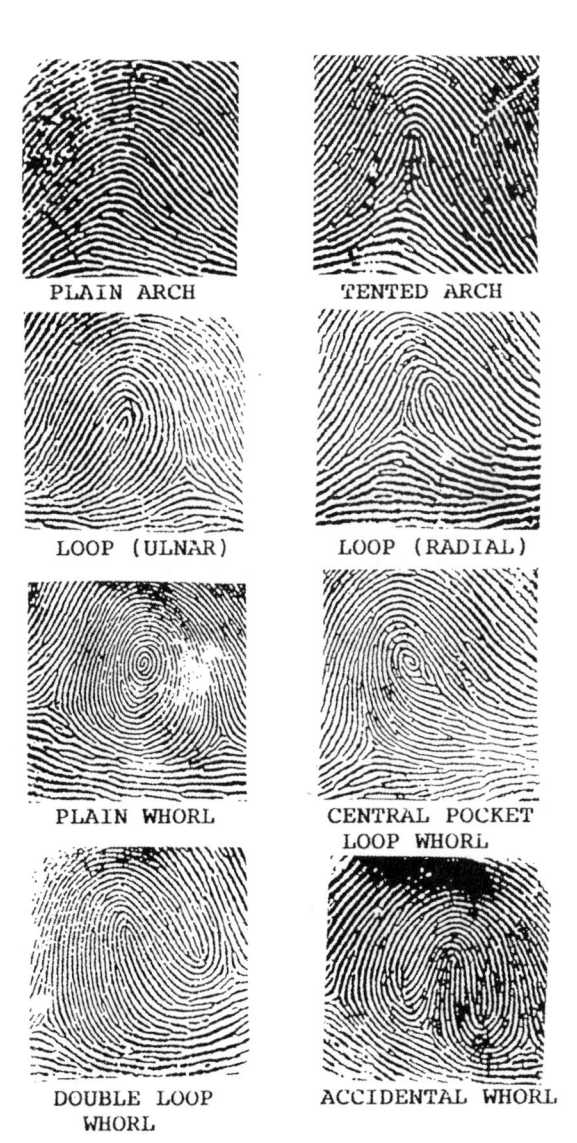

TYPES OF FINGERPRINTS FOUND AT CRIME SCENES
Visible prints are ridge impressions created by the transfer of a colored substance such as paint, ink, blood, or other substances from a finger or hand to a smooth surface. Visible prints are also referred to as "patent" prints, not to be confused with latent prints (which are invisible—see below).

Plastic (three-dimensional) prints are three-dimensional ridge impressions made when the finger or hand is pressed into a soft material like wax, clay, putty, milk chocolate bars, or other soft surfaces.

▲ The basic types of fingerprint patterns.

47

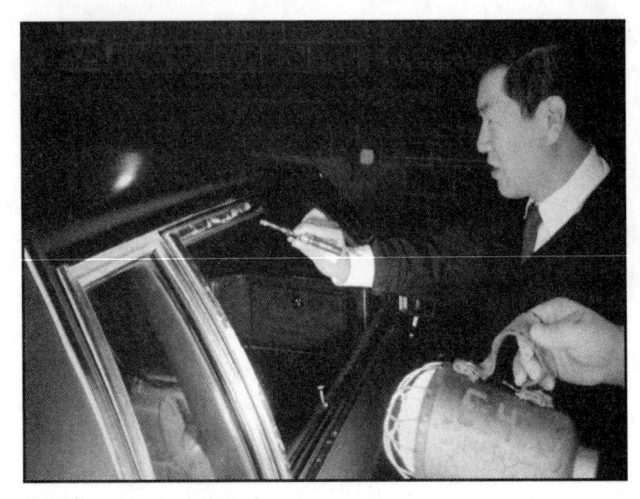

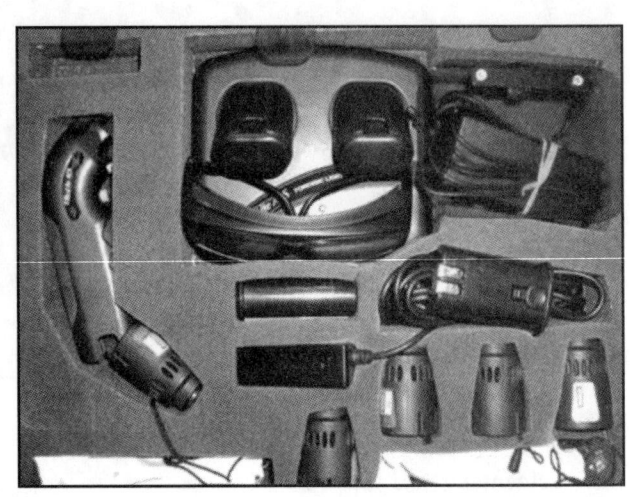

Forensic Science Today

▲ A fingerprint examiner using a Super Glue fuming wand to search a vehicle for latent fingerprints. (left)

▶ ALS forensic light source is useful equipment for visualizing latent prints at a crime scene. (right)

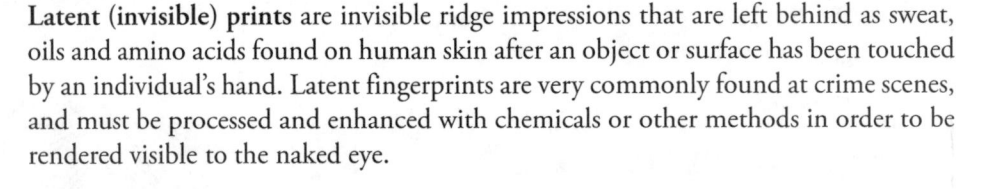

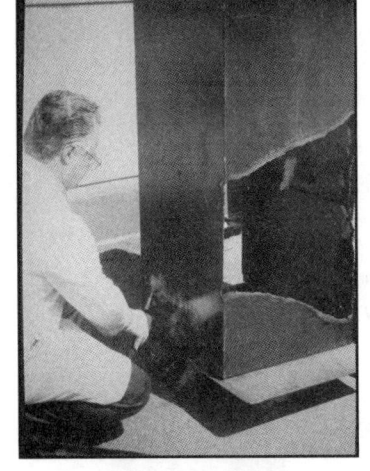

▲ Using super glue wand on large items.

Latent (invisible) prints are invisible ridge impressions that are left behind as sweat, oils and amino acids found on human skin after an object or surface has been touched by an individual's hand. Latent fingerprints are very commonly found at crime scenes, and must be processed and enhanced with chemicals or other methods in order to be rendered visible to the naked eye.

HOW TO FIND AND EXAMINE LATENT FINGERPRINTS AT THE CRIME SCENE

Dusting for prints. This is the traditional method of lifting fingerprints from an object so that it can be used for evidence. The dust used can be black powder, magnetic powder or fluorescent powders that come in several colors. First, the latent print is dusted using one of these powders and a fibrous or magnetic fingerprint brush. The developed prints are numbered with a grease pencil or marker, and photographed. Then the print is lifted from the surface onto a piece of sticky transparent tape, which is then stuck onto a piece of plastic fingerprint lifter of a contrasting color. All prints must be labeled and photographed showing the location of the print and the order in which it was taken.

Using the Super Glue wand to "fume" for prints. The Super Glue fuming method applies cyanoacrylate fumes to the body or object in order to bring out latent prints. This is a portable fuming system previously only used in the forensic laboratory. Now, it is available to crime scene investigators on the crime scene. The Super Glue wand allows for fuming of larger items and items that can't be removed from the crime scene and brought in to the lab for further processing. This method is often superior to the old-fashioned dusting method. Authors of this book—George Taft and Dr. Henry Lee—and others were instrumental in assisting with development of the Super Glue wand.

Using an alternative light source (ALS) to view prints. Latent fingerprints can also be viewed by using portable alternative light sources with appropriate filters, or together with fluorescent chemicals and powders. When the chemicals and powders are applied to the object containing the prints, the light source is shined onto the surface and the prints will then glow eerily. This is also often shown on television crime shows.

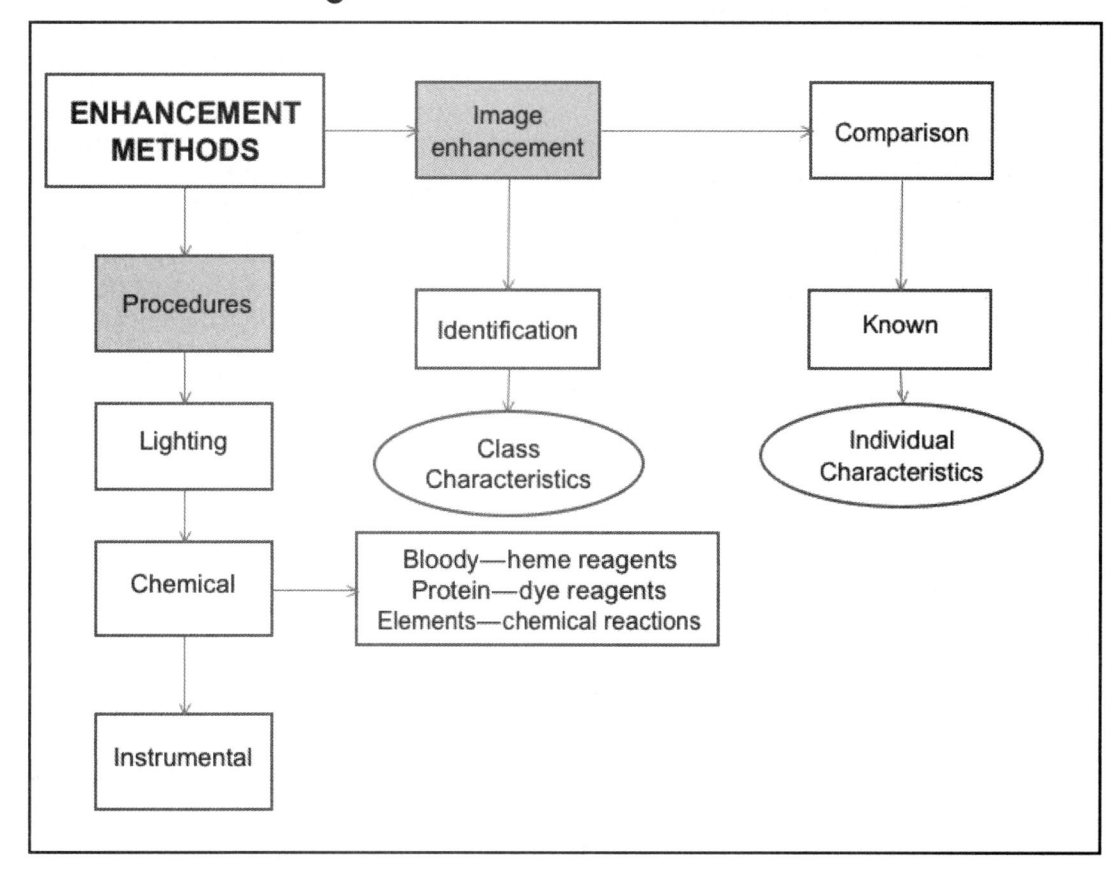

Logic Tree: Enhancement Methods

```
ENHANCEMENT METHODS
├── Procedures
│   ├── Lighting
│   ├── Chemical ──→ Bloody—heme reagents
│   │                Protein—dye reagents
│   │                Elements—chemical reactions
│   └── Instrumental
├── Image enhancement
│   └── Identification
│       └── Class Characteristics
└── Comparison
    └── Known
        └── Individual Characteristics
```

EXAMINING LATENT FINGERPRINTS IN THE FORENSIC LABORATORY

Latent fingerprints submitted to the lab may be analyzed in a variety of ways. These include the same type of fingerprint dusting techniques, which might be done in this case when the prints are not dusted at the crime scene but rather submitted to the lab for processing. There are several traditional methods used to chemically develop prints. Iodine fuming, ninhydrin and silver nitrate are used to develop prints on porous surfaces. Iodine interacts with lipids and fatty acids in the skin yielding a yellow-brown print. Ninhydrin and other reagents react with amino acids and peptides from body secretions resulting in a violet color print. Silver nitrate reacts with salt in the latent print to produce a gray print when silver chloride is formed. Fluorescent chemicals produce highly fluorescent patterns in reaction to amines from body secretions. Gentian violet or physical developer is used to develop skin cells in the shape of fingerprints left on the sticky surface of tape. Super Glue fuming mentioned above forms a polymer coating over the fingerprint. Finally, there are other alternative methods such as x-rays, vacuum metal deposition, and alternative light illumination.

Prints are compared by examining the lifted print and the prints on a suspect's ten print card. This must be done manually by a qualified fingerprint examiner. Currently there is no accepted standard number of points on a lifted print that must match points on a suspect's prints in order to say that a lifted print came from a suspect. Some agencies require eight points to match to say that fingerprints came from the suspect. The FBI requires twelve points.

Forensic Science Today

Table 1
METHODS OF VISUALIZING LATENT FINGERPRINTS

Methods	Principle	Applicable Surfaces
Powder dusting	Adherence of inert material to fingerprint residues.	Smooth, nonporous surfaces such as metals, glass, plastics, tile, and finished woods.
Oblique lighting and photography	Natural residues of fingerprints.	Smooth and nonreflective surfaces.
Laser or alternate light source	Fluorescent material in fingerprint residues.	Smooth, nonporous, or lightly porous surfaces such as plastic and paper.
Physical developer	Salts in fingerprint residues.	Smooth surfaces.
Crystal (Gentian) violet	Fatty acids and lipids.	Sticky side of tapes.
Iodine fuming	Chemical interaction with fatty acids and lipids.	Smooth surfaces such as paper and human skin.
Ninhydrin	Chemical interaction with amino acids, peptides, and proteins of residues.	Absorbent surfaces such as paper and cardboard.
Super glue	Formation of polymers with amino acids and water.	A wide variety of smooth, nonporous, and slightly porous surfaces including human skin.
Small particle reagent	Fatty acids and lipids.	Nonporous and damp surfaces.
X-ray	Adherence of lead to residues.	Smooth surfaces such as plastic films, polyethylene, and paper.
Vacuum coating	Adherence of gold, silver, or cadmium to residues.	Heat stable nonporous surfaces.
TMB (Trimethyl Benzidene)	Heme (a part of hemoglobin-from red blood cells).	Bloody fingerprint.

▶ Sixth-grade drawing of fingerprints.

Chapter 4
Fingerprint Evidence

CASE PROFILE
Superglue Wand Brings the Lab to the Crime Scene

Hartford, CT: Thirty-year old Irma Horvath, a waitress at the University Club of Hartford, reported to work for her usual 9:00 A.M. shift. The mother of a two-year old, she changed into her black and white uniform in the basement locker room and then began making sandwiches and the morning coffee for the club's members.

At 9:30 A.M. a coworker noticed the coffee percolating, but didn't see Irma anywhere. A few hours later the manager called police, and while members were served lunch, the chef quietly led police through the building. This preliminary search turned up nothing.

At 6:00 P.M., Dr. Henry Lee as well as the State Police and a K-9 unit were called, but club officials wouldn't let them search the fancy "Blue Room" because a special dinner party was underway. Finally, at 10:30 P.M. after the diners had left, the police dogs were allowed in and Irma Horvath's body was found stuffed behind a ventilation duct in a storage area, about 15 feet from the party room that had just been vacated. She had been stabbed with a fork and strangled with a hair dryer cord. Her clothes were found nearby. A button with threadlike material that had been attached to her dress was also found, which indicated to Dr. Lee that she had been forced to the attic.

At the scene, Dr. Lee was asked to field test a new (1994) forensic instrument called a "magic wand." The handheld wand emitting Superglue fumes was waved over Irma's body, picking up the bloody fingerprints on her back. This allowed investigators to bring the forensic lab to the crime scene, cutting down on wasted time by not having to wait until the body is taken to the medical examiner's office to discover latent fingerprints.

While the club stayed closed, detectives started interviewing the other employees, including chef Julio Rodriguez. He told investigators that he was a friend of the victim. But when the club reopened a couple weeks later, Rodriguez quit, saying he found a new job. Six weeks later, the new chef discovered bloodstained pants in the basement. They belonged to Rodriguez, and the DNA analysis showed that the blood on them belonged to Irma Horvath.

Ten days later, as police moved to serve a warrant on Rodriguez, Hartford detectives were called to his apartment. He had fatally stabbed his sister-in-law and had taken off. After a citywide search, Rodriguez was found in a bar. He was arrested and confessed to the murders, claiming he suffered from blackouts. With overwhelming forensic evidence pointing to him, the former Marine and married father of three pled guilty in exchange for the prosecutors' recommendation of 110 years in prison.

TECHNICAL DEFINITIONS

AFIS: Automated Fingerprint Identification System—a computerized database of fingerprints from people around the world.

Latent Fingerprints: Fingerprints that are not visible to the naked eye.

Patent Fingerprints: Fingerprints that are clearly visible.

Plastic Fingerprints: Fingerprints left when a person presses a finger into a soft substance.

Ridge Impressions: The pattern left behind by an individual's fingerprint; another way of saying "fingerprint."

Chapter 5

Trace Evidence

☞ CHAPTER OBJECTIVES

- ✔ Understand the nature of types of glass.
- ✔ Discuss how glass evidence can provide the missing key to crime.
- ✔ Describe how to collect glass evidence from crime scenes and known standards.
- ✔ Explain how glass samples are analyzed in the forensic laboratory.
- ✔ Understand the nature of soil, dirt, and dust evidence.
- ✔ Describe how to carry out documentation and collection of soil, dirt, and dust evidence.
- ✔ Explain how soil, dirt, and dust evidence is analyzed in the forensic laboratory.
- ✔ Understand the nature of hair and fiber evidence, and where such evidence might be found.
- ✔ Explain the different components of a hair.
- ✔ Discuss how fiber evidence can provide the missing key to a crime.
- ✔ Describe how to collect hair and fiber evidence from crime scenes and known standards.
- ✔ Explain how hair and fiber samples are analyzed in the forensic laboratory.
- ✔ Understand when glass, soil, hair and fiber can have individual characteristics.
- ✔ Define the following vocabulary words and phrases: *density, refractive index, soil, sieve, mineralogical evidence, root, cuticle, medulla, cortex,* and *known hair standard*.

WHAT IS TRACE EVIDENCE?

Trace evidence, sometimes also referred to as "transfer" evidence, usually means a minute quantity of material such as glass, soil, hair and fiber left behind at a crime scene. These materials can easily transfer or be deposited from one surface to another. Trace evidence is collected by crime scene technicians and scientifically analyzed in the forensic laboratory. When the origin of trace evidence is established with certainty, it can be used to link the suspect, victim, and/or crime scene with one another.

GLASS EVIDENCE

From broken bottles to broken windows, glass is found at many crime scenes. Glass can play an interesting role in determining the sequence of events during a crime. Pieces of glass can be taken away from a crime scene by falling into the cuff of the pants of someone at a crime scene. They can then be deposited at the carrier's home or in their car.

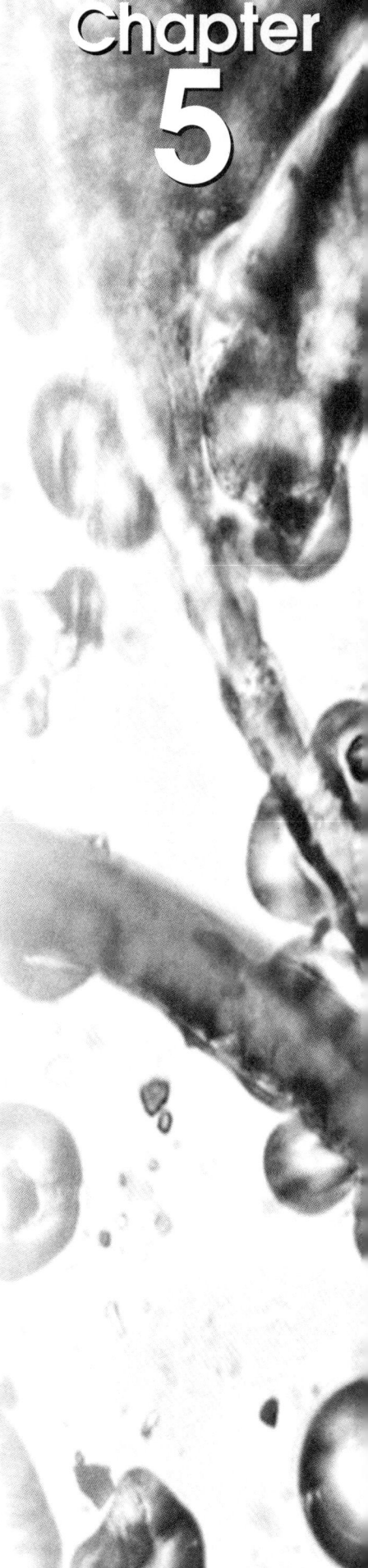

KEY TERMS

Soil
Sieve
Mineralogical
Root
Cuticle
Medulla
Cortex
Telogen hair
Anagen hair
Catagen hair

NATURE OF GLASS EVIDENCE

The basic component of glass is sand, silicon dioxide. The molecular structure of glass differs slightly for different types of glass. Soda-lime glass is sand that has had sodium carbonate and calcium oxide added to the silicon dioxide. When the sand is heated to melting temperature, the molecules form new bonds between the silicon dioxide molecules producing glass, a very crystalline structure. A few examples of the more common types of glass and their differences are as follows:

- Soda-lime glass—sodium carbonate and calcium oxide added.
- Tempered glass—has been produced to fracture into very small pieces.
- Windshield glass—two pieces of soda-lime glass with a sheet of plastic between the layers to allow the "sheet" of glass to flex upon impact.
- Heat resistant glass—boron oxide added to the sand to produce glass that resists fracturing due to rapid temperature changes.
- Leaded glass—lead added to the sand to create a more rigid glass with sharper edges.

DOCUMENTING AND COLLECTING GLASS EVIDENCE

Care should be taken in the collection of glass evidence to prevent breakage. The glass should be carefully wrapped in cotton and placed in a rigid container such as a small box or film canister. If glass is broken at a crime scene, investigators should collect all of the glass to allow for comparison of any pieces discovered on suspects or at other scenes. The containers should be properly sealed and labeled.

ANALYZING GLASS EVIDENCE

Generally glass cannot be individualized. There can be slight differences in density between glass of the same type, soda-lime glass, for instance. Forensic scientists can render an opinion as to the significance of similarity between a known sample and a questioned sample. Their opinion can be determined by examining the glass using several techniques.

The density of the glass can be determined using the formula below. Density is the amount of mass per unit volume of a substance.

$$\text{Density of glass} = \text{Mass of glass}/\text{Volume of the glass}$$

Each type of glass has a range of densities in which it can fall. Density can be found by using the flotation method, in which glass is placed in a column of liquid of varying densities. The glass will be suspended in the liquid that has a density equal to that of the glass.

The forensic scientist can also determine the refractive index of the glass. Refractive index is the ratio of the speed light that travels through a vacuum to the speed it travels through that substance. Refractive index is determined by comparing the unknown glass to a liquid of known refractive index. Glass has the same index of refraction as the known substance if it disappears in the known substance. These properties only allow for the identification of type of glass.

Chapter 5
Trace Evidence

Although large pieces of glass left behind at a crime scene are no longer considered trace evidence, they are equally important for investigations. Sometimes, even more conclusive results might be achieved when forensic scientists can physically match two or more pieces of glass from known and unknown sources. If the pieces fit like two puzzle pieces, then they were originally part of the same sheet or bigger piece of glass. This is an individual characteristic, allowing the scientist to say that the two pieces came from the same source (for example, a window pane).

Additionally, the fracturing patterns in glass allow scientists to determine information about the sequence of events that occurred during the fracturing. For example, if two shots are fired from a gun and the two bullets hit a glass window, it can be determined which bullet hit the glass first. This determination can be made because the second bullet is limited by the radial fracturing that already occurred due to the first bullet. That is, there will be a point or points on the fracture lines of the first shot where lines from the second fracture are stopped by the first fracture's lines. (See Figure 1.)

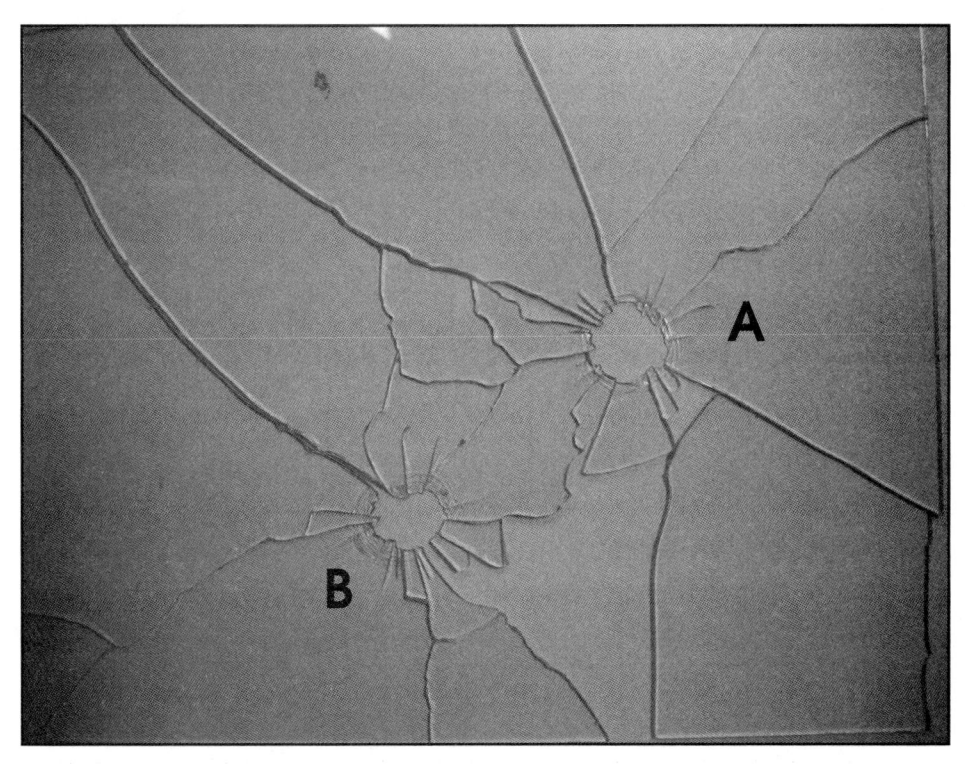

◀ Figure 1. Two bullet holes founds on a windshield in a homicide case. The radial fractures indicate Hole A impacted the windshield first, and that Hole B is from the second shot.

One can also determine which side of the glass that the bullet entered and exited. When the bullet presses against the entry side, the glass stretches slightly before breaking. The exit side is stretched more than the entry side causing a concaved area of glass to be forced out of the glass on the exit side. Therefore, the exit side has a larger opening and is rougher than the entry side. The fracture on the exit side has conchoidal fracture lines; that is, the fractures look like concentric circles. (See Figure 2.)

▶ Figure 2. Radial and concentric fractures shown in a close-up view of a bullet hole on a windshield.

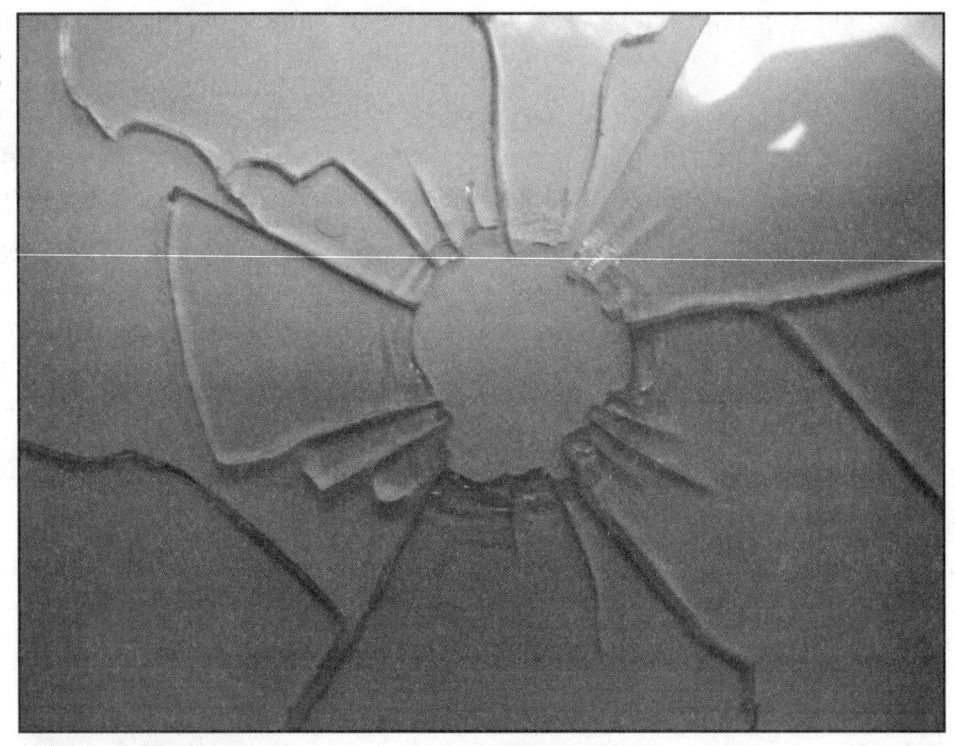

SOIL EVIDENCE

Since most people carry particles of soil on their clothing, shoes, and bodies in the form of dirt or dust, most crime scenes also contain soil in one of several forms. Soil is usually a mixture of clay, dust, sand, rocks, plant and insect particles, and other trace material. It can play an important role in investigating both indoor and outdoor crime scenes. Soil can be left at a crime scene in the form of imprints or particles that fall off an individual; or it can be taken away from the crime scene as it is picked up by an individual who then leaves the scene. It can also be brought from a primary crime scene (such as a murder in the woods), and deposited at a secondary crime scene (the trunk of the car where the body was stored). Soil is usually found on shoes and clothing, in or on automobiles, or on tools or weapons.

NATURE OF SOIL EVIDENCE

Soil is made up of many different types of materials. It contains organic matter from plants and animals, inorganic matter from rock-derived mineral grains, and human-made materials such as bits of glass, paint chips, concrete, asphalt, plastic, fibers, rubber, and other trace materials.

DOCUMENTING AND COLLECTING SOIL EVIDENCE

Before collecting samples of soil evidence, crime scene investigators should properly document both the crime scene and the individual item containing the soil evidence. This should be done through photographs, sketches, notes, and videotape. Next, if the soil sample is to be taken directly from the ground (and not as trace evidence found on another object), investigators should collect approximately four tablespoons of soil from each of several different locations in the crime scene. The samples must be placed in individual containers, sealed, and labeled. Soil should never be packaged in plastic bags, paper envelopes, or other containers with openings from which the soil could escape.

Chapter 5
Trace Evidence

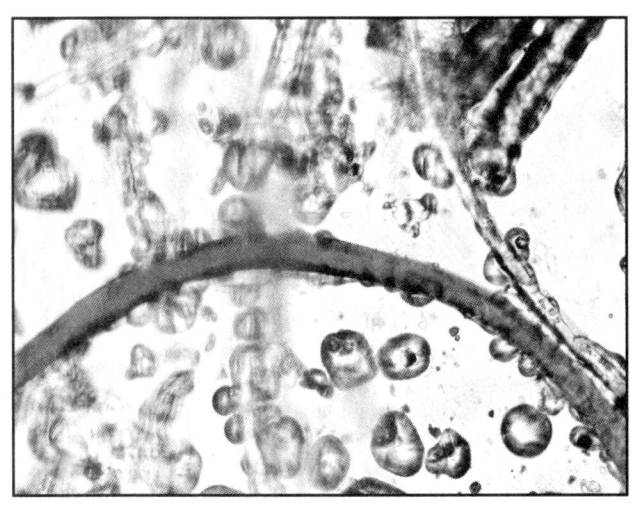

▲ Fibers and trace particles from a shooting crime scene. (Courtesy of Robert O'Brien of Forensic Lab.)

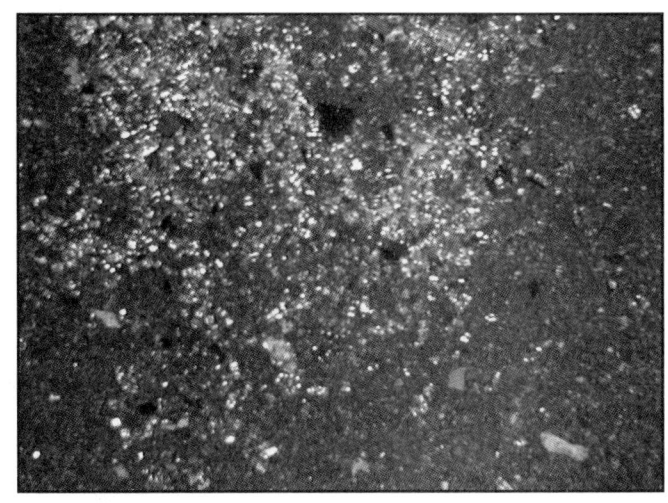

▲ Trace evidence removed from tape and examined under a microscope.

▼ Multi-layer paint chip found on a victim's clothing in a hit and run case.

▼ Trace evidence being examined by Dr. Lee at Alaska Crime Lab.

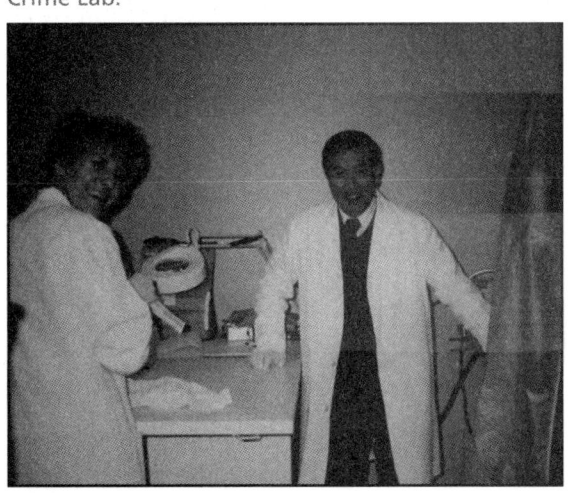

If the soil sample is to be taken from an impression such as a shoe print or tire track in dried mud, then a cast should first be made of the impression (See Chapter 11, Imprints and Impressions). Next, soil should be collected from each impression; and additional samples should be taken from the four compass points surrounding the impression at a radius of a few feet. Another collection should also be taken at the same four compass points but further away, at a radius of 25 feet. This is to establish the individuality of the area of interest.

57

Finally, if soil is present on the clothing of a victim or suspect, the clothing should be removed and packed with paper between any folds or layers to avoid transfer of the soil to another surface. The clothing should then be wrapped in paper or placed into bags so that no soil evidence is lost. Again, items of clothing, including shoes, should never be packaged in plastic bags, since the static electricity created by friction with the plastic could cause some vital evidence to stick to the plastic and be lost.

ANALYZING SOIL EVIDENCE IN THE FORENSIC LABORATORY

It is very difficult to individualize a source of soil due to the wide variety of components in soil and the endless types of soil mixtures. Usually, forensic scientists can only render an opinion as to the significance of similarity between a known sample and a questioned sample. In order to come to this opinion, scientists use a variety of methods to analyze soil in the lab.

The first step in laboratory analysis of soil is to dry the soil so that it can be studied more carefully. Soil is dried in the lab by spreading it out on clean paper and placing it in an oven with a temperature setting of 25 degrees Celsius. Using a microscope, soil examiners then look at both questioned and known soil samples at a low magnification to see if they can identify any of the materials present in the soil. The examiner notes the physical characteristics of the soil, such as texture and color, and any trace materials and human-made components that may be present. Some trace materials, such as metal filings, fibers, and paint, can then be isolated and analyzed separately to aid in both the soil analysis and the investigation of the crime as a whole.

After the initial comparison of soil samples, the examiner next separates the soil components by passing the soil mixture through a series of sieves. This will help determine particle size and distribution. The examiner then determines what percentage of the total weight of the soil sample is held in each sieve. Then, the examiner analyzes the physical, mineralogical, morphological, and chemical characteristics of each portion.

Additionally, soil examiners can gather information about the density of individual particles contained in the soil, and the distribution of such particles within each soil sample for comparison purposes. By using a polarizing light microscope or petrographic microscope, the examiner can further compare and identify the mineral components of the soil. Finally, examiners can use atomic absorption spectroscopy, emission spectroscopy, differential thermal analysis, and other types of instrumental analysis to analyze other organic and inorganic components of the soil.

OTHER TYPES OF MINERALOGICAL EVIDENCE

There are many other types of forensic evidence in this general category, such as rock, mineral, crystals, plastic, asphalt, and so on. The collection and examination of these types of evidence are generally the same as those described for soil evidence above.

Chapter 5
Trace Evidence

CASE PROFILE
TRACE EVIDENCE LINKS SUSPECT TO VICTIM AND CRIME SCENE

Newington, CT: Twenty-four year old Mary Carrington was the manager in the family owned restaurant, Mr. Steak. She never liked being the last one in the restaurant, but this night she stayed behind and closed up after the last diners left. When Mary didn't call the security company to notify them to turn on the alarm system, they called the Newington, Connecticut police to ask them to check the restaurant. The police drove by and found Mary's car in the lot, the lights on and the restaurant locked. They immediately called Mary's father, the owner of the restaurant. When he arrived and unlocked the doors, he found his daughter in the manager's office, dead of what was first thought to be a gunshot wound to the head and later changed to death by blunt object.

That night, Chief Forensic Scientist Dr. Henry Lee was at home nearby, trying to shake a fever of 104. Committed to investigating the murder of Mary Carrington, he was taken by stretcher to the restaurant at midnight, where crime scene investigators joined the lone Newington detective in gathering trace evidence. They found a drink on the table, a cigarette in the ashtray, and change on the floor. An opened telephone book sat on a table near the victim. Was Mary waiting for someone?

But police discovered that the safe was open and $8,000 was missing. It looked like a robbery gone wrong, something that didn't happen very often in this small town. The neighborhood merchants were shocked and the restaurant employees could not understand why Mary had altered her routine and stayed behind, alone in the restaurant after hours. Everyone knew she hated to be alone in the building late at night.

The crime squad stayed most of the day and gathered a great deal of trace evidence. Dr. Lee saw footprints of a work boot in white powder. What were the footprints made of? The white powder on the restaurant floor, later identified as cockroach insecticide, gave them a great picture of the killer's tracks. The phone book provided more evidence – blood spatter and hair evidence - that seemed to point to a phone listing that Mary might have called. They found the number of a social club where a motorcycle club was having a party.

Police zeroed in on several suspects, including Mary's boyfriend, Walter Dalie—a motorcycle enthusiast with an impressive criminal record. Dalie had worked at the same restaurant and Mary had bought him a motorcycle and a leather zippered Harley Davidson jacket. Detectives gathered Dalie's clothes and brought them to the Forensic Lab. The lab assistants said the clothes were clean but they were instructed to look closer—especially at the metal zipper. Clothes can be washed but it is really hard to carefully clean a zipper. Under higher magnification, the forensic team found brain tissue trapped in the metal teeth. They also found other trace evidence, including some roach powder, which put Dalie at the scene.

But without a murder weapon, the case would be tough to make. Detectives searched the roof and storm drains of the building but came up with nothing. They moved to Dalie's backyard. They found no weapon but they did find a freshly dug hole – the hiding place for the money taken from the restaurant. There was now enough evidence to convince Dalie to plead guilty, even without a weapon. Dr. Lee believes Dalie kicked his girlfriend Mary Carrington to death.

Forensic Science Today

HAIR AND FIBER EVIDENCE

Hair and fibers are among the most common types of trace and transfer evidence that investigators find at a crime scene. Hair and fibers are transferred to shoes, clothing, and other physical evidence whenever an individual or object comes into contact with a source of hair, fabric, or other fibers. Thus, a suspect will often leave hair and fibers from his own body and clothing at a crime scene, and he will also take some with him from the scene when he leaves. Forensic scientists may study hair and fibers left behind at a crime scene or on a victim's body to help identify the victim and the victim's assailant, to determine where a murder or assault may have occurred, or to prove or disprove an alibi. This type of evidence can also be found at many other types of crime scenes, such as cases of pedestrian-motor vehicle accidents and sexual assault.

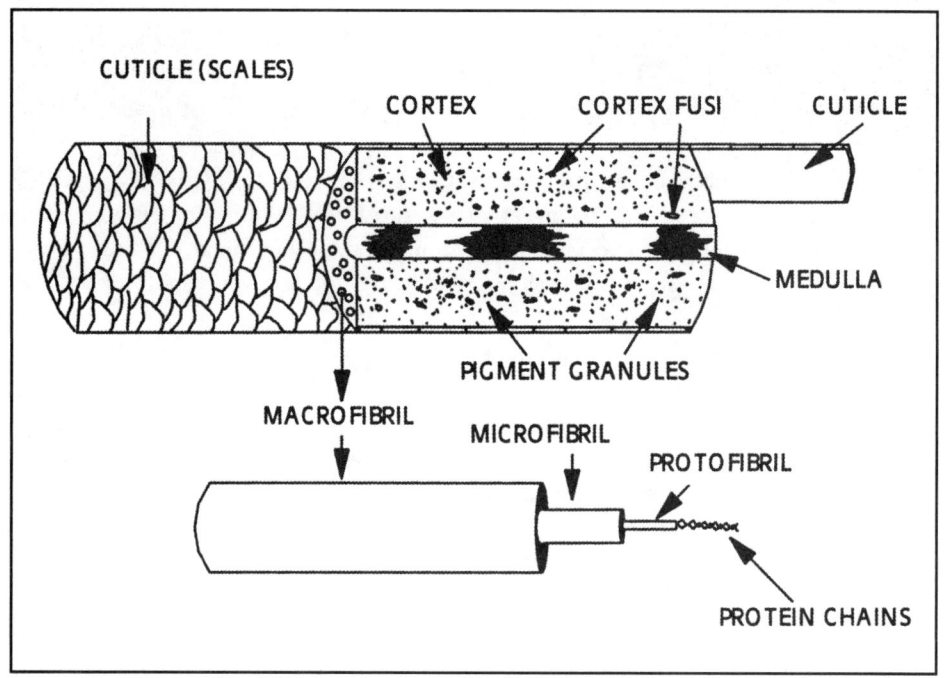

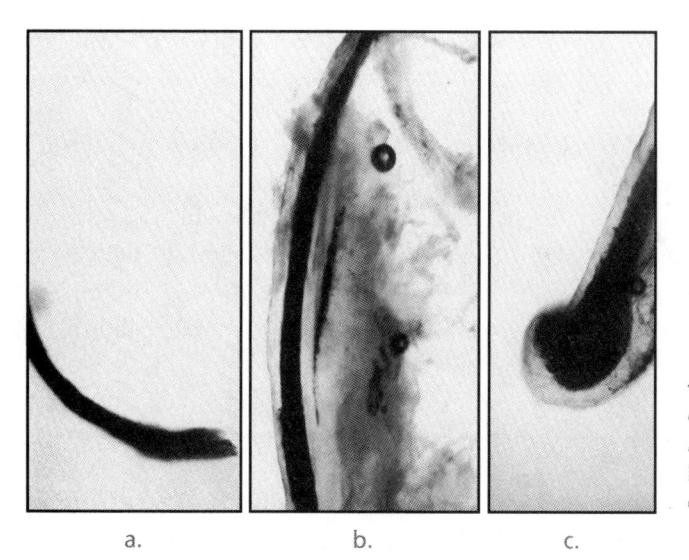

▲ Cutaway diagram of the structural components of a human hair. (top)

◀ Three growth phases of hair:
a.) Telogen,
b.) Anagen, and
c.) Catagen.

Chapter 5
Trace Evidence

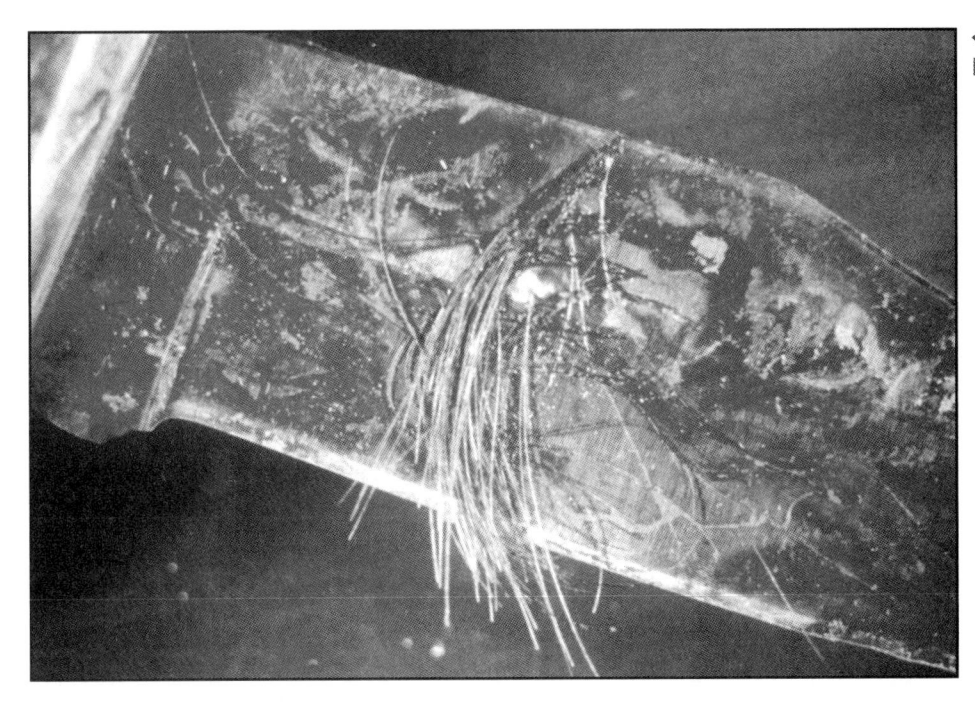

◀ Hairs found on a bloody knife at a homicide scene.

CHARACTERISTICS OF HAIR EVIDENCE

Hair is created inside the hair follicle and becomes a non-living substance as it comes to the surface of the skin. Certain characteristic areas are developed during the formation and growth of hair. These areas are useful for comparing hair samples.

When you look at a hair that has been pulled out of a person's head, you will see two parts—the *root* and the *shaft*. Until a hair is pulled out of the head, the root is always embedded in the follicle, beneath the epidermis. One way of telling whether a hair belongs to an individual is through DNA typing of the root. Also, the shape and pattern of the root may vary across species. The shaft, on the other hand, is that portion of the hair you can see as it grows out of the body.

When you view a cross-section of hair under a microscope, you'll see that there are three main components as you go from the outside of the hair in toward the center of the shaft:

1. **Cuticle.** This is a layer of scales covering the hair shaft, and can come in a variety of patterns. These patterns can also help identify whether the hair is human; and if so, how it compares to the hair of other persons. The scales come in three basic patterns—coronal, spinous, and imbricate—and always point down toward the tip of the hair. Human hair has imbricate cuticle.

2. **Cortex.** Going in one layer from the cuticle toward the center of the hair is the cortex. This is the main section of the hair, and is made up of elongated cells. The cortex may also have inclusions present, such as cortical fusi (small air spaces), pigment granules (which give hair its color), and ovoid bodies (solid structures that are spherical or oval in shape).

3. **Medulla.** This is the central core of the hair, which also comes in several different forms. Sometimes the medulla is translucent, meaning that it allows light to pass

through it; and sometimes it is opaque, meaning that no light can pass through. The medulla can either be continuous, intermittent, or fragmented. Again, this also helps determine whether the hair is human or belongs to a different species. The simplest way to determine if hair is human or not is to use the medullary index. If the diameter of the hair is less than one-third of the diameter of the shaft of the hair, it is human hair. Human hair can have fragmented medulla or an absent medulla.

Moreover, at any given time the hair follicle may be "resting" (telogen hair), actively growing (anagen hair), or transitioning out of growth (catagen hair).

CHARACTERISTICS OF FIBER EVIDENCE

Fibers are either natural—made from plants, minerals, or animals—or synthetic, such as nylon or polyester. Fibers can also be derived or regenerated, meaning that they come from reprocessed natural materials. How much fiber is transferred from its source depends upon factors such as the type of fiber, amount of loose fiber available, weave of the fabric, the mode of physical contact, the degree of contact, and various environmental factors.

After forensic investigators collect fibers from a crime scene, the next step is to identify the source of the fiber before linking it to the origin, person, or crime scene. This is very difficult as today's world is flooded with fibers from many different sources. For example, carpeting is only one source of fiber, yet in the U.S. alone millions of square yards of carpeting are sold. Because of the pervasive nature of fibers in our world, fiber evidence is most likely to be only one of many types of evidence that come together to make it more likely than not that the suspect, victim, and/or crime scene are somehow linked.

Sometimes, however, fiber evidence can provide the vital missing key to a crime. If a fiber is only produced in a small amount for a limited period of time, or comes in a very unusual color, or a piece of fabric was found, then it will be easier to trace to a unique source. Therefore, the first step in fiber analysis is to determine how common or uncommon it may be. Investigators must ask themselves: Who manufactured this fiber? When was it manufactured? How much of it was manufactured that year? Where was it sold?

For the Record

The Indus Valley was one of the birthplaces for cotton. Remnants found at Mohenjo-Daro prove that cotton was produced there as early as the Second Millennium B.C. The skill of tinting with the use of mordants (chemical agent to fix color) was even mentioned in the Old Testament, "more enduring than the dyed colours of India". Cotton soon circled the world and was grown by American Indians as early as 700 B.C.

COLLECTING HAIR AND FIBER EVIDENCE FROM A CRIME SCENE

Hair and fibers can either be picked up with gloved fingertips or with clean tweezers or forceps. Using tweezers or forceps is preferable since it prevents contamination. Hair and fibers can also be collected through a vacuum fitted with a special hair and fiber collection attachment in the hose. The filter must be changed with each use. Another method is referred to as a "tape lift," which uses clear cellophane tape to pick up the evidence, which can then be stuck to a clean glass slide to protect it until later examination. The tape lift is preferable as this method only collects the last evidence deposited.

Once hair and fibers are collected, they must be packaged properly so as not to contaminate or destroy the evidence in transit to the forensic laboratory. Vacuum sweepings should be kept in their filters in separate containers and labeled properly. The containers must be sealed well, with no openings through which small particles may be lost. Plastic bags should not be used. Alternatively, individual fibers that are picked up can be placed in a piece of paper folded in a "druggist's fold" (see Chapter 14, Chemical Evidence, for a diagram of the druggist's fold), which in turn is sealed in a paper envelope and labeled accordingly.

COLLECTING KNOWN HAIR STANDARDS FROM A SUSPECT OR VICTIM

Forensic investigators must also collect hair from suspects, victims, and witnesses in order to compare the hairs with those found at a crime scene. This can be used to link individuals to the crime scene. The hair collected from a person's body for comparison is called the "known standard" because the source of the hair is known, just like known standards are collected for fingerprints, impressions, and other types of physical evidence. In order to properly collect hair samples, the investigator must first comb the sample area of the head to remove all loose hairs. These hairs should then be packaged and labeled. Next, the investigator uses tweezers to pull out hair samples to include the root. The root of the hair is very important because it contains vital DNA and genetic material. Hair samples should be taken from various parts of the head, and a minimum of five to ten hairs from each location should be taken.

EXAMINING AND COMPARING HAIR SAMPLES IN THE FORENSIC LAB

First, hairs are examined to determine if they are actually hairs, and not some other type of fiber. If the hair is of animal origin, the scientist can determine what kind of animal the hair came from. If the hair is human, the next step is to study it under a microscope to look at the surface of the hair to see if any trace material such as blood, hair dye, or grease is present. Microscopic examination also helps narrow the pool of potential donors by examining properties such as medulla pattern, color, reflectivity, root type, racial origin, the part of the body the hair came from (somatic origin), and genetic markers. There are several types of genetic markers found in a hair, and by studying these markers scientists in earlier days determined the ABO blood group and the sex of the donor. Today, forensic labs determine the unique characteristics of the donor as determined by their nuclear DNA (with hair root), or Mitochondrial-DNA (with hair shaft).

Forensic Science Today

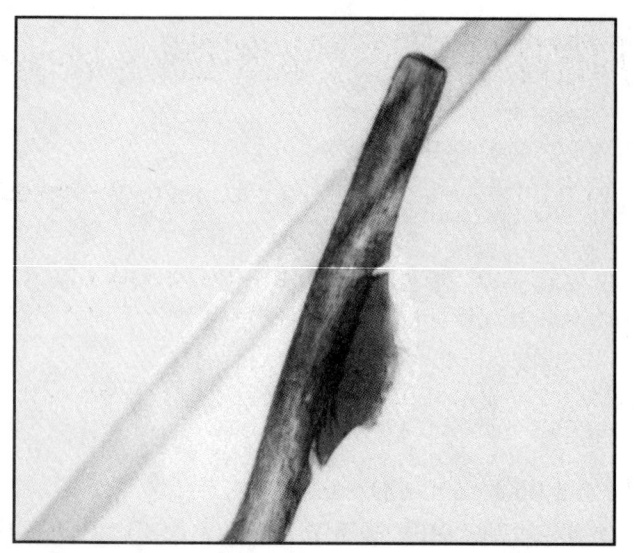

▲ Photomicrograph of hair fragment with blood crust. (left)

▶ Photomicrograph comparison of fibers (right): (a) red fiber found in a victim's fingernail, (b) the known fiber sample removed from the suspect's sweater.

EXAMINING FIBER EVIDENCE IN THE FORENSIC LAB

Forensic scientists use several different methods of examining fiber evidence in the lab. First, the examiner can physically match two or more pieces of fabric that had been torn apart during the crime by fitting the two pieces together. If the pieces fit together like pieces of a puzzle (edges match up), then the pieces were originally a part of one whole. Second, fibers are examined under the microscope to determine characteristics such as color, shape, texture, pattern, twist, cross-sectional appearance, and surface characteristics. Third, polarized light microscopy helps identify the general class that a fiber belongs to. Fourth, examiners measure physical characteristics of fibers such as their density, melting point, ash formation, and tensile strength. Fifth, scientists use chemical reagents to test the solubility of a fiber. Sixth, and finally, other various types of instrumental methods are used to determine the chemical composition of fibers. Some of these instrumental methods, including infrared spectroscopy, pyrolysis gas chromatography, scanning electron microscopes, atomic absorption, and dye comparison, have replaced the more traditional techniques.

OTHER TYPES OF POLYMER EVIDENCE

There are many other types of polymeric trace evidence that can be found at a crime scene or on a victim or suspect's body. These types of evidence are also valuable in forensic investigation. Polymeric trace includes plastic, glue, grease, plastic bags, small pieces of artificial fingernails, buttons, parts of plastic eyeglass lenses, automobile parts, polymer ropes and string, and similar materials. Even the plastic emblems sometimes found on shirts have been used to solve crimes. The collection and examination of polymeric trace evidence are generally the same as fiber evidence.

Chapter 5
Trace Evidence

TECHNICAL DEFINITIONS

Density: The amount of mass per unit volume of a substance.

Refractive Index: The ratio of the speed that light travels through a vacuum to the speed it travels through a substance.

Soil: A mixture of both organic and inorganic matter. It may contain a wide variety of substances derived from minerals, plant matter, glass, paint chips, plastic, and other trace materials.

Sieve: A tool used by forensic scientists to strain out and separate the various components of soil.

Mineralogical: Having to do with the mineral properties of a substance.

Root: The bulb of the hair found in the hair follicle, the only part of the hair containing nuclear DNA.

Cuticle: A layer of scales covering the hair shaft.

Medulla: The central core of the hair.

Cortex: The primary portion of the hair surrounding the medulla, covered by the cuticle.

Known Hair Standard: Samples of hair collected from a person's head to compare with hairs found at a crime scene.

Chapter 6

Drug Evidence

👉 CHAPTER OBJECTIVES

- ✔ Discuss the nature and different types of drug evidence.
- ✔ Describe how blood and urine samples are taken for drug testing.
- ✔ Describe how drug evidence is collected.
- ✔ Explain how clandestine drug laboratories are investigated.
- ✔ Understand how drug evidence is analyzed in the forensic laboratory.
- ✔ Discuss the issues involved in drug-facilitated rape, particularly how to prevent and survive such attacks.
- ✔ Define the following vocabulary words and phrases: *clandestine drug lab, controlled substance, synthetic drug, drug-facilitated rape (DFR), rohypnol,* and *Gamma Hydroxybutyrate (GHB)*.

Drugs come in many forms—both legal (as prescribed "controlled substances") and illegal. Illegal drugs will obviously be seized as evidence of criminal activity based on the fact that even possession of such substances is against the law. Controlled substances, or legal drugs which are available only through prescription, can also be abused by those with addictions or misused with criminal intent by those selling such substances on the black market.

Most drugs and controlled substances are seized by law enforcement agencies during the course of drug investigations, or "incident to" (as part of) an arrest for other reasons. While most chemical analysis of drug evidence occurs in the forensic laboratory, police officers may also conduct field tests on the suspected drugs to gain a preliminary idea of what they have seized. Once the suspected drug is submitted to the forensic laboratory, forensic *chemists* conduct chemical tests as a guide toward what type of quantitative and qualitative analyses need to be conducted.

TYPES OF DRUG EVIDENCE

Drugs may be defined as any type of substance—whether natural or synthetic—which is used to produce certain psychological or physiological effects on the body. Drug evidence may be classified into the following categories: opiates or narcotics (drugs that reduce pain and other sensation and lead to a sleepy state), stimulants (drugs that stimulate the sympathetic nervous system and cause agitation, high blood pressure, increased heart rate, and rapid breathing), hallucinogens (drugs that alter one's perceptions and may lead to disturbing hallucinations), and depressants (drugs—including alcohol—that depress the central nervous system and result in loss of motor coordination and reduced respiration). Some examples of drugs in these categories are found in Table 1.

Forensic Science Today

KEY TERMS

Chemist
Controlled Substance
Clandestine Drug Lab
Drug-Facilitated Rape (DFR)
Rohypnol
Gamma Hydroxybutyrate (GHB)

COLLECTING BLOOD AND URINE SAMPLES FOR DRUG EVIDENCE

People may be tested for drugs for a variety of reasons; and whether it is for participation in sports, workplace safety, as a term of probation, or as part of an arrest, certain procedures must always be followed to prevent contamination or deterioration of the biological sample.

When testing for drugs in the blood, blood samples must be drawn by a licensed physician, nurse, or medical technician; and disinfectant must be used to clean the skin where the sample will be drawn. Two tubes of blood (10-cc each) are drawn and collected into gray-topped tubes containing sodium fluoride as a preservative. The tubes are then sealed, labeled, and kept in the proper chain of custody.

Logic Tree: Drugs and Poisoning

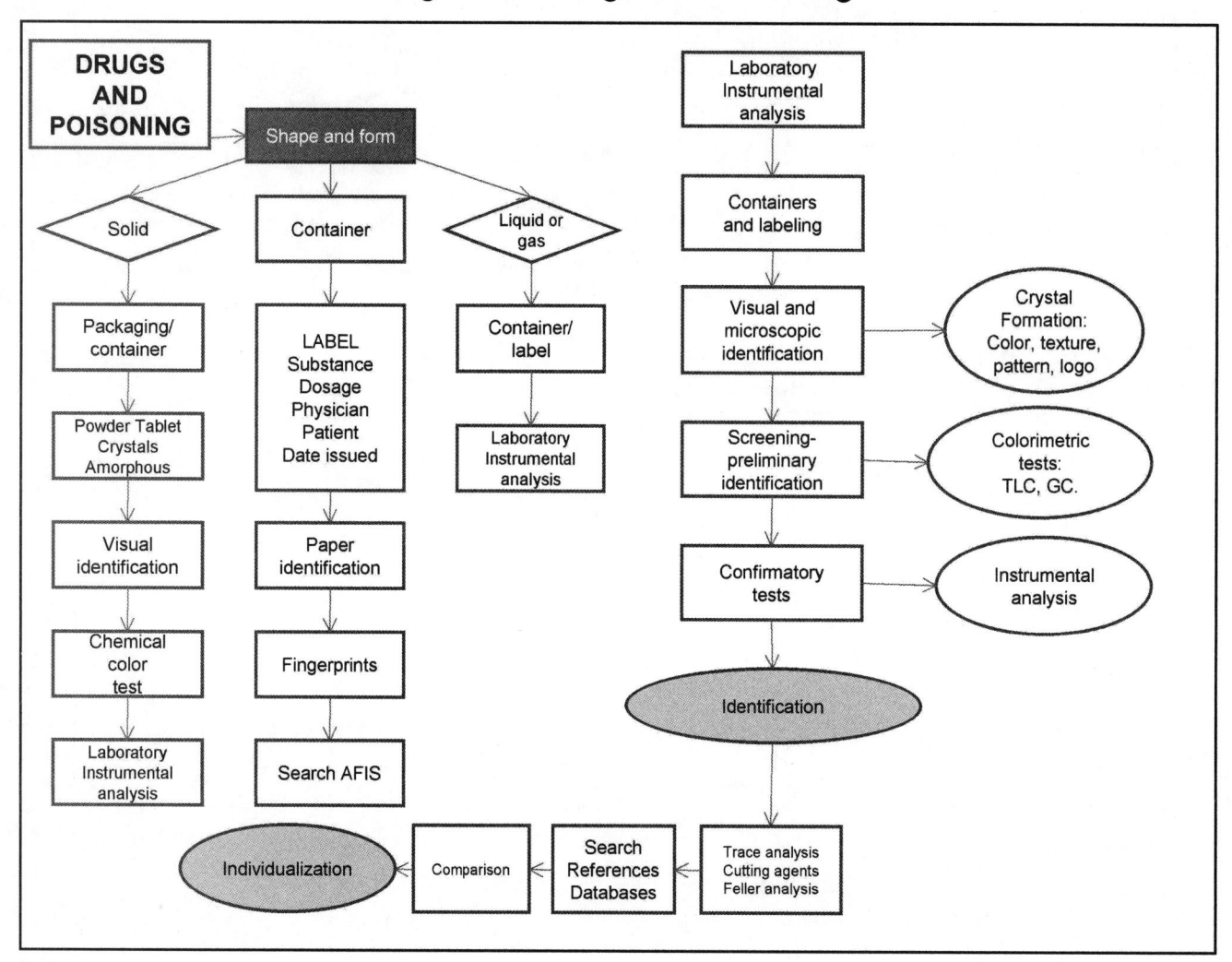

When testing for drugs in the urine, the collection of the urine sample must be witnessed by an authority to make sure that the person giving the sample does not dilute the sample with water from the sink or toilet (or with someone else's urine!). The urine sample should be collected in a special unused plastic screw-top jar, the same type that is used by hospitals. Again, the container should be properly sealed, labeled, and kept in the chain of custody.

COLLECTING OTHER FORMS OF DRUG EVIDENCE

When a plant or plant material is suspected to contain illegal drugs, it must be submitted to the forensic laboratory for chemical analysis. The plant parts should all be collected in a paper evidence bag and allowed to air dry before sealing the bag, in order to prevent the growth of biological organisms. If the materials are already in bags or containers, they must then be placed into a clean evidence bag to preserve the fingerprints on the original bag or container. The evidence bag must then be sealed and labeled properly.

For powders, tablets, and capsules, any containers they are kept in should similarly be placed into evidence bags in order to preserve any fingerprints on the container. All drug packages and containers must be counted, marked, and described on the inventory sheet, and placed into evidence bags that are labeled and sealed properly.

Liquid forms of drugs should be kept in their original containers. Such containers should be placed in a security evidence container to protect against breakage, contamination, or damage.

When collecting drug paraphernalia such as needles, syringes, spoons, cigarettes, foils, pipes, papers, and other materials, investigators must exercise extreme caution not to get in contact with biohazardous materials which can lead to serious infection if puncture or exposure to broken skin occurs. Crime Laboratory employees should be made aware of evidence containing needles/syringes so that they may avoid a potential biohazard. Each type of paraphernalia has an appropriate container that must be used, sealed, and labeled.

Alcohol + Marijuana = Recipe for Disaster

Combining alcohol and marijuana is a particularly dangerous choice. Not only does this combination make you more vulnerable and less likely to make good decisions, the use of marijuana with alcohol increases your risk of alcohol poisoning. Under normal circumstances you have a built-in protection against drinking too much alcohol. Your body will protect you against this excess by causing you to expel the contents of your stomach—you throw up. Marijuana reduces your ability to feel nauseated, thus, reducing your protection against alcohol poisoning. What is alcohol poisoning? Alcohol poisoning occurs when you have taken in too much alcohol for your liver to process. Alcohol poisoning can easily kill you.

Forensic Science Today

INVESTIGATING CLANDESTINE DRUG LABORATORIES

We may all have an image from television and movies of what a clandestine (secret) drug laboratory looks like; but in truth there are a wide variety of places and settings in which illicit drugs are made. When investigating clandestine drug labs, investigators must use utmost caution and care not only in following safety guidelines but also in preserving evidence.

What types of evidence might be found at illicit drug labs? Anything from the substances used to produce precursors to the drugs and the drugs that are made there, to computers, money, records, documents, and fingerprints. These can all be of value and seized from this type of crime scene. Investigators must follow safety precautions against both chemical hazards and biological hazards. Security should be maintained until a qualified chemist reaches the scene to dismantle and shut down laboratory operations. Nothing should be disposed of or destroyed by pouring it into water. Nothing should be tasted or smelled; and gloves should be worn when handling any liquid and powder forms of chemicals.

EXAMINING DRUG EVIDENCE IN THE FORENSIC LABORATORY

Forensic chemists use many different methods to analyze drugs in the forensic laboratory. This is due to the wide variety of different drug substances and chemical components that make up the drug. Some of the techniques used in the lab are similar to analyzing other types of chemical evidence.

First, all drugs submitted to the lab are macroscopically examined, weighed, and catalogued. The forensic scientist must carefully describe the package and overall characteristics of the suspected drug sample. Any color, shape, or identifying markings of pills or capsules should also be recorded for comparison to known standards.

THE FORENSIC DRUG ANALYST

Forensic drug analysts use analytical chemistry to identify the presence and quantity of controlled substances. The drug chemist analyzes unknown powders, liquids, vegetative materials (plants), pills, capsules, and other forms of drug dosages. The drug analyst must not only identify the active ingredient, but also determine the exact amount present. Forensic drug analysts are also called upon to assist with processing materials from clandestine (secret and illegal) laboratories and to develop important investigative information about the drugs found in these labs, such as the country of origin or other sources.

Forensic chemists will also conduct chemical tests on the suspected drug, often using color tests as screening agents. Certain drugs react with certain chemical reagents to give characteristic color changes or precipitates. To conduct such a test, the chemist places a small portion of the suspected drug powder on a clean microscope slide and adds a drop of the appropriate chemical reagent. The chemist places the slide on the stage of the microscope and observes the drop under the microscope field. Specific chemical reagent solutions cause characteristic microscopic crystals to form when added to a drug. These microscopic crystals can then be identified by their characteristic shapes.

Some examples of color tests used to give a preliminary identification of a drug are:

- Duquenois-Levine test- used in the identification of marijuana.
- Marquis test- used in the identification of opiates and amphetamines.
- Scott test- used in the identification of cocaine.

According to the law, each drug must be identified by a test that allows for the identification of that drug as only that drug. In order to accomplish this goal, most forensic laboratories identify most drugs using a combination of the gas chromatograph and mass spectrometer (gc/ms). One exception to this final test is marijuana. Marijuana is identified botanically; that is, the parts of the plant known to be common only to marijuana must be seen by the forensic scientist on the questioned sample. Finally, as with other chemical substances, forensic scientists can use a variety of instrumental techniques to further analyze drugs. These techniques include thin layer chromatography, infrared spectroscopy, high-performance liquid chromatography and gas chromatography/mass spectrometry.

▶ A variety of pills seized by police.

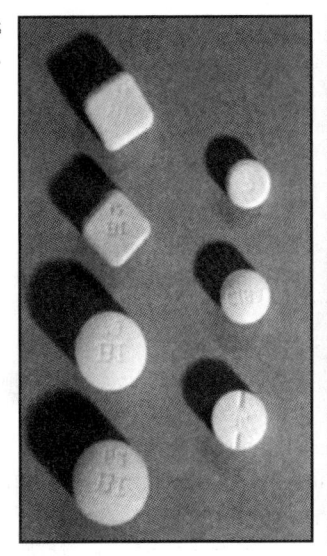

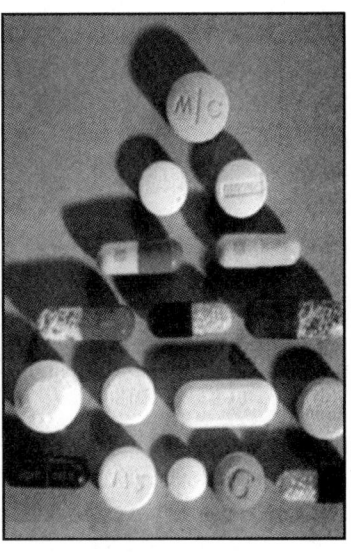

Forensic Science Today

Table 1
Commonly Used Drug Screening Reagents

Drug	Reagent	Color Reaction
Amphetamine	Mandelin Marquis	gray-green red-brown
Barbiturate	Dille-Koppanyi	purple
Cocaine	Cobalt Thiocyanate Mandelin	blue precipitate orange
Codeine	Mandelin Marquis	green purple
Heroin	Mandelin Marquis Nitric Acid	brown purple yellow-green
LSD	Marquis Erlich Van Urk	dark purple pink-purple blue-purple
Marijuana	Modified Duquenois	blue
Methadone	Cobalt Thiocyanate Marquis	blue yellow-pink
Morphine	Mandelin Marquis Nitric Acid	orange-brown purple orange
Psilocin and Psilocybin	Weber test	red to blue
Benzodiazepines	Janovsky	orange

◀ A variety of opiate products.

72

Chapter 6
Drug Evidence

WARNING!

DRUGS AND SEXUAL VIOLENCE: WHAT YOU DON'T KNOW *CAN HURT YOU!*

Every two minutes, somewhere in America, someone is sexually assaulted. One in four women will be the victim of rape or attempted rape in their lifetimes. Boys and young men are also victimized by sexual violence. Alcohol and drugs are often involved in rape and violence, whether they are ingested by the perpetrator, victim, or both.

ALCOHOL AND DRUGS MAKE YOU MORE VULNERABLE TO SEXUAL VIOLENCE. HERE'S HOW:

Ingesting any kind of drug – including alcohol – impairs your judgment and renders you less able to protect yourself against rape and other types of sexual assault. Most sexual violence is committed by a person known to the victim, such as a relative, friend, boyfriend, family friend, co-worker or other acquaintance. Ingesting alcohol and other drugs makes it more difficult to see the warning signs of an impending attack, which can then seem to come "out of the blue" by someone you thought you could trust.

Rape, while always a crime of violence, does not always occur with the use of weapons or brute force. Coercion or trickery is also used to gain submission of a victim by someone she knows and trusts. When a young woman or man becomes impaired through the use of alcohol and other drugs, it becomes much easier for the perpetrator to make an act of sexual violence appear to the victim and others to be consensual when in fact it is not.

Over 80% of rapes against women are perpetrated against a victim under the age of twenty-four. This is the same period in a young person's life when she is more likely to be testing the new waters of adulthood and may be experimenting with alcohol and other drugs. Add to this the fact that over 80% of all sexual assaults are committed by someone known to the victim, and it becomes clear that you are never completely safe if you consume alcohol and other drugs even with people you know, trust, and would never guess to be capable of rape.

PREDATORY DRUGS—BE ON THE LOOKOUT

You may have heard of cases where a young woman's drink is "spiked" with a "date-rape drug" and she is then sexually attacked. In truth, the term "date-rape drug" is a misnomer because quite often the victim whose drink is spiked is not on a date with her attacker at all. A more accurate term is *Drug-Facilitated Rape (DFR)*.

- **The nature of DFR drugs.** Such predatory drugs make one incapable of resisting sexual attacks—they are odorless, tasteless, induce memory problems (up to twelve hours after occurrence), and are quickly metabolized. Worse yet, DFR drugs induce a state of amnesia so the victim often has difficulty identifying the rapist or the events surrounding the assault. However, convictions can still be obtained in these types of cases so it is important to report the crime to police as soon as possible.
- **Why does the attacker use drugs?** Because it is easier to gain control of the victim, and the victim cannot defend herself. The victim may also have little memory of the assault. Moreover, to outsiders, the attacker may look like a "rescuer" who is simply taking home a "drunken friend."
- **How do DFR drugs work?** They are considered to be anesthetic agents, hypnotics, and depressants of the central nervous system. The drug is usually added to beverages, and the effects begin within 30 minutes of ingestion. First, the victim will begin to feel lethargic, that is, very tired and unenergetic. She may find it difficult to move around. The condition then progresses to the victim blacking out and suffering from amnesia. By this time, it is too late for the victim. Unless she is with friends who are looking out for her, it will appear to all intents and purposes that the rapist is just a nice guy who is taking care of a girl who drank too much.
- **Rohypnol is one of the most common forms of Drug-Facilitated Rape drugs.** It renders the user mentally and physically incapacitated; and its effects are heightened when used in combination with alcohol.

Forensic Science Today

- **Gamma Hydroxybutyrate (GHB) is the other most common type of DFR drug.** Some people also voluntarily take GHB in low doses in order to produce an alcohol-like intoxication. However, in higher doses it makes one extremely vulnerable to sexual predators. Some of its side effects include topical irritation (skin-mouth-eyes), nausea, vomiting, incontinence (loss of urinary control), seizures, loss of consciousness, liver damage, kidney failure, respiratory depression, and death. Moreover, it is physically addictive. Why risk it?

- **How do other drugs impact rape?** The most common drugs found present in samples from sexual assault victims include alcohol, marijuana/hash, cocaine, and benzodiazepines. While drugs such as alcohol and marijuana are not "secretly" administered to the victim in the same way as DFR drugs are, ingesting any kind of drug—including alcohol—impairs the victim's judgment and renders him or her less able to protect against rape. This is especially so when alcohol and other drugs are mixed, which worsens the effect of the drugs and doubles or triples the victim's vulnerability.

- **How do I protect myself against Drug-Facilitated Rape?** Awareness is the first step. Young women must be very careful not to abuse any types of drugs; and they should establish a "buddy system" with their female friends before going out to a party or club. Never leave a friend alone even though you may want to leave the situation; and never let your female friends leave you alone and vulnerable in such circumstances.

What should I do if I have been sexually assaulted?

All types of sexual attacks should be reported to the police as soon as possible. Your safety is your first priority—get yourself to a safe place such as your home or a friend's home; or you can call a taxi wherever you are and they should take you to the hospital for free when you explain it is an emergency. Do not shower, wash yourself, or change your clothes; doing so could destroy vital evidence. You have several options—some victims may feel comfortable calling the police right away, some prefer to tell their parents first. The best resource to call on may be your local rape crisis center, which may be able to send a victim advocate to meet you at your house, the hospital, or the police station to support and guide you through the process of reporting the incident. To find the number of the nearest rape crisis center, call a free 24-hour national hotline at **1-800-656-HOPE**. For more information check out the Rape, Abuse, and Incest National Network at **www.rainn.org**.

Victims of rape often needlessly feel shame and blame themselves for an act of violence that was not their fault. Due to the nature of acquaintance rape, it is often easier to blame one's self and to dismiss the attacker's actions as something other than rape. But this should not stop a victim from gathering her courage and reporting the attack to police. It is the duty of police officers to investigate all reported rapes. Many police stations, especially substations near college campuses, have specially trained female liaison officers who will provide support as the victim files the report.

Chapter 6
Drug Evidence

TECHNICAL DEFINITIONS

Clandestine Drug Lab: An illicit, secret laboratory where drugs are illegally made.

Controlled Substance: A drug available by prescription only, illegal if possessed in unlawful quantities or by unauthorized persons.

Synthetic Drug: A drug that is made in a laboratory, and does not occur in nature.

Drug-Facilitated Rape (DFR): Sexual violence committed with the aid of drugs given to the victim by the perpetrator. DFR drugs create an amnesia-like state in the victim.

Rohypnol: A common DFR drug used by sexual predators to facilitate rape.

Gamma Hydroxybutyrate (GHB): A common DFR drug used by sexual predators to facilitate rape.

Forensic Serology

Chapter 7

CHAPTER OBJECTIVES

✔ Understand the history of serological evidence.
✔ Identify the different chemical reagents used to conduct field tests on suspected bloodstains.
✔ Identify the different ways of collecting liquid blood evidence from a crime scene.
✔ Identify the different ways of collecting blood evidence from bloodstained materials found at a crime scene.
✔ Explain the steps taken to examine and analyze blood evidence in the forensic laboratory.
✔ Define the following vocabulary words and phrases: *serology, blood type, false positive, Phenolphthalein, Leucomalachite Green, Tetramethylbenzidine*, and *Luminol*.

Criminal investigators may encounter two different types of blood evidence at a crime scene: (1) liquid blood and/or (2) dried bloodstains. In this chapter we will learn about collecting and analyzing liquid blood and the science of serology, and in Chapter 9 we will study bloodstain patterns.

Forensic serology involves the study of body fluids and stains remaining at a crime scene, particularly blood. Crime scene personnel may come across blood evidence in many different types of crimes, most often in crimes of violence such as homicide, assault, and sexual assault. However, blood evidence can also be found at non-violent crime scenes such as burglaries (where a burglar may have cut himself on broken glass, for instance). Thus, if blood at a crime scene contains DNA matching that of a suspect, the blood can be an important piece of evidence linking the suspect to the scene. Likewise, blood containing the victim's DNA that is found on a suspect's clothing can also prove an important link between the suspect, victim, and crime.

▶ Hair, blood, and sperm are common types of serological evidence found in rape cases.

Forensic Science Today

KEY TERMS

Serology
Blood Type
False Positive
Phenolphthalein
Leucomalachite Green
Tetramethylbenzidine
Luminol

Prior to 1990 and the subsequent surge in DNA testing, forensic serologists used blood typing to link suspects, crime scenes, and victims. Today, however, most forensic laboratories no longer use this technique, preferring to use the newer and more accurate DNA testing methods instead. (See Chapter 8 for more information on DNA evidence and testing).

Crime scene technicians should always remember to document any blood patterns that may have resulted from the crime, prior to collecting blood evidence. They must also follow rigorous safety protocols since blood is considered a biohazardous material that can spread infection and disease such as hepatitis and HIV.

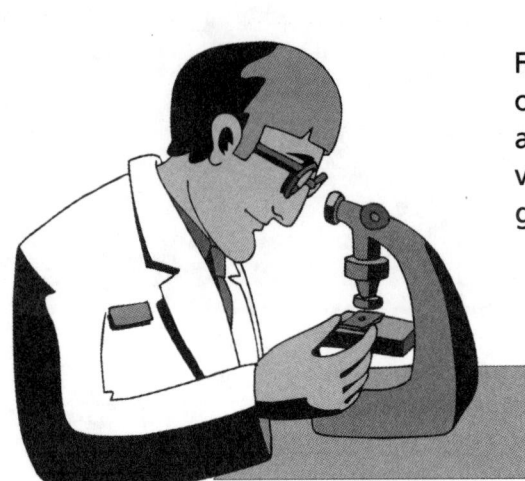

The Forensic Serologist

Forensic serologists apply the principles and techniques of biochemistry, serology, immunology, hematology, and molecular biology to the identification and individualization of blood and other body fluids. Serologists ask the following questions about their sample:

1. Is this the stain of a body fluid?
2. What type of body fluid was it?
3. What species did the stain come from?
4. Does the stain have the same blood types or DNA profile as a certain known sample?
5. Can we reconstruct the evidence to determine where it came from?

FIRST STEP: CONDUCT FIELD TESTS TO DETERMINE WHETHER BLOOD IS PRESENT AT THE CRIME SCENE

Bloodstains at crime scenes are not always directly visible to the naked eye. If blood is present, it may have come into contact with other substances; or the perpetrator may have tried cleaning it up. Therefore, crime scene investigators must conduct chemical screening tests to determine if blood is present at the crime scene.

There are four common chemical reagent field tests that are used at crime scenes to determine the possible presence of blood. These tests work best to discover areas where an attempt has been made to clean up the blood. Because they are such sensitive tests, they will work even after the suspect has mopped or wiped the area to remove any blood. Additionally, bloody shoeprints developed with chemical reagents have been known to identify the make of the shoe that made the print.

1. **Kastle-Meyer (Phenolphthalein) Test.** A cotton swab is rubbed against the suspected bloodstain; and then a drop of Phenolphthalin (reduced Phenolphthalein) plus a drop of hydrogen peroxide (3 percent) is added to the swab. If the stain contains blood, the swab will turn pink-to-red within 15 seconds.

2. **Leucomalachite green.** This test is conducted in the same manner as the Phenolphthalein test above but without the peroxide. With this test, a positive blood result will immediately turn the swab into a greenish-blue color.
3. **Tetramethylbenzidine.** This test is also conducted like the Leucomalachite test, except a positive reaction is indicated by an intense blue color.
4. **Luminol.** Luminol is also a field test that you may have seen on one of the forensic investigation television shows. It involves spraying the chemical onto the suspected bloodstain and viewing it in total darkness. If blood is present, it will begin to luminesce (glow in the dark) within 5 seconds.
5. **Fluorescin.** This is a chemical used for a presumptive test for blood. It gives a fluorescence reaction (glowing yellow-green) in the presence of Heme-like molecules.

It is important to remember that these field tests are not conclusive proof of blood, since false positives can occur (meaning that the test shows the presence of blood when there is actually none). Rather, they indicate to crime scene investigators that blood is *most likely* present, and the stains should be submitted to the forensic laboratory for further analysis and testing to confirm that blood is indeed present. Thus, while a negative field test definitely does mean that blood is *not* present at the scene; a positive field test only means that further steps need to be taken to *confirm* the presence of blood using a high standard so that the evidence will be legally admissible in court. Additionally, if a positive reaction occurs between the chemical agent and the suspected bloodstain, it does not tell us what kind of blood it is (such as whether the blood is from a human, cow, or dog), only that it is indeed blood.

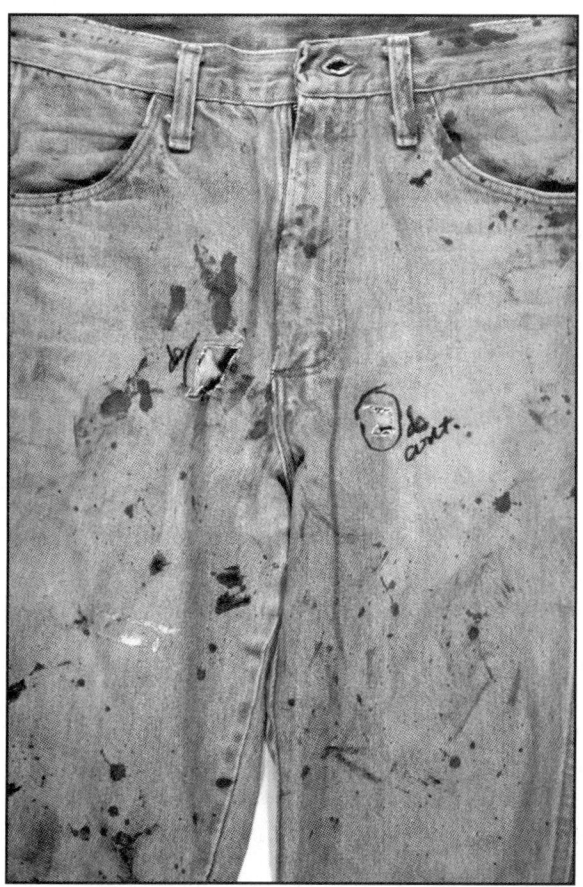

◀ Blood stains found on a suspect's jeans.

Forensic Science Today

CASE PROFILE
Blood Evidence Remains in Garage Even After Clean Up

Bridgeport, CT: One afternoon, Bridgeport police got a frantic call from Tammy Thief about her missing boyfriend, Alex Palmieri. Tammy expressed to police that she was extremely fearful that something terrible happened to Alex. Tammy told investigators that Alex had a job cashing checks for Thomas E. Marra, Jr. who might be responsible for Alex's disappearance.

The Bridgeport police questioned Tammy about why she believed that Thomas Marra may have harmed Alex. She explained that she overheard Alex crying and talking to his father about how he had to plea bargain to get out of a desperate situation involving illegal check cashing that he had been doing for Marra. To begin the search, the police needed a physical description of Alex Palmieri. Tammy described a grey jumpsuit that Alex had been last seen wearing along with a unique and quirky pair of sneakers he had just gotten that week.

Investigators then questioned Thomas Marra and soon discovered that he was working as a paid informant for the FBI. When questioned about the disappearance of Alex Palmieri, Thomas Marra told police that he had no idea what could have happened to him.

In the months following Alex's disappearance, several other associates of Marra's also turned up missing. Police were now more than just suspicious of Marra, so they started questioning his associates.

When detectives questioned Nicholas Byers, a close friend of Marra's, he advised police that Marra had told him that he had killed Palmieri and dumped his body in the water near a local boat dock. Byers also told investigators that after Marra killed Palmieri he went to Palmieri's home to get access to his bank account and withdrew $20,000. This information was not enough to charge Marra or even get a search warrant for Marra's house. To charge Marra, the police desperately needed to locate Alex Palmieri's body. For six to eight months the police searched the waters with dive teams and sonar equipment, yet they found nothing.

Nearly a year later, a woman and her daughter were walking along the water and found an odd looking sneaker laying on the beach. Upon closer inspection, the shoe contained a sock with some bone fragments. When the woman reported the shoe to the police, an investigator recalled the description that Tammy Thief had given to the police about Palmieri's shoes. Tammy positively identified the left shoe as the same shoe that Palmieri had been last seen wearing.

Now armed with some evidence of a dead body, the Bridgeport police called on forensic investigator Dr. Henry Lee to examine the evidence and help solve the disappearance of Palmieri. With the help of Marra's ex-wife, the Bridgeport police obtained a warrant to search Thomas Marr's garage, the location where Palmieri usually met with Marra.

Dr. Lee led the forensic team that surveyed the garage. He theorized that if Palmieri was killed in the garage, blood splatter would more than likely have been deposited on the walls and maybe even up onto the rafters. Although the garage had been recently painted, Dr. Lee sprayed chemicals onto the walls of the garage to expose any trace of blood evidence. The chemicals revealed tiny traces of blood splatter. After collecting samples, Dr. Lee scraped shards of wood from in between the floorboards. The wood samples also tested positive for traces of blood evidence.

continued on next page...

Chapter 7
Forensic Serology

The blood evidence collected from Marra's garage gave enough probable cause for a warrant to search Marra's van, which may have been used to transport Palmieri's body to the water. Dr. Lee examined the van and, after taking it apart screw by screw, he found traces of blood evidence.

Finally, Dr. Lee compared DNA from the bone fragments found in the shoe with the blood evidence collected from the van and garage. It matched. Lee could now place Palmieri in Marra's garage and van.

Police charged Thomas Marra with the first-degree murder of Alex Palmieri. Marra said he had nothing to do with Palmieri's death or the disappearance of his other associates. At the trial Dr. Lee explained the forensic evidence to the jury. After deliberation, the jury found Thomas Marra guilty of first-degree murder. Thomas Marra will spend the rest of his life in prison with no possibility of parole.

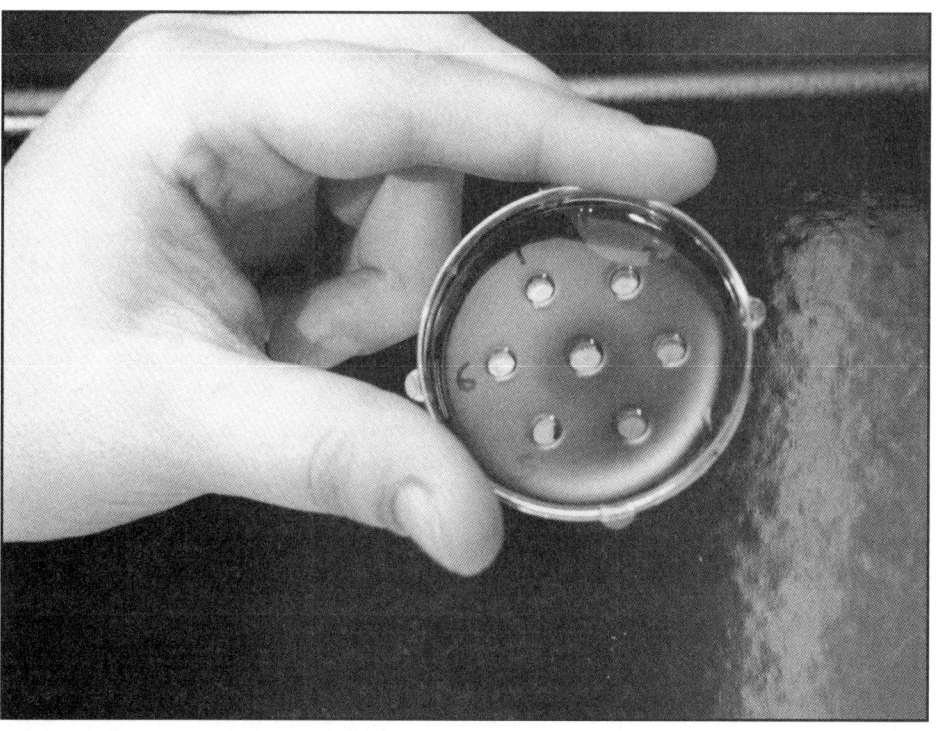

▲ Double immuno-diffusion plate used for species identification of blood stains.

COLLECTING LIQUID BLOOD EVIDENCE AT THE CRIME SCENE

If liquid blood is present at a crime scene, every effort should be made to collect a sample before it coagulates. Table 1 on the following page lists the different ways of collecting liquid blood at a crime scene.

COLLECTING EVIDENCE FROM BLOODSTAINS AT A CRIME SCENE

In general, if a bloodstain is still wet on a piece of clothing, the material should be allowed to dry undisturbed (without heating or blowing fans) before the item is packaged in clean paper. This type of evidence should never be placed in plastic or airtight containers, since retained moisture will speed degeneration of the blood in the stain, often making it useless for analysis. If a bloodstain has already dried, the entire piece of material should be transported to the lab as soon as possible. If the stain, however, is on something solid and nonabsorbent (like a cement wall), it can be collected either by scraping it onto a piece of paper, or transferring it onto cotton swabs that have been moistened with saline solution or distilled water. Table 2 on the next page lists the different ways of collecting bloodstains at a crime scene.

EXAMINING AND ANALYZING BLOOD IN THE FORENSIC LABORATORY

After blood is tested and collected from the crime scene, it is transported to the forensic laboratory where it must undergo a series of tests for various purposes.

First, the sample is tested to identify and confirm that it is indeed blood. Laboratory personnel conduct chemical screening tests using the same reagents as those used in the field tests. They also perform confirmatory tests, such as microcrystal tests and immunological tests to confirm that the sample is blood.

Blood can also be tested to determine the species of animal from which it originated. One such test is the immunological precipitin test, where an antiserum for a particular species causes a visible hazy white area to develop at the interface between the suspected blood solution and the test reagent.

As noted earlier in this chapter, prior to 1990 forensic serologists also determined the type of the blood contained in the bloodstain. Two different tests—which detect ABO antibodies in the serum or which detect blood group antigens—can determine whether the blood is type A, type B, type O, or type AB. Additionally, there is a series of additional tests that scientists conduct on blood in the forensic laboratory. These tests might include typing of red blood cell isoenzymes and serum proteins, typing of the serum protein Gm and Km factors, or typing of genetic variants of red cell hemoglobin through electrophoresis or isoelectric focusing. Nowadays, however, DNA analysis is the method of choice in forensic laboratories.

Chapter 7
Forensic Serology

Table 1
Methods Of Collecting Whole Blood at Crime Scenes

Blood Condition	Collection Mode
Fresh, liquid blood	Use hypodermic syringe or pipette; put blood into EDTA vial. Use a clean cotton swab and air dry.
Fresh, wet, thick, clotting	Collect as above.
Whole blood from a living person	Have person drawing blood collect two tubes: one with EDTA anticoagulant, the other without any anticoagulant, and refrigerate. Transport to lab immediately. Keep if overnight storage is necessary.

Table 2
Methods Of Collecting Bloodstains at Crime Scenes

Blood Condition	Collection Mode
Crusts of dried blood	1. Scrape into clean vial. 2. Scrape into paper, fold, then place in an envelope. 3. Tape lift from surface. Note: Collect material from surrounding area as a control.
Stained knives, rocks	Submit the item without sampling.
Upholstery, rugs (fabric)	Cut out section and submit. Note: An unstained area should be submitted as a control.
Stains on walls: small stains	1. Moisten unused cotton and swab stained area gently until swab has uniform deep red or brown color and air dry. Note: Also collect a control sample. 2. Tape lift
Very small stains	Same technique. Use swab or tape lift.
Large stains	Use swab.
Clothing	Air dry at room temperature; keep out of direct sunlight; put each item in a separate bag and staple shut—never use plastic bags.

Forensic Science Today

COMPONENTS OF HUMAN BLOOD

Cellular
- Red Blood Cells (RBC)
- White Blood Cells (WBC)
- Platelets

Plasma
Serum—plasma-clotting agents
- Antibodies
- Enzymes
- Proteins

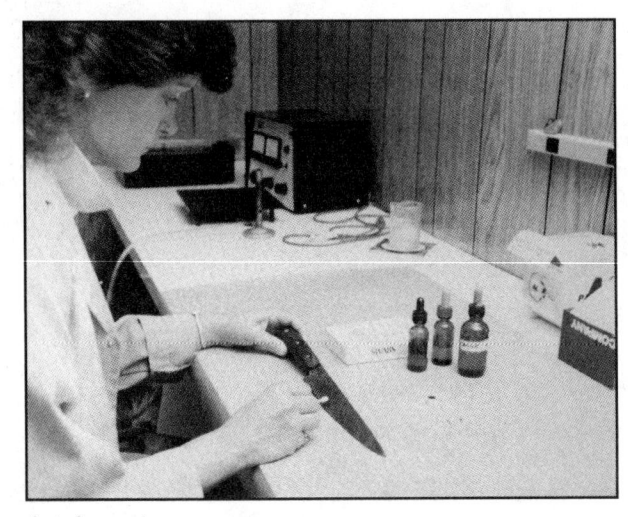

▲ A forensic serologist examines a bloody knife from a homicide scene.

Table 3
Serological Typing Results

	ABO	Rh	ADA	PGM	AK
victim	O	+	1	2-1	1
suspect	B	+	1	1	2-1
jean	H	+	1	2-1	1
shirt	H	+	1	2-1	1
stain	H	+	1	2-1	1

Serological typing results show blood stains found on a suspect's clothing could have originated from the victim and could not have been contributed by the suspect.

TECHNICAL DEFINITIONS

Serology: The study of fluids remaining after coagulation, particularly in blood.

Blood Type: Also known as Blood Group, the system of classifying human blood according to the presence or absence of genetically determined antigens upon which immunological compatibility is based. This system has largely been replaced by DNA testing in most forensic laboratories.

False positive: A false field test result indicating that blood is present at a scene when in actuality no blood is present.

Phenolphthalein: A chemical reagent that is used to field test at crime scenes for the presence of blood, which turns pinkish-red within 15 seconds if blood is present.

Leucomalachite Green: A chemical reagent that is used to field test at crime scenes for the presence of blood, which immediately turns a greenish-blue color if blood is present.

Tetramethylbenzidine: A chemical reagent that is used to field test at crime scenes for the presence of blood, which turns an intense blue color if blood is present.

Luminol: A chemical reagent that is used to field test at crime scenes for the presence of blood, which must be applied to suspected bloodstains in total darkness, and makes bloodstains luminesce if blood is present.

Chapter 8

DNA Evidence

CHAPTER OBJECTIVES

✔ Understand the structure and function of DNA, including the roles of nucleotides, tandem repeat sequences, and variable number tandem repeats (VNTR).
✔ Explain the difference between nuclear and mitochondrial DNA, and how they are used differently in forensic science.
✔ Give examples of how DNA evidence can help solve crimes.
✔ Give examples of where DNA evidence may be found at a crime scene.
✔ Describe the steps in the proper collection and preservation of DNA evidence.
✔ Explain the four stages of DNA laboratory analysis, namely: extraction, amplification, analysis, comparison.
✔ Understand how the CODIS DNA database is used.
✔ Define the following vocabulary words and definitions: *nuclear DNA, mitochondrial DNA, tandem repeat sequence, variable number tandem repeat (VNTR), Polymerase Chain Reaction (PCR), Restriction Fragment Length Polymorphism (RFLP), Short Tandem Repeat (STR) markers*, and *Combined DNA Index System (CODIS)*.

The use of DNA evidence in solving crime is exploding around the world; and the availability of new and improved technology makes DNA analysis more critical than ever in the field of forensic science. Analysis of DNA evidence is extremely valuable in a wide variety of cases, such as homicide, sexual assault, physical assault, hit-and-run incidents, missing person investigations, identification of human remains, and paternity cases.

TYPES OF DNA DATABASES

- Sex Offender
- Felony Offender
- Felony Arrest
- All Type Arrest

Total 1,407,000 DNA profiles in the U.S. CODIS database and 64,500 in the forensic database.

▶ The structure of DNA, found in the nucleus of a cell, resembles a double-helix.
(Photo courtesy of Bruce Bowdle, FBI Laboratory.)

Forensic Science Today

KEY TERMS

DNA
Nuclear DNA
Mitochondrial DNA
Tandem Repeat Sequence
Variable Number Tandem Repeat (VNTR)
Polymerase Chain Reaction (PCR)
Restricted Fragment Length Polymorphism (RFLP)
Short Tandem Repeat (STR) Markers
Combined DNA Index System (CODIS)

DNA is a unique genetic identifier found in the human body. Half of an individual's DNA is inherited from the mother, and the other half from the father. Each person's DNA is unique because it is inherited from both parents in a completely random fashion. Thus, like the fingerprint, a person's DNA is unlike that of any other human, except in the case of identical twins. Moreover, an individual's DNA is the same in every nucleated cell of the body. Not only does DNA vary from person to person, this variation can also actually be measured in a scientific and legally admissible manner. Therefore, DNA analysis is the most accurate way to determine the identity of the person who left behind evidence at the scene of a crime in the form of blood, body fluids, hair roots, or bones.

WHAT IS DNA?

DNA is an abbreviation for Deoxyribonucleic Acid, which is composed of a sequence of chemicals—known as nucleotides—that contain an individual's unique genetic material. DNA contains two strands of these nucleotide chains, which together twist around each other to form a double helix.

The four different nucleotide subunits that are the building blocks of DNA are arranged in a specific sequence that contributes to one's genetic makeup. Therefore, a gene can be defined as a region of DNA that controls a specific hereditary characteristic. Each nucleotide has a letter abbreviation that corresponds to the first letter of the chemical name of the nucleotide. These four abbreviations are G (guanine), A (adenine), T (thymine), and C (cytosine). They also link up in certain ways: One way to think of it is that A and T are like "sisters;" and G and C are like "brothers." Thus, A and T are always paired, and G and C are always paired.

The four nucleotides are always lined up together within the two strands of the DNA double helix, but they occur in different orders and repeat in an almost endless variety of patterns. These different patterns sometimes repeat themselves in different sequences, and these repeating sequences are known as *tandem repeat sequences*.

Every individual has a different number of copies of their of tandem repeat sequences. For instance, someone might have three copies of the repeat sequence AGTC. Thus, this portion of DNA sequence would look like this: AGTCAGTCAGTC. These repeat sequences are one of the reasons why each individual's genome is unique. What is a genome? It is the totality of genetic information belonging to a cell or an organism; in other words, the DNA that carries this genetic information.

Another term to be familiar with is *Variable Number Tandem Repeats*, also referred to by its abbreviation, VNTR. This refers to the *length* of tandem repeats in your DNA. The chances of anyone else in the world having the same VNTR results, even if replicated under laboratory conditions, are less than 1 in 1,000,000. This is why knowledge of VNTR is so useful in forensic science.

Chapter 8
DNA Evidence

COMMONLY USED DNA TERMINOLOGY
- **DNA**—Deoxyribonucleic Acid
- **RFLP**—Restriction Fragment Length Polymorphism
- **PCR**—Polymerase Chain Reaction
- **STR**—Short Tandem Repeats
- **Y STR**—Y Chromosome Short Tandem Repeats
- **mtDNA**—Mitochondrial DNA
- **SNPs**—Single Nucleotide Polymorphism

WHERE IS DNA FOUND IN THE CELL?

DNA is found both in the nucleus and mitochondria of all human cells, except red blood cells (which have no nucleus). *Nuclear DNA*—DNA in the nucleus of the cell—looks and acts differently than *mitochondrial DNA*, which is found in the cell's mitochondria. Nuclear DNA is in the form of *chromosomes*, which is a very long DNA molecule containing the nucleotide sequences that code for genes. Humans have 46 chromosomes in 23 pairs. These chromosomes are made up of DNA that codes for thousands of genes. Nuclear DNA is inherited from both parents. In addition to non-sex determining chromosomes, which are known as autosomes, humans also have what are known as sex chromosomes. These consist of either an X chromosome or a Y chromosome. Females have two X chromosomes, and males have one X chromosome (inherited from their mother) and one Y chromosome (inherited from their father). Thus, forensic scientists can also determine the sex of the person whose DNA has been found at a crime scene.

While every cell contains only one nucleus, it can contain over 10,000 mitochondria cytoplasma. Mitochondrial DNA, unlike that found in the nucleus, is smaller than nuclear DNA, and circular in shape. Mitochondrial DNA is only inherited from a person's mother. There are many mitochondria within a cell, and their function is to convert energy into a form that can be utilized by the cell. The DNA within the mitochondria produces enzymatic proteins that function in this energy conversion process.

Both nuclear DNA and mitochondrial DNA have pros and cons in relation to forensic identification. Nuclear DNA is more informative and can tell us more about the person to whom the DNA belongs. Mitochondrial DNA, however, is useful because it can last longer than nuclear DNA, and can be more easily extracted from degraded samples such as ancient bones. This is because there are thousands of mitochondria in each cell (contrasted to only one nucleus), and mitochondrial DNA is smaller and circular in shape.

Both nuclear and mitochondrial DNA have had interesting uses in addition to solving crimes. Nuclear DNA has been used to determine the descendants of Thomas Jefferson, to determine paternity (the father of a child) in child custody/support cases, and to analyze the genetic qualities of Egyptian mummies. Mitochondrial DNA, on the other hand, has been used to test Peruvian mummies and the charred remains of Tsar Nicholas and his family, Russian royals who were murdered during the Russian revolution of the early 20th century.

WHAT CAN DNA TESTING DO FOR INVESTIGATION OF A CRIME?

DNA analysis is useful for a number of different purposes in solving crimes. First, DNA can identify the source of biological evidence found at a crime scene. It can also tell us who is definitely *not* the source of such evidence; and can then redirect the investigation in a new direction. DNA evidence can also aid investigations by doing the following:

- identifying an unknown victim's remains,
- linking serial crimes,
- identifying the number of assailants,
- identifying additional victims,
- identifying possible suspects whose DNA has been stored in a convicted felon DNA database, and
- exonerating people who have been wrongfully convicted of a crime they did not commit (see the section later in this chapter on The Innocence Project).

In addition, DNA analysis can tell us the following:

- whether the sample contains DNA from more than one person,
- the minimum number of contributors of a sample,
- the gender of the contributor if the sample is from a single person, or whether a male has contributed to a sample containing a mixture of DNA from more than one person, and
- whether there is a biological relationship between sample donors.

WHERE CAN DNA EVIDENCE BE FOUND?

DNA evidence can be found in a remarkable number of places. Here are just a few examples:

- on the back of a licked stamp or envelope,
- in a piece of used chewing gum,
- on a cigarette butt,
- on the sweatband of a baseball cap,
- in semen stains found on a sexual assault victim's underwear,
- in blood trapped under the handle grips of a pocket knife,
- in a saliva stain from a bite mark on a victim,
- in fecal matter or vomited materials,
- in the skin of a perpetrator found under the victim's fingernails, or in the skin of the victim found under a perpetrator's fingernails, and
- in body tissue found on the bumper of a vehicle involved in a hit-and-run incident.

In addition, forensic scientists collect known samples of DNA from victims and suspects by taking a blood sample, by collecting strands of hair containing the root bulb, and through swabs taken from the inside of the mouth.

Chapter 8
DNA Evidence

Logic Tree: Rape/Sexual Assault

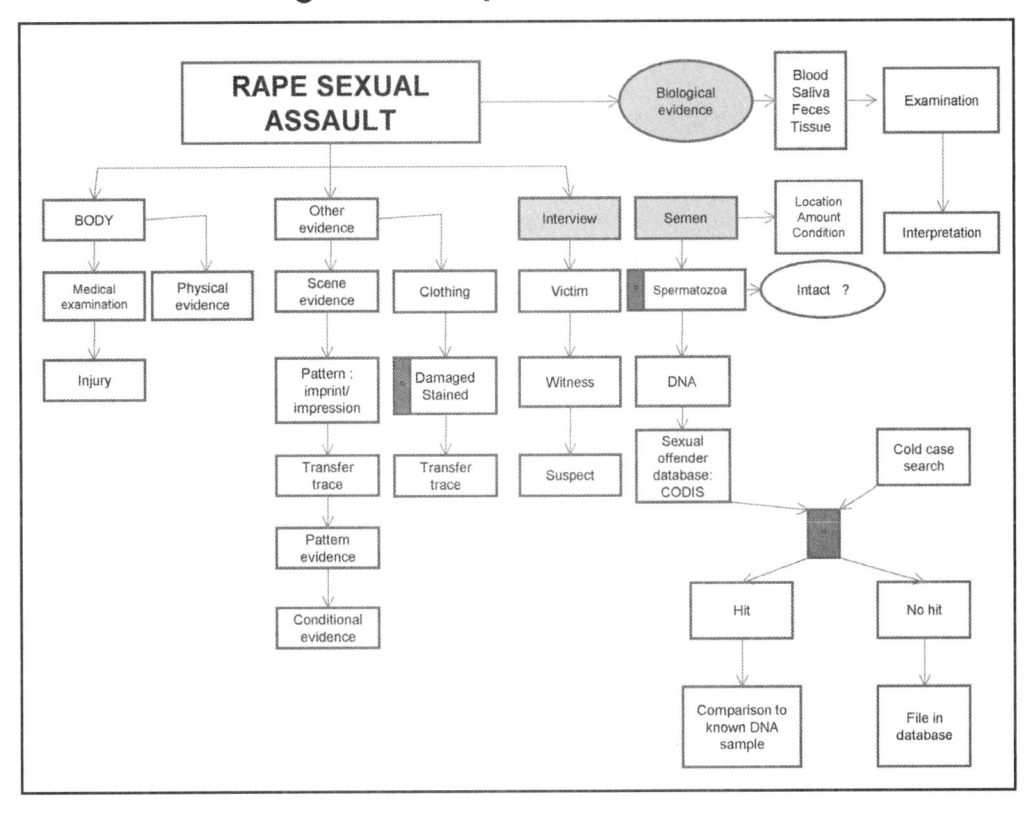

COLLECTING AND PRESERVING DNA EVIDENCE

The ability to successfully analyze DNA recovered from biological evidence at a crime scene depends on what kinds of specimens are collected, and the methods used to preserve them. Additionally, in order for DNA evidence to meet the legal and scientific requirements for admissibility in a court of law, it must be properly documented, collected, packaged, preserved, and kept within the chain of custody.

Therefore, the first step in collecting DNA evidence is thorough documentation following the guidelines applicable to all physical evidence at crime scenes. No evidence should ever be disturbed until its original condition and position have been recorded through photography, video, notes, or sketches. All specimens should be labeled properly, including the case number, item number, location, date, time, and the name of the person collecting the evidence. In addition, all patterns in bloodstains or body fluids should be documented prior to collection of samples for DNA analysis since, as we will learn in Chapter 9, bloodstain patterns themselves can tell us a great deal about what happened during the course of a crime.

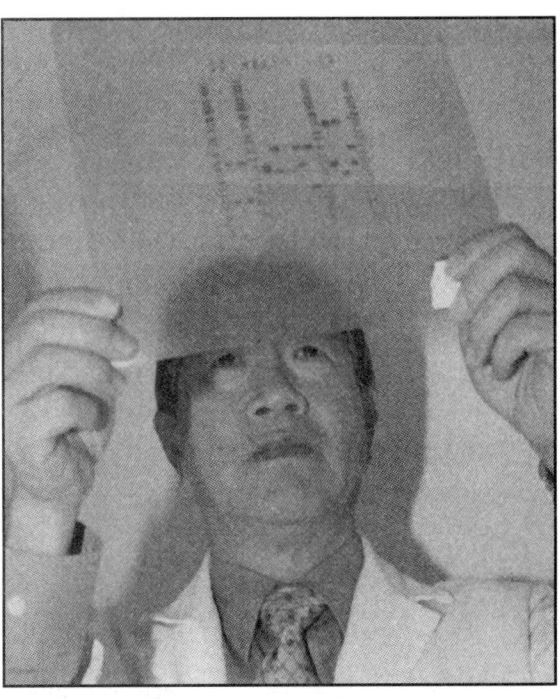

▲ Dr. Lee examining a DNA RFLP autored from a rape case.

When collecting biological evidence, gloves should always be worn both to protect the collector from biohazards and to prevent contamination of the evidence. When DNA is to be extracted from a liquid (such as wet blood or other body fluids), the liquid should be collected with a clean syringe or disposable pipette and transferred to a clean test tube. Samples found on snow or in water should be collected immediately to prevent the evidence being further diluted by the water or melting snow. If the evidence is present as a wet stain, the stain should be air-dried before packaging, in order to prevent bacterial growth and sample deterioration. If DNA is to be extracted from a dried stain on a small piece of evidence—such as an item of clothing or small object—the entire piece of evidence should be carefully transported to the forensic laboratory for testing. If, however, the dried stain is found on an immovable object such as a wall, the stain should be cut out of the object, or swabbed after it is photographed and noted.

Sometimes DNA evidence will be extracted from body tissue, organs, or bone. In this case, each item of tissue or bone can be picked up with a clean pair of forceps—or with the hand if gloves are worn—and put into a clean container. The forceps should be carefully cleaned after each use, though it is preferable to use disposable forceps for each specimen. If no forceps are available, disposable gloves should be changed between the collection of each item. Once the evidence is in the container, the container should be sealed, labeled, and put into a freezer as soon as possible. The evidence should be kept cold during transport to the laboratory.

Finally, DNA can also be found in the root of the human hair. When collecting hair evidence, each hair or group of hairs should be picked up using a clean pair of forceps, and packaged separately in a paper fold. Hairs that are mixed with blood, tissue, or other body fluids should be air-dried prior to being placed in a clean container, which is then properly labeled, sealed, and stored in a freezer.

ANALYZING DNA EVIDENCE AT THE FORENSIC LABORATORY

Forensic biologists use molecular biology to extract and analyze DNA from blood, semen, bones, body tissue, hair roots, saliva, fecal matter, and any other material that may contain DNA. DNA analysis works well with many difficult stains, even where the sample is limited, or where the sample is aged and the biological materials have partially degraded, or are part of a mixture. The ability to separate cellular material of spermatozoa from other types of cellular material may lead to identification of the suspect's DNA profile in sexual assault cases.

There are four steps in the examination of DNA evidence at the forensic laboratory. Put simply, they are:

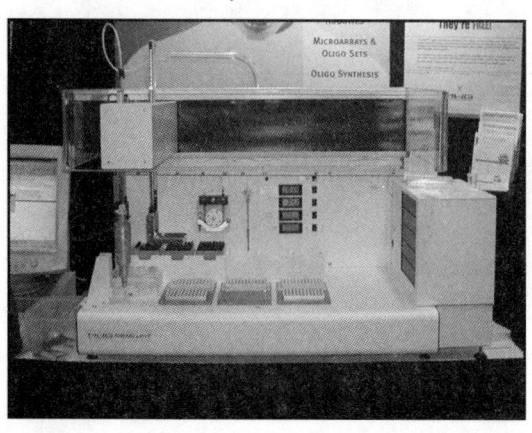

▼ DNA extraction experiment.

1. **Extraction.** DNA is extracted from the biological sample found at a crime scene.
2. **Amplification.** DNA is amplified to replicate (make copies of) the number of DNA molecules (strands) present, thereby making it easier to analyze.
3. **Analysis.** The amplified DNA is analyzed using various technical methods.
4. **Comparison.** The analyzed DNA is compared to the DNA from known standard samples, as well as to DNA profiles maintained in state and national DNA databases.

These four steps are explained in detail below.

Chapter 8
DNA Evidence

Extraction and Amplification

After a biological sample is transported from a crime scene to the forensic laboratory, the DNA is extracted from the sample so that it may be analyzed. Next, after the DNA is extracted from the biological sample, it must then be amplified. Think of the amplifier in a car stereo: it takes a small sound and makes it louder. Likewise, DNA amplification takes the very minute (small) portions of DNA available in biological evidence and makes more of it by increasing the number of copies of the molecules present so that further testing can be done.

This method of amplification is called a *Polymerase Chain Reaction*, also referred to by its abbreviation, PCR. PCR analysis is done in a machine called a *thermal cycler*, also known as a "PCR machine." In PCR analysis, the DNA molecule is placed in a mixture of nucleotides and DNA replication enzymes that cause the DNA to replicate—copy—itself. The DNA replication enzyme, which is known as "DNA polymerase," is extracted from bacteria that live in thermal pools and can survive in extremely high temperatures.

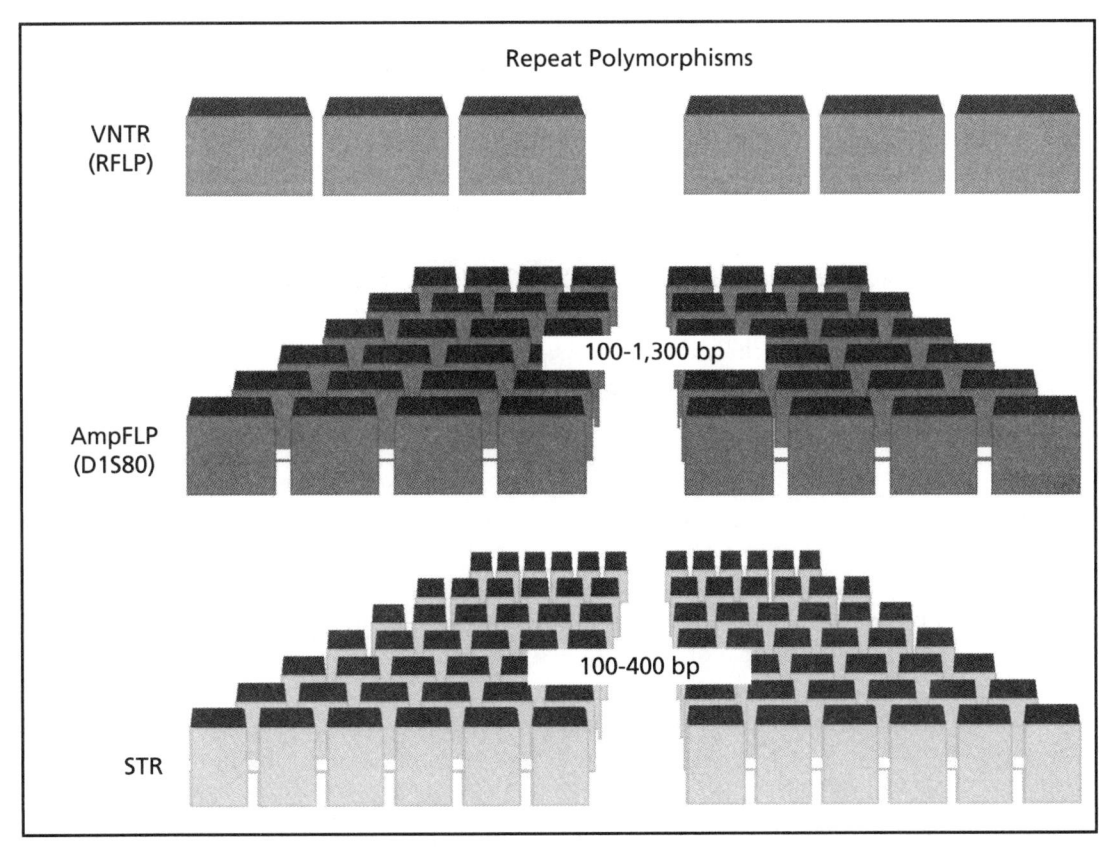

Courtesy of Dr. Carl Ladd and Dr. Michael Bourke, DNA section CSP Lab.

Once the PCR machine is started, it first separates the double-stranded DNA helix into its two separate strands using high temperatures. The process of separating DNA strands is called *denaturation*. For each cycle of amplification, the thermal cycler raises and lowers the temperature to allow denaturation. When the temperature is raised, the strands separate, causing the nucleotides and DNA replication enzymes that are present in the reaction mixture to allow the DNA to replicate, hence copying itself.

Now, each of the single strands has doubled, so that there are twice as many strands (four) than in the beginning (two). Then, the new double strands are separated again, and allowed to replicate themselves as well, so again there are twice as many strands of DNA (eight) than before (four). This process is repeated over and over again (28 to 30 times) until there are a large enough number of copies of DNA to complete the necessary lab testing and analysis of the evidence.

Analysis

The ways in which DNA evidence is analyzed in the lab are greatly expanding and improving with time. Even as we learn about the methods of analysis, some of these methods are gradually being phased out and replaced with newer and even better methods. However, it is still important to have a good understanding of these methods in order to make the best use of new technology as it arrives.

One earlier type of DNA analysis is called *Restriction Fragment Length Polymorphism* analysis, or RFLP. In this type of analysis, chemicals called "restriction enzymes" are used to cut the DNA into specific pieces, or fragments. The different lengths of DNA are separated via a process known as gel electrophoresis and then stained with a chemical called Ethidium Bromide. This allows the analyst to visualize the separated DNA under ultraviolet light. The resulting fragments will be of different sizes, based on the unique genetic characteristics of the individual to whom the DNA belongs. So, RFLP refers to the variations in the length of these DNA fragments.

Remember what we learned at the beginning of this chapter about VNTR—or Variable Number of Tandem Repeats. This refers to the unique number of repeating nucleic acid chains each individual will have. VNTR is also the cause of the variations in the sizes of DNA fragments that are created when a restriction enzyme is used to cut a piece of DNA. And this is what forensic scientists use to differentiate the DNA of one person from that of another.

Comparison

The most rapidly developing area in forensic applications of DNA technology is the analysis of *Short Tandem Repeat* (STR) markers. An STR marker refers to a specific part of DNA that is different for different people. While a single STR marker is not enough to conclusively differentiate individuals, a dozen or more STR markers can comprise a genetic profile that can virtually identify any person. These STR markers are what scientists use to create DNA profiles to be maintained in DNA databases. When all of the STR markers in a biological sample match all the STR markers in a sample given by a person or found in a database, a match has been made.

Chapter 8
DNA Evidence

CASE PROFILE
Identical Twins Both Confess to the Same Murder: Who Really Did It?

New Orleans, LA: Twenty-five year old Damon Green decided to visit New Orleans with his girlfriend in the spring of 1988. They checked into Room 202 at the Evergreen Plaza Inn and spent time taking in the sites, going to jazz clubs and gambling in local casinos. On April 4th, the couple took a cab downtown, but Damon Green returned to their hotel room alone. Around 12:30 A.M., the motel manager received a complaint about people arguing in Room 202. No other guests had complained, so the manager dismissed the incident as a fight. Later that night a motel guest walked past Room 202 and noticed that the door was ajar. From the hallway, the guest saw Damon Green's body lying face down. Jefferson Parish Police quickly arrived at the scene. Damon Green was pronounced dead at the scene.

While crime scene investigators were collecting evidence, an informant contacted Jefferson Parish Police. The informant told detectives that his roommate, James Ranna, had come home with another individual named Neel Thomas. Neel had blood on his person and both James and Neel were missing clothing. Police officers found Neel Thomas at his home and took him to the police station for questioning. After being interrogated by police for several hours, he confessed to murdering Damon Green with James Ranna. Neel Thomas was then arrested and charged with murder.

A dramatic turn of events came two days later when Neel's identical twin brother, Arthur Thomas, left an audio-taped confession to the murder of Damon Green and then committed suicide. Arthur Thomas said on the tape that he was the only one responsible for the murder and that his brother Neel had nothing to do with it. Had the informant confused an innocent Neel Thomas with his twin brother Arthur?

Unfortunately, the tape gave very few details of the murder and left investigators with a lot of questions. Baffled, the district attorney contacted Dr. Henry Lee, the noted forensic scientist, to review the case.

After Dr. Lee arrived in New Orleans, he went to the crime scene at the Evergreen Plaza Inn to analyze the evidence. Dr. Lee reviewed the crime scene photographs as he moved through the motel room. He immediately noticed that there were signs of a struggle. The furniture had been tossed and the room appeared ransacked or searched, so robbery was a possible motive. A bloodstain on the floor appeared to be a footprint. Transfer smears along the lower part of the wall were also visible. Dr. Lee noted that a hairdryer cord had been wrapped around Damon Green's neck and he had also sustained a single gunshot wound. However, a second 9mm casing was found underneath Green's body. This indicated that two shots had been fired, though only one bullet had hit Green's body.

The scientific facts in the case told police that at least two people had attacked Damon Green. The attack started near the bed. Green was hit at least two times, causing medium velocity blood spatter on the mattress. Green then tried to get away by crawling down on the floor. This caused the transfer smears on the lower portion of the wall. A telephone cord was then wrapped around Green's neck. Someone attempted to strangle Green before shooting him once in the back of the head. While the motel room had been fumed with Ninhydrant, it only produced a few latent fingerprints from the bathroom area. Those prints matched Green's girlfriend. No other prints were recovered. Detectives seized cigarette butts from the scene to test for DNA, but Dr. Lee pointed out that the twins were identical. Therefore, the twins would share the same profile.

continued on next page...

After the district attorney reviewed Dr. Lee's analysis, he noted that the scientific facts in the case didn't match up with Neel Thomas' confession and account of what had happened. Perhaps Arthur Thomas – the twin who had committed suicide – was responsible for the murder after all. Two bullet casings were recovered, but Neel said he fired his gun only once. No physical evidence linked Neel Thomas to the crime scene. When Thomas later recanted his statement, detectives went back to him to conduct a polygraph test.

When asked about his inconsistent statements, Neel Thomas admitted that his brother, Arthur Thomas, had indeed committed the murder and had given him some of the details. Because Neel didn't actually commit the crime, the details of his post-interrogation confession were wrong. Prosecutors now believed that Neel Thomas tried to protect his twin brother by confessing to a crime he didn't commit. When the true murderer, Arthur Thomas, saw that his twin brother had been arrested for a crime Arthur had committed, he felt remorse and committed suicide. "Because there is so much reasonable doubt as to Neel Thomas' guilt, justice dictates that we not go forward with the trial," the prosecutor said. "Neel Thomas may have been involved in some way, but we have no proof. We're obligated to follow the evidence."

James Ranna later confessed to his part in the killing and said that Neel Thomas had nothing to do with Green's murder. Ranna was later sentenced to five years, as part of a plea deal, for accessory to manslaughter. The murder charges against Neel Thomas were dropped.

DNA STR Population Genetics

Locus	Most	Least	Locus	Most	Least
D3S1358	0.1140	6.05 x -6	D13S317	0.1696	1.28 x -3
VWA	0.1166	2.98 x -5	D7S820	0.1174	6.05 x -6
FGA	0.0655	1.04 x -4	CSF1PO	0.1594	6.05 x -6
D8S1179	0.0137	1.04 x -4	TPOX	0.2762	6.05 x -6
D21S11	0.0841	6.50 x -6	THO1	0.1384	6.05 x -6
D18S51	0.0651	4.11 x -5	D16S539	0.1847	6.15 x -6
D5S818	0.2903	6.05 x -6			
Combined	6.027 x -12	3.597 x -62	Combined: 1 in.	1.66 x 10 ft. 11	2.78 x 10~61

CODIS: THE NEW DNA DATABASE

Just as law enforcement officials around the world can track down suspects by matching their fingerprints in a database, so too can police in many countries match a suspect's DNA profile with DNA maintained in a database. The name of this database in the U.S. is *CODIS*, which stands for Combined DNA Index System. CODIS contains STR DNA profiles of thirteen core locations within human DNA that were chosen by the FBI and other laboratories to standardize testing. These thirteen DNA locations were chosen based on their high variation in the number of tandem repeats.

The CODIS database contains over 1.4 million DNA profiles from convicted offenders. It also contains close to 65,000 DNA profiles collected from crime scenes where no suspect has been found. CODIS serves two major functions:

1. **Identify Suspects.** Police can compare DNA profiles from unknown crime scene evidence to convicted offender DNA profiles obtained locally and nationally.
2. **Link Cases.** Police can also compare unknown DNA profiles collected as evidence from various crime scenes.

A recent case in Arizona shows the usefulness of the CODIS database in keeping violent criminals off the streets. A convicted felon was in prison for aggravated DUI. When he only had 54 days left to serve of his sentence, his DNA profile was fed into Arizona's DNA database. Bingo—police got three hits, matching his DNA with DNA found in three unsolved sexual assault cases. The three sexual assault cases went to trial, and the prisoner was convicted of all three, resulting in an additional 31 to 57 years in prison. Imagine what would have happened prior to CODIS: this violent, repeat sexual offender would have been walking the streets again in 54 days.

While the CODIS database in most U.S. states only keep DNA samples of *convicted* felons, there is also a movement in some states to expand the number of people from whom DNA can be attained. For example, Virginia—which has the largest database in the U.S.—keeps DNA profiles of all people who were *arrested* of violent crimes, regardless of whether the arrestee was eventually convicted of the crime. As a result, police get 45 hits on the database per month. While defense attorneys argue that this may violate the civil rights of people who are arrested (and after all, not everyone who is arrested for a crime has actually committed one), the fact remains that the number of "hits" matching a suspect to a crime increases exponentially as the scope of the database is broadened. If DNA databases were restricted to only violent felons, we would lose 85 percent of all hits. The DNA database of Great Britain—the most comprehensive one in the world—contains 1.5 million DNA samples and results in 1600 hits *per week*!

The downside of widening the scope of DNA databases is that it also exponentially increases the amount of work that must be done in the lab. This creates the need to prioritize cases, with the result that some cases can wait for months or years before being tested. For example, in New York City alone, there are over 16,000 rape kits—kits containing valuable DNA evidence of rapists—sitting unanalyzed. That's 16,000 rape victims waiting for justice, and potentially 16,000 violent predators walking the streets. This is just one reason why we are bringing forensic science to young people—the field is in desperate need of dedicated, skilled, and highly trained individuals to work in the labs as well as on the streets solving crime.

Forensic Science Today

DNA EVIDENCE: IS IT PERFECT?

While DNA analysis is the best way to identify suspects and link individuals to a crime, it is, nonetheless, still subject to human fallibility. Just as police investigators can ruin a case by doing a shoddy job with otherwise good evidence, so too can DNA evidence lose its value if the people working with it fail to maintain rigorous standards. For one thing, there are simply not enough resources—meaning people and money—to keep up with the huge backlog of testing that advances in DNA technology have brought about. For another, problems can occur in forensic laboratories that can also result in a miscarriage of justice if the evidence is mishandled. In order to keep up with the demand for quality DNA testing that will be admissible in court, forensic laboratories could benefit from accreditation processes, peer review, and proficiency exams of its employees. It is the interaction of science with human behavior that can make or break the value of DNA evidence in court; and it is the duty of those working in the world of justice to address the myriad of human-caused weaknesses in the system.

THE INNOCENCE PROJECT

DNA technology is now being used not only to convict the guilty, but also to set free the wrongfully convicted—those people who were sent to prison for crimes they did not commit. Since the advent of post-conviction DNA testing of prisoners, over 220 people have been exonerated—proven innocent—of crimes for which they have been sent to prison in the United States. Of these, at least seventeen had been on death row. There is no way to know if any innocent people have been executed.

The Innocence Project is an organization formed in 1992 by a law professor and other lawyers to use DNA testing to prove the innocence of wrongfully convicted inmates. At the Project, law students, lawyers, and volunteers work together to assist prisoners whose long-standing claims of innocence can be proved or disproved through DNA testing. Over the past thirteen years, the Innocence Project has spawned a national civil rights movement and a network of over 25 similar projects across the country and even around the world. This movement brings together Democrats and Republicans, prosecutors and defense attorneys, and advocates for both prisoners and victims to free the innocent and track down the true perpetrators of crimes. For more information go to www.innocenceproject.org.

Chapter 8
DNA Evidence

Sample Case Analysis

THE LIFE OF A CASE FROM COLLECTION TO COURT

Imagine you are a forensic scientist working in a crime lab specializing in analysis of DNA evidence. Here's what a typical case scenario might look like. It's 10:00 A.M. and the lab receives a call from Detective Alice Jones. She will be bringing in evidence samples from a crime scene today. The case samples arrive. What do you do?

First, unpack the evidence and record the following:
- How was the evidence delivered?
- Who sent or delivered the evidence?
- Exactly what was received?
- Who received the evidence?
- Who unpacked the evidence?
- Where was the evidence stored?
- What temperature was the evidence stored at?
- What is the evidence?
- In what condition is the evidence?

Why must such particular detail be noted with each piece of evidence? In Chapter 3, we learned about the "chain of custody." As we learned, it is vital to keep track of evidence at every stage as it changes hands, so that it will be admissible in court.

Now, imagine that you received the following evidence from Detective Jones:

Item #1 (Q-1): Bloodstain sample of unknown origin from the wall near the light switch

Item #2 (Q-2): Men's running shoe of unknown origin with bloodstain on left side of right shoe, recovered from the suspect's house

Item #3 (Q-3): Bloodstain sample of unknown origin from the bathroom floor

Item #4 (K-1): Known standard blood sample of the suspect, taken while suspect was in custody

Item #5 (K-2): Known standard blood sample from a female victim, taken while victim was in hospital

Item #6 (K-3): Known standard blood sample from a male victim, taken while victim was in hospital

The next thing you do is separate the known samples (labeled with the letter "K") from the questioned evidence samples (labeled with the letter "Q") collected from the crime scene. Then, for both sets of evidence, you record all the steps you take in the procedures used, as well as who performed each procedure. For each set of evidence (known samples and questioned samples from the crime scene), you take the following steps:

- Extract DNA from each sample.
- Estimate amount of DNA recovered of good quality from each sample.
- Using PCR analysis, amplify the correct amount of sample for the appropriate STRs.
- Run samples on the instrument which separates the DNA fragments.

Now, you must analyze and report your results:

- Use computer software to visualize and evaluate the results from the samples and controls.
- Compare results for crime scene evidence with the results for known samples.
- Draw your conclusions and write up your report.

In this imaginary scenario, here are the conclusions you may have made for each of the evidence samples collected from the crime scene:

Item #1 (Q-1)—Bloodstain from wall near light switch—results:

- Mixture of male and female DNA.
- Female DNA is consistent with the female victim.
- Male victim is excluded from this evidence.
- Suspect is excluded from this evidence.

Therefore, the blood on the wall near the light switch came from the female victim and also from an unknown male. (Who was the unknown male?)

continued on next page...

Item #2 (Q-2)—Men's running shoe with bloodstain on the left side of the right shoe, recovered from the suspect's house—results:

- Blood came from a single source: female.
- DNA from the bloodstain matches that of the female victim's DNA profile.
- DNA recovered from the sweat on the inside of the shoe is consistent with the suspect.

Therefore, the running shoe had been worn by the suspect, and the blood on it came from the female victim.

Item #3 (Q-3)—Bloodstain from the bathroom floor—results:

- Blood came from a single source: male.
- DNA profile matches that of the suspect.

Therefore, the bloodstain on the bathroom floor came from the suspect, who must have been injured in some way during the course of the crime.

Now, things are not looking very good for the suspect, are they? After all, the female victim's blood has been found on his shoe; and his own blood has been found at the scene of the crime. *But: what about the blood from the unknown male found on the wall near the light switch?* There may have been a second perpetrator in this crime that got away. How can we find out?

This is where CODIS, the DNA database, comes in. Using the blood sample from the wall, you run the DNA profile of the unknown male in the CODIS database. First, you run the DNA against the database containing DNA profiles of other convicted felons. Imagine you hit a dead-end here, and the DNA from the unknown male does not match any convicted felons in the database. The next step would be to run the DNA profile against other DNA profiles also found in other crimes, for which no match to a person has been found. Good news—you get a hit, and you see that the same unknown male was also involved in two other unsolved crimes: one burglary, and one sexual assault.

Now where does this take us? Well, things are still not looking very good for the suspect, since he is as yet the only suspect who can be definitively linked to the crime. On another level, this case is now linked with two other cases—a burglary and a sexual assault—which may (or may not) help Detective Jones find the "second suspect."

Several weeks go by, and now the case finally goes to court. In this case, the suspect is found guilty because his blood was found at the crime scene (on the bathroom floor); and the blood of one of the victims was found on his shoe.

This leaves the following question unanswered: "Who was the unidentified male?" In this case, the "second suspect" escapes the long arm of the law, while his co-perpetrator pays the full price for them both. However, the DNA of the unidentified male is now preserved forever in a DNA database that can link him to any crime committed anywhere in the world. Sooner or later, justice will come knocking on his door.

Chapter 8
DNA Evidence

TECHNICAL DEFINITIONS

Nuclear DNA: A unique genetic identifier found in cells of the human body, composed of two strands of inherited material forming a double helix.

Mitochondrial DNA: The DNA found in the mitochondria of the cell, inherited from the mother only.

Tandem Repeat Sequence: The repeating sequences of nucleic acids comprising DNA.

Variable Number Tandem Repeat (VNTR): The number of tandem repeat sequences in an individual's DNA.

Polymerase Chain Reaction (PCR): The method used to amplify (copy) DNA samples in small quantities so that they may be further analyzed.

Restriction Fragment Length Polymorphism (RFLP): The variations in lengths of DNA fragments that are created by adding a special "restriction" enzyme, which cuts the DNA.

Short Tandem Repeat (STR) Markers: A specific portion of variation DNA that is compared, together with other specific portions, with other DNA samples.

Combined DNA Index System (CODIS): The name of the database in which DNA profiles of convicted felons as well as DNA found at unsolved crime scenes are kept for comparison.

Chapter 9

Bloodstain Patterns

CHAPTER OBJECTIVES

✔ Understand the nature of bloodstain pattern evidence.
✔ Understand that the surface tension in blood contributes to the shape of blood stains.
✔ Understand that Newton's Laws of Physics are used to interpret patterns of blood.
✔ Discuss what bloodstain patterns can tell us about a crime.
✔ Explain the considerations of examining, measuring, and analyzing dripping blood patterns, impact/spatter blood patterns, and special blood patterns.
✔ Define the following vocabulary words and phrases: *blood spatter, target surface, terminal velocity, directionality, angle of impact, point of origin*, and *point of convergence*.

Bloodstain patterns can reveal many facts about crimes where blood has been shed. Forensic scientists study the patterns of bloodstains left behind at a crime scene in order to reconstruct the sequence of events; and the patterns of bloodstains can sometimes be even more useful than the DNA or serological information found in the blood itself. While DNA can tell us about the "who" of a crime, bloodstain patterns can tell us the "what, how, when and where."

By examining the size, shape, location and distribution patterns of bloodstains at a crime scene, criminal investigators can determine such information as where the victim was located at the time of the injury, in which direction the weapon was pointed, and other vital facts.

WHAT DO BLOODSTAIN PATTERNS TELL US ABOUT THE CRIME?

By carefully documenting and examining bloodstain patterns at a crime scene, investigators are able to reconstruct what happened during and after the crime. This must be based on detailed measurements and analysis of blood patterns. As with all pattern evidence, blood patterns must be interpreted with extreme care. Blood pattern examiners must also conduct controlled laboratory experiments to verify the interpretations they have made of the blood patterns.

Forensic Science Today

KEY TERMS

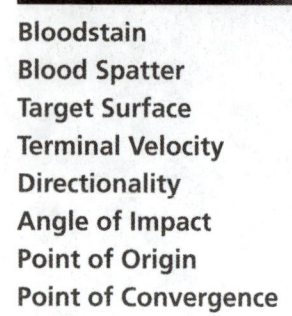

- Bloodstain
- Blood Spatter
- Target Surface
- Terminal Velocity
- Directionality
- Angle of Impact
- Point of Origin
- Point of Convergence

So, what can we learn from examining blood patterns? Here are some examples of what blood patterns at a crime scene can tell us:

1. The approximate *distance* between the blood source and the surface upon which the blood landed (also known as the *target surface*).
2. The approximate *speed* of the blood droplets at impact. This will help determine what type of injury caused the blood droplets. For example, blood spatter from a gunshot wound will travel at a much faster speed than blood dripping slowly from a cut hand.
3. The approximate *direction* the blood droplets were traveling in when they impacted the target surface. This can help determine, for instance, whether a shooter was pointing a gun from above the victim's head or at chest level.
4. The *position of the victim*.
5. The *position of the perpetrator*.
6. The *direction* in which a *weapon* may have been swinging.
7. The *direction* in which blood may have *trailed* after impact; i.e., blood dripping down a wall after impact with the target surface.
8. Whether blood was then *wiped* or *smeared* by a person or another object.
9. The estimated amount of blood and the *time* that has elapsed since the blood was deposited upon the target surface.
10. The *movement of* the blood between focal points.

Finally, bloodstain patterns can help investigators determine the sequence of events during a crime. For example, a crime scene investigator can determine from bloodstains that the perpetrator shot the victim twice, then dragged the victim to another room, while his own hand was bleeding from a knife used in self-defense by the victim. Subsequently the perpetrator attempted to clean up the blood at the scene. All the blood patterns from these three actions (shooting, dragging, bleeding hand and cleaning up) will be different.

EXAMINING, MEASURING, AND ANALYZING BLOODSTAIN PATTERNS

Once crime scene investigators have confirmed the presence of blood, they can then carefully examine the bloodstain patterns in order to reconstruct the sequence of events of the crime. There are three main categories of bloodstains:

1. Dripping blood patterns
2. Impact/spatter blood patterns
3. Other blood patterns

Determining which category a particular bloodstain falls into will enable investigators to carry out a more detailed analysis.

Dripping Blood Patterns

Blood that has dripped on to a surface will look very different than high-velocity impact spatters. When blood drips down, it has a viscosity four times greater than water. Under normal conditions, a drop of blood has an average volume of roughly 0.03 to 0.06 ml (or 20 drops per milliliter). Drops of blood have a spherical shape due to the surface tension of liquid blood; and when released from the source blood drops oscillate somewhat during free fall.

The *diameter* of a blood drop stain can tell us a great deal about how far the drop traveled before hitting the target surface. While the size and shape of a blood drop stain can vary based on different types of target surfaces (porous or nonporous, smooth or rough, etc.), in general it will produce a circular shape upon a 90-degree impact. The further the distance the blood drop has traveled, the larger the stain will be, until the point at which the blood drop has reached *terminal velocity*. Terminal velocity is the constant speed reached by a falling object when the downward gravitational force is equal to the frictional air resistance of the substance through which it is falling. A free-falling blood drop has a terminal velocity ranging from 20 to 25 feet per second. If a drop of blood falls from a distance any greater than that producing terminal velocity, it will continue to leave the same-sized bloodstain upon impact.

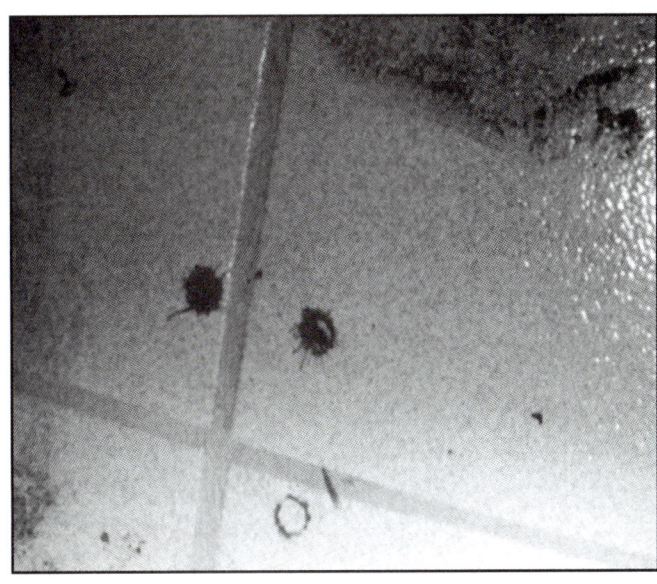

▲ Vertical low-velocity blood drops found at a crime scene.

Investigators can also determine the *directionality* of a drop of free-falling blood. This is arrived at by measuring the angle of impact and ascertaining the direction in which the blood was traveling. The *angle of impact* of a free-falling drop of blood is measured by examining the *shape* of the drop and using a trigonometric formula. The more elongated or elliptical in shape the bloodstain, the more acute the impact angle.

By measuring the length and width of the bloodstain, and determining the trigonometric relationship between the ratio of the long axis and the short axis, you can calculate the angle of impact. The trigonometric formula is as follows:

$$\text{Sine of Impact Angle} = \text{Bloodstain Width} / \text{Bloodstain Length}$$
$$\text{Sine}^{-1}(\text{width}/\text{length}) = \text{Impact Angle}$$

In general, the "tail" of the bloodstain will point in the direction of the traveling blood drop. This can also help determine the origin of the droplet.

Crime scene investigators also look carefully at *blood trails* formed by free-falling blood drops. Blood trails are formed whenever an actively dripping source of blood is moved, such as a bleeding victim who is running across the floor to another room. Blood trail patterns can tell us the direction the blood source was traveling in as well as the relative traveling speed. Investigators look closely not only at each individual blood drop in a trail, but also the distance between the drops. The faster the speed of the moving blood source, the greater will be the distance between blood drops, and the more elongated those blood drops will be. Additionally, the distance between the blood drops can be related to the rate of bleeding.

Impact/Spatter Blood Patterns
Sometimes blood does not simply fall freely through the air and land upon a surface; rather, it will move with greater velocity and spatter across the target. This is usually what happens when a victim is shot by a gun close to a wall or hit with a baseball bat. The energy and force of the bullet driving through the victim's body will cause the

blood to move with greater velocity, and to be broken into smaller droplets of blood than normal. By the same token, the swinging of a baseball bat will also create greater energy and transfer to the blood spatter.

The energy that creates blood spatter will come from either an internal or external force. Internal force comes from the body's own blood circulation system, so that a knife wound to the carotid artery of a living person will cause a strong spurting motion of blood, resulting in a spray of blood upon the target surface. External force refers to the force that created the bleeding, or that was exerted on the blood source itself. Examples of this would include the gun shot wound described above, or perhaps a heavy object dropped into a pool of blood causing impact spatter on the walls nearby.

Blood spatters can be loosely grouped into three categories based on how fast the blood drops were traveling when they hit the target surface:

- *Low-velocity impact spatter* where most blood drops remain full-sized, are not broken into small drops, and drop diameter is 4 mm or more.
- *Medium-velocity impact spatter* where most drops are broken into smaller droplets of 2 to 4 mm, resulting from a force of greater than 25 ft/sec (terminal velocity). The most common types of medium velocity impact spatters are:
    - *Projected blood.* Sometimes blood will gush onto a surface—such as when an artery is opened or blood spatter will project on the surface upon an external impact force. This pattern will have sharp edges and be in the direction on the surface in which the blood has projected.
    - *Repetitive dripping pattern*. This is what is created by blood dripping down into a pool of blood and creating small satellites of blood spatter around the pool.
    - *Cast-off bloodstain pattern.* This is created when an already bloody object (such as a bloody weapon or body part) is swung through the air, casting off droplets of blood as it arcs through the air. The resulting pattern is linear in nature and the individual blood drops are of differing shapes. This is due to the changes in impact angle.
- *High-velocity impact spatter* where blood droplets have been highly spattered and measure less than 2 mm, usually associated with forces of greater than 100 ft/sec and produced by high energy impacts such as explosions, high-speed collisions, and gunshot wounds.

Crime scene investigators must also determine the *point of origin* of the blood spatter in order to reconstruct the series of events in a crime involving bloodshed. This will allow investigators to reconstruct the sequence of events, the type of force that was used to create the blood spatter, and the relative position of people or objects near the impacted surface. The first step in determining the point of origin is to determine the *point of convergence*. This refers to the point where all lines coming from the individual blood spatter stains come together. Using long pieces of string or graphic tape, crime scene investigators create a line through the center, lengthwise, of a representative sample of each blood impact spatter. Next, investigators must determine the angle of impact of each bloodstain. Then the axis line, as represented by the piece of string or tape, will be set using a protractor. Once the point of convergence of these lines is determined, investigators can then measure the distance of the point of convergence from the floor, walls, and ceilings. This will lead to the point of origin in three-dimensional space.

Chapter 9
Bloodstain Patterns

◀ Blood spatters found on the wall of a homicide scene. Point of convergence was determined by using the string reconstruction.

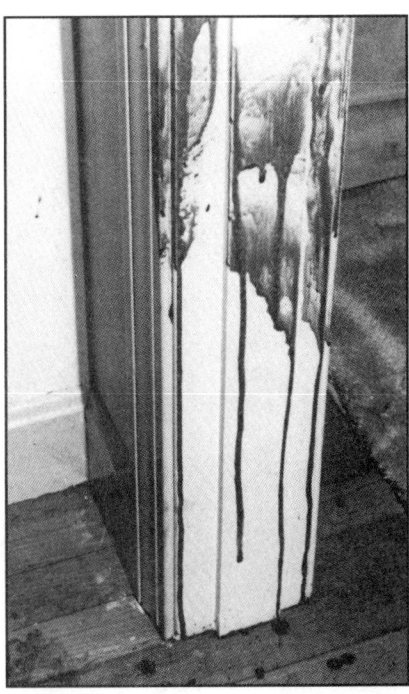

◀ Medium-velocity castoff bloodstains found at a crime scene. (left)

◀ Contact-transfer bloodstains, blood smears, blood dripping, and blood flow patterns were noted in another area of the same apartment. (right)

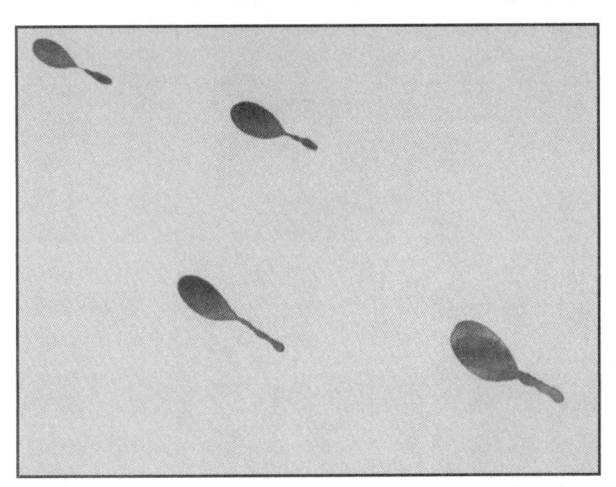

▲ Blood stain pattern showing the impact direction.

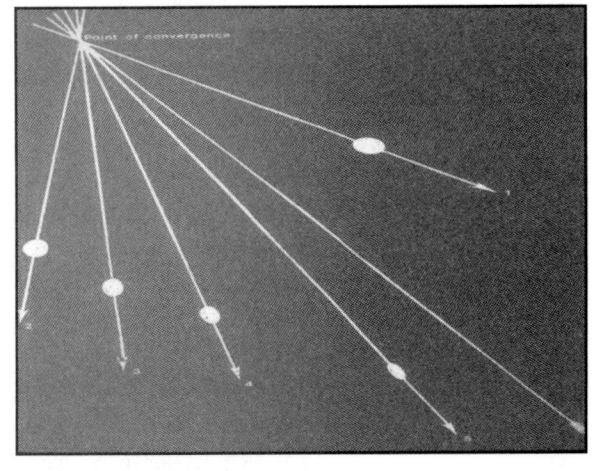

▲ Reconstruction of the point of convergence of the blood stains.

105

CASE PROFILE
Blood Stain Evidence: The Case of Kenneth Mathison

Hilo, HI: On November 26, 1992, a couple driving on Highway 11 in Hawaii was approached at 8:40 P.M. by a man asking them to call the police. The man was later identified as Sergeant Kenneth Mathison. The couple drove to a local store and called 911. They returned to the scene, lit a flare, and left when the police arrived. At about the same time, two retired police officers drove by and saw Sgt. Mathison holding the body of a woman, rocking her back and forth, in the rear of a van. The woman was later identified as Mathison's wife, Yvonne. The van was pointed the wrong way and its driver's side was tilted against some bushes at a forty-five degree angle. The retired officers did not stop, thinking it was an accident and help was on the way.

The first police officers on the scene discovered that Yvonne Mathison was dead. They found a large amount of blood in the van and on the bodies and clothing of both Yvonne and Sgt. Mathison. Police also observed a pair of extensively damaged woman's eyeglasses in the rear of the van. One lens from the eyeglasses was discovered in the front of the vehicle, under the passenger's seat.

Investigators found Yvonne's badly damaged Seiko watch hanging from the signal switch. The watch showed evidence of forced contact that could have been caused by friction against a very hard surface such as concrete. Subsequent laboratory examination indicated that the Seiko was damaged in close proximity to Yvonne's head. Fragments of her hair were discovered trapped in its cracked crystal. The watch's time and date showed 8:56 P.M. November 26.

Autopsy results indicated that Yvonne had a badly broken jaw, two broken fingers on her left hand, a fractured rib, multiple lacerations and abrasions on her head and arms. Forensic pathologists also found that the abrasions on her arms could have been caused by rope burns.

Sgt. Mathison gave a statement indicating that Yvonne was driving the van that night when Mathison told her that he was named in a paternity suit with another woman. She turned white and leapt out of the vehicle before he was aware of her intentions. Mathison stated that he had difficulty moving to the driver's seat and regaining control of the car. He turned the car around in the dark and rainy night to desperately search for her. During the search, he accidentally struck Yvonne and ran her over. Mathison claimed his van went out of control and swerved up into the muddy ditch in the opposite direction. He said that he found his wife lying in the road and picked her up and tried to reenter the van from the driver's side. According to his statement, Mathison cradled his wife's body and tried to stop oncoming traffic for help.

Yvonne Mathison was a maternity nurse at the Hilo Medical Center. She died a day later, the evening of November 27, 1992 at the age of 52. Kenneth Mathison, a sergeant in the Hilo police force, was ten years younger than his wife. In addition to his police work, he had success in recent years as a contractor and real estate investor. The Mathisons had a turbulent relationship, and had been married to each other twice. The first marriage ended in divorce several years earlier. The second marriage was in trouble.

The van was subsequently examined and large amount of bloodstains were found inside the vehicle. The following bloodstain patterns were observed:

- Contact smears and hair swipes were found on the sun visor
- Hundreds of medium velocity type blood spatter were observed on the instrumental pane
- Medium velocity impact spatters were located on the inner surface of the van

continued on next page…

- Overhead cast off type blood spatters were found on the roof
- Vertical low velocity blood drops were found on the dry wall on the floor. Some of those bloodstains were underneath a variety of soil stains
- Two bloody hand prints were noticed on the inside wall of the van

The bloodstain patterns found inside the van were inconsistent with the statement provided by Sgt. Kenneth Mathison. From these bloodstain patterns investigators reconstructed the sequence of events showing that Sgt. Kenneth Mathison had murdered his wife. He was subsequently convicted for the crime.

Other Blood Patterns

Not all bloodstains will fit neatly into either a dropped or impacted blood category. Crime scenes are as unique as the individuals who commit them (and the individuals against whom they are committed), and as a result, crime scene investigators will come across many different patterns of bloodstains depending on the circumstances. A few such variations are described below:

- **Blood pools.** Pools of blood created by blood flowing from wounds can tell investigators how much blood was shed, and the type of injury that was inflicted on the victim.
- **Contact-transfer bloodstain patterns.** Blood that has formed a pool can also be transferred onto other surfaces through contact with the surface, either directly or through an intermediate object. An example of this would be a person stepping through a pool of blood and leaving bloody footprints. Contact-transfer patterns can also be formed by other sources of blood, such as blood dripping on to a person's hand that then touches a doorknob, or a bloody hammer being dropped on the ground. Usually the contact-transfer pattern is in the shape of the object that transferred the blood.
- **Wipe/swipe patterns.** These are additional types of contact-transfer bloodstain patterns. When a clean object wipes across a bloody surface, this is referred to as a "wipe." When a bloody object wipes across a clean surface, it is referred to as a "swipe." This can give important clues about movement of the transferring objects and the direction in which they were traveling.
- **Blood imprint patterns.** These include bloody shoe prints, hand prints, footprints, fingerprints, weapon and tool prints.

THE SCIENCE USED TO INTERPRET THE PATTERNS

Blood behaves as any other fluid would. Its movements obey Newton's Laws of Motion. When in motion it continues in that same motion until a force acts on it. Gravity pulls it down toward the earth. The force applied to it is directly proportional to its mass and the acceleration it achieves. The shape that the droplets of blood take can be related to the actions that occur when the droplets are moving.

As a person moves, any fluid (e.g., blood) found on that person is moving at the same speed as that person. When gravity causes the fluid to fall off of the person, the fluid continues to move in the same direction with the same speed as it was when on the person until a force is applied to it. Air resistance and gravity slow the fluid down and pull it down toward the earth. As these forces are being applied to the fluid, the shape of the droplet of fluid changes. The droplet will change to a spherical shape. The faster the person was moving the more elongated the shape. The speed of movement of the person from whom the fluid is falling is proportional to the length of the stain.

The direction of movement of the bleeding person can be determined. All substances resist the change in movement. This resistance results in the change of the shape of the drops of fluid. When a person is moving fast enough to cause the elongated drops to have tails on them, the pointed end of the stain points in the direction the person was moving.

As a fluid falls its velocity increases due to the acceleration caused by gravity. The greater the height it falls from the greater the kinetic energy (K.E. = $^1/_2 mxv^2$) it has. The greater the kinetic energy the more force will be applied to the surface it strikes. The greater the force applied the more disruption to the spherical shape there will be. If the fluid is water or a fluid that has a high percentage of water, the surface tension of the fluid can be disrupted if enough force is applied to it. In other words, blood falling from a short distance will produce a very round, bubble-like shape upon impact with a flat surface. The further it falls the more likely that the bubble will be flat and the greater the chance that the blood will produce tiny side splatters outside of the main circle of blood.

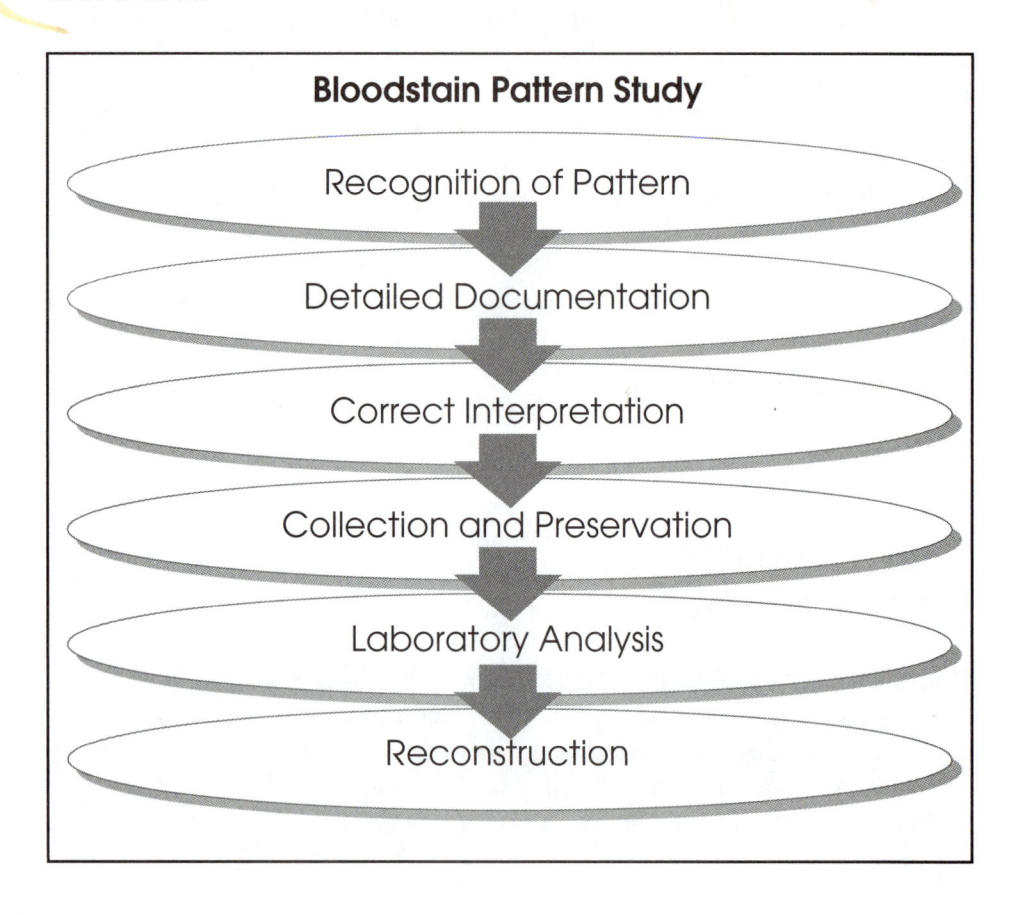

Bloodstain Pattern Study

Recognition of Pattern
↓
Detailed Documentation
↓
Correct Interpretation
↓
Collection and Preservation
↓
Laboratory Analysis
↓
Reconstruction

Chapter 9
Bloodstain Patterns

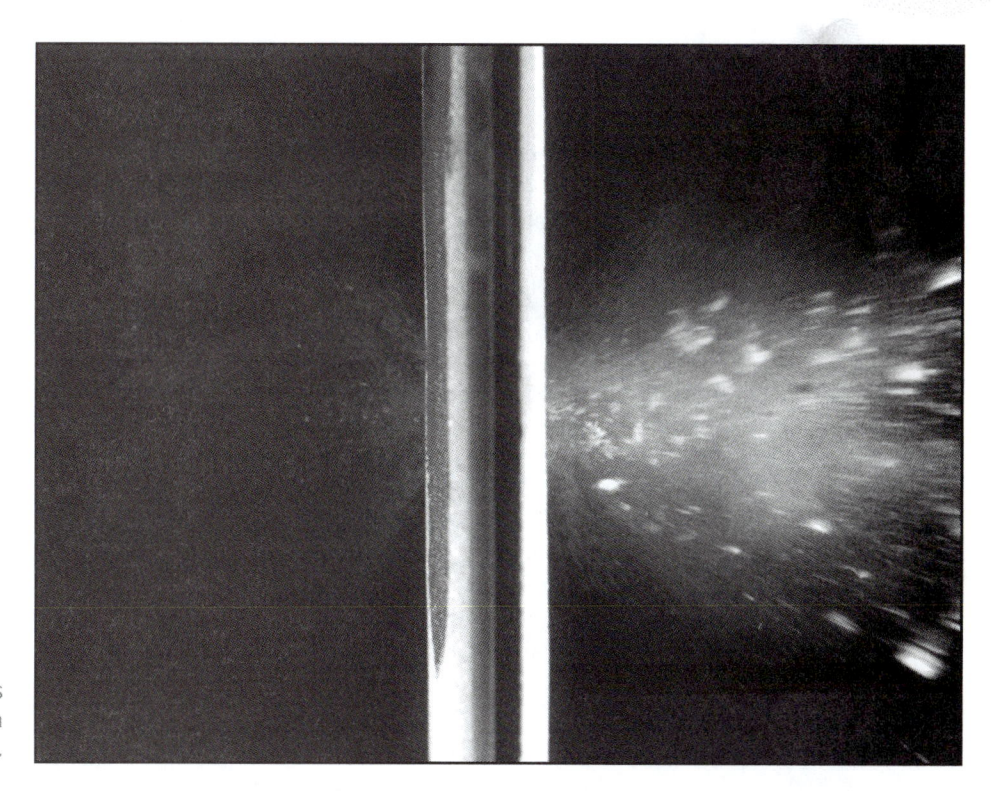

▶ High-velocity blood spatters commonly produced by a gun shot.

🔍 TECHNICAL DEFINITIONS

Blood Spatter: A bloodstain caused by medium- or high-velocity force impact of blood upon the target surface.

Target Surface: The surface upon which drops of blood land.

Terminal Velocity: The constant speed reached by a falling object when the downward gravitational force is equal to the frictional air resistance of the substance through which it is falling.

Directionality: The direction in which a drop of blood had been traveling prior to hitting the target surface and creating a bloodstain.

Angle of Impact: The angle at which a drop of blood approaches and hits the target surface.

Point of Origin: The place where a source of blood released blood drops into the air prior to the blood hitting the target surface.

Point of Convergence: The point at which all lines coming from individual bloodstains come together, used to measure the point of origin.

109

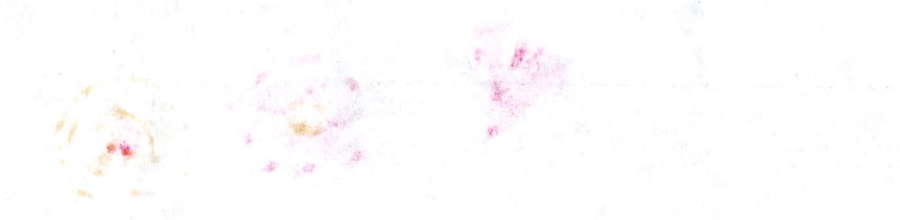

Chapter 10

Questioned Documents

CHAPTER OBJECTIVES

✔ Understand the role that document examiners play in solving crime.
✔ Identify several types of documents that can be used as evidence.
✔ Discuss the methods of how document evidence is collected and preserved.
✔ Explain at least five things that documents can tell investigators about a crime that has been committed.
✔ Define the following vocabulary words and phrases: *document examiner*, *forgery*, *known sample*, *questioned document*, and *handwriting analysis*.

Document evidence provides a multitude of clues to many different crimes. Document examiners study both the contents and the physical form of documents in order to determine the age, source, and authenticity of a document. Documents that are questionable in terms of authenticity or origin are called *questioned documents*. Usually, documents are most important when solving white collar crimes, such as fraud, bad check writing, or forged wills. However, document examination also comes in handy for solving homicides, sexual assaults, blackmail, and especially kidnapping when a ransom note has been left behind (such as in the Jon Benét Ramsey case discussed in the Case Profile at the end of this chapter).

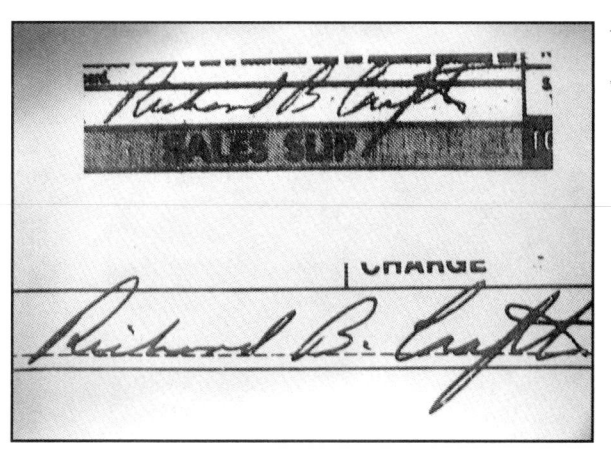

◀ A comparison of signatures used as evidence in the Woodchipper Murder case, profiled in Chapter 3.

Forensic Science Today

KEY TERMS
- Document Examiner
- Questioned Document
- Known Sample
- Forgery
- Handwriting Analysis

TYPES OF DOCUMENTS THAT CAN BE USED AS EVIDENCE

Virtually any written, printed, or typed material can be considered document evidence for solving crimes. Even writing on a wall or piece of furniture is considered a document! Here are some examples of the different types of documents that are submitted to forensic laboratories for analysis:

- Credit card slips
- Checks
- Wills
- Business records
- Shredded documents
- Ransom notes
- Robbery notes
- Suicide notes
- Lottery tickets
- Letters
- Computer printouts
- Extortion or blackmail notes

COLLECTING AND PRESERVING DOCUMENT EVIDENCE

Collecting questioned documents for laboratory analysis. Crime scene investigators must be very careful in handling questioned documents, especially to prevent their own fingerprints or DNA from getting on the sample. Using forceps or tongs, the examiner places the document in a clear plastic envelope that is then sealed and initialed. The document should not be folded or marked in any way.

Collecting handwriting samples (exemplars) from individuals. When a suspect, victim, or other individual is brought in to the police station to give a writing sample, it is very important that stress be minimized and the individual is as comfortable as possible in order to get the best sample. The conditions under which the document was made should be duplicated as closely as possible. These conditions include the type of paper, type of surface, whether a pen, pencil, or printer was used, and the context of the written message. The investigator should then read out loud the contents of the text of the known sample, and ask the subject to write out the message by hand. The subject should not be allowed to look at the original document, otherwise he or she may try to copy it. As soon as the subject has written a sample, it should be removed immediately; and he or she should be asked to write the sample several times if possible. This will help ensure that the subject is writing naturally and has given up any attempts to disguise his or her sample. Additionally, it is a good idea to ask the subject to repeat the exercise using the hand not normally used to write with, since sometimes perpetrators of crimes leave notes behind using their non-writing hand as a means of disguising their identity.

Collecting authentic handwriting samples from public records. True samples of handwriting can be found in a wide variety of public records. Some of these records include voter registration, applications for government jobs, marriage licenses, hunting licenses, birth certificates, credit card applications, hospital admissions records, military records, driver's license applications, and many more. A list of these authentic writing samples can be found in Table 1 on the following pages.

Chapter 10
Questioned Documents

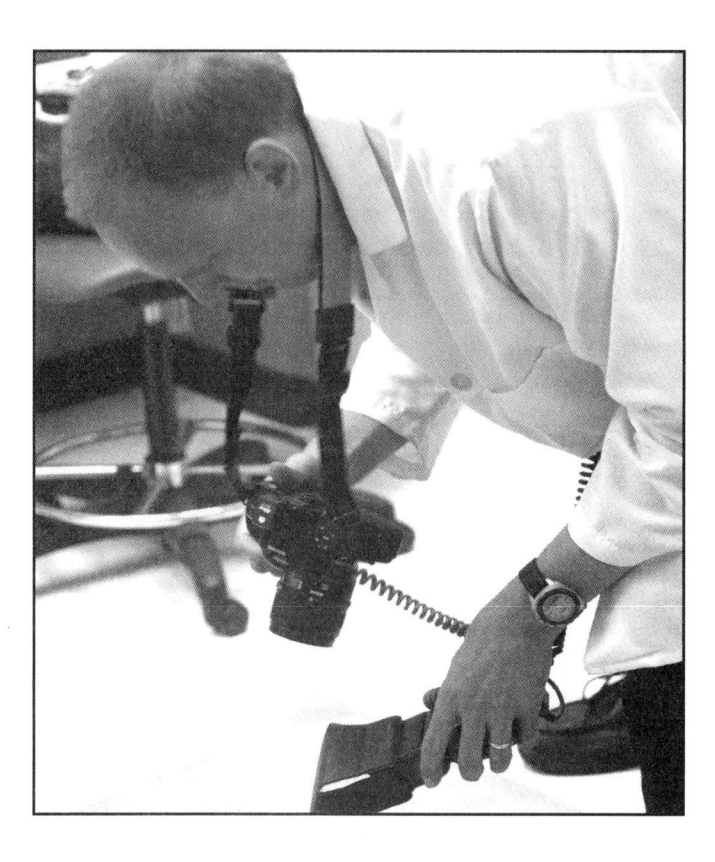

◀ A document examiner conducts an examination and photographs a questioned document.

Table 1
SOURCES OF AUTHENTIC HANDWRITING SAMPLES

City Records	Building department permits.
	City auditor: canceled checks and office records.
	City clerk: licenses for peddler, tavern, special permits, and so on, and voter registration lists.
	Personnel department: civil service applications.
	Permits for dogs, building, parks and other city facilities, and applications.
County Records	County clerk: civil service applications, claims for services or merchandise, fishing, hunting, and marriage licenses.
	Department of taxation: state income tax returns.
	Purchasing department: bids and contracts.
	Register of deeds: deeds, birth certificates, public assistance applications, and ID card applications.
	Selective service (local board): registrations and appeals.
Department Store Records	Credit applications.
	Receipts for merchandise.
	Signed sales checks.
	Merchandise delivery records.
Drug Store Records	Register for exempt narcotics or poisons.
	Signed prescriptions.

Table 1 (continued)

Hospital Records	Admission and release forms.
	Consent forms.
	Checks and payment records.
	Communications and letters.
Library Records	Applications for cards.
	Check out and reserve slips.
Education Documents	Applications for entrance.
	Athletic contests.
	Daily assignments.
	Examination and research papers.
	Fraternity or sorority records.
	Receipt for school supplies, i.e., laboratory and athletic gear.
	Registration cards and forms.
	Federal and state loan and grant applications.
	Scholarship applications.
Federal Records	Customs documents: immigration and naturalization records.
	Department of Justice (FBI): fingerprint cards, applications, and records.
	National fraudulent check file, checkwriter standards file, safety paper standards file, rubber stamp and printing standards file, and typing standards file.
	Military records.
	Patent office applications.
	Post Office department: P.O. box application, registered and special delivery receipts.
	Social Security administration: applications for numbers and benefits.
	United States Treasury: canceled payroll checks.
	Veterans administration: application for benefits for veterans and widows.
Social, Recreational, and Fraternal Documents	Civic organizations, clubs such as sports, luncheon, and so on, lodges, nonprofit groups, political groups, PTA organizations, and religious organizations.
Financial Documents	Canceled bank checks.
	Contracts and related correspondence.
	Credit applications, i.e., department store.
	Deeds.
	Deposit slips.
	Expense accounts.
	Insurance documents, including health and accident.
	Lease agreements.
	Loan company records.

Table 1 (continued)

	Microfilm bank records.
	Pension applications and checks.
	Promissory notes.
	Safety deposit vault register and applications.
	Bankruptcy proceedings, cash received slips, and withdrawal slips.
	Title company documents.
	Account applications.
	Stock certificates.
Real Estate Records	Property listing agreements.
	Purchasing contracts.
Public Utility Records (Corporate Documents)	Applications for service: cable television, internet, electricity, garbage removal, gas, telephone, and water.
	Book of accounts.
	Invoices.
	Minutes.
	Original telegram messages.
	Reports to intra and interstate commerce agencies.
Miscellaneous Documents	Administrator and estate.
	Airplane logs.
	Answers to decoy letters.
	Architect plans.
	Asylums.
	Auctions.
	Bail bonds.
	Building after-hour registers.
	Close associates.
	Complaint bureaus.
	Copyright applications.
	Death certificates.
	Decoys, delivery receipts, and return receipts for registered mail.
	Exchanges.
	Express company, cartage, and mover's receipts.
	Express records and receipts.
	Furniture contracts.
	Guardian.
	Janitors (wastepaper).
	Legal papers.
	Messenger receipts.

Table 1 (continued)

	Neighbors.
	Newspaper reporters.
	Notaries.
	Partner memos and records.
	Permit to open mail.
	Railroad passes.
	Rent receipts to tenants.
Military Documents	Bases and stations: Air Force, Army, National Guard, Navy, and Marines.
	General service related papers: tax exemption filings, lien, real estate, pension, medical, educational.
	Record depots (ex-service men).
	Selective Service (draft board) records.
	Leave records.
	Hospital records.
Motor Vehicle Documents	Applications for registration.
	Court documents relating to accidents.
	Credit card applications and invoices based thereon.
	Hotel and motel registration and reservations upon routes of travel, collected from credit purchases.
	Installment contracts on vehicle purchases.
	Insurance papers.
	Operator's and chauffeur's licenses and applications.
	Orders for service.
	Accident reports.
	Report of loss or theft.
	Tickets and fines.
On the Person	Contents of wallet, i.e., signed ID cards and photographs.
	Letters and postcards.
	Notebooks.
	Passports.
	Permits.
Personal Documents	Autograph albums.
	Automobile repair work order receipts.
	Back of photographs.
	Bank account books.
	Birth and baptismal certificates and records.
	Book contracts.
	Books in general.
	Canceled checks.

Table 1 (continued)

	Check stubs.
	Correspondence and postcards.
	Diaries.
	Family Bible.
	Greeting cards.
	Hospital and medical records.
	Insurance policies.
	Labeling on cans, bottles, and so on, in kitchen and workshops.
Household Documents	Memoranda about home and office, such as a note to a delivery person.
	Military service records.
	Pages of photograph albums.
	Passports.
	Personal notebooks.
	Prescriptions.
	Receipts, i.e., movers, credit, and rent.
	School yearbooks.
	Telephone and correspondence listings.
	Wills.
Police Department Records and General Criminal Documents	Complaints and reports to police, sheriff's departments, and district attorney's.
	Arrest records including fingerprint cards.
	Court of claims.
	Court clerk.
	Exemplars obtained incident to booking procedures.
	Jail and penitentiary records.
	Jury records.
	Juvenile court records.
	Parole and probation reports.
	Receipts for returned property.
	Writings obtained by other agencies in prior investigations.
Relatives	Letters of all types.
	Greeting and gift cards.
State Records	Conservation files: boat, fishing, and hunting licenses.
	Corrections files: probation and parole reports.
	Incorporation documents filed with state agencies.
	Motor vehicle files: driver's and title.
	Personnel files: Civil Service applications and examinations.
	Secretary of State: applications for notary public.
	State Treasurer: canceled checks.
	Taxation files: beverage and cigarette tax applications.

Table 1 (continued)

Vocational Documents	Account books.
	Applications for employments.
	Applications for professional and vocational licenses.
	Canceled payroll checks.
	Civil service papers.
	Client checks.
	Credit union paperwork.
	Employment bureau and personnel office papers.
	Labor union documents.
	Order blanks.
	Professional rolls.
	Public examinations, i.e., Civil Service.
	Receipt books.
	Receipted bills.
	Receipts for pay.
	Reports and surveys.
	Secretary.
	Stenographic and clerical memoranda.
	Time cards.
	Vacation and petty cash requests.
	Withholding exemption forms.

WHAT CAN DOCUMENTS TELL US ABOUT A CRIME THAT HAS BEEN COMMITTED?

Document evidence can be valuable in solving a variety of crimes. Here are some of the many ways in which documents are analyzed by forensic document examiners:

Handwriting comparisons. Handwriting analysts carefully examine the handwriting found on a document to determine if it was really written by the person who claims to have written it. Handwriting experts do this by comparing the handwriting on the seized evidence with a known sample of the true and genuine handwriting belonging to the person by whom the document was supposedly written. Document examiners obtain known samples of authentic handwriting from public records, such as those listed in Table 1 above.

Determining what types of materials were used in a document. Using various chemical, microscopic, and instrumental techniques, forensic document examiners can also discover what type of ink, pencil, pen, printers, computers, and paper were used in a questioned document. They can even determine which manufacturer created the inks, papers, and printing devices used to make the document. This can provide an important lead for tracking down the source of the document, and can provide a link between a suspect and the crime.

Chapter 10
Questioned Documents

Determining whether the document has been altered in any way. Documents can be altered by chemical or physical removal of information. For example, information can be removed from a check by chemically washing the check or by scraping information from the check. Using a variety of techniques, light sources, and filters, forensic document examiners can tell whether a document has been altered, such as a check whose amount has been changed. Some of these techniques include microscopic examination; reflective or luminescent illumination, and photography; infrared and ultraviolet examination; and laser and other alternate light source illumination.

Reconstructing invisible or damaged messages. Sometimes documents can contain important evidence of writing or printing even when it is not visible to the naked eye. One such example is a piece of paper that has been underneath another piece of paper upon which a message has been written. Although the ink may not have penetrated to the bottom piece of paper, by using special technology document examiners can discover what had been written on the top piece. They do this by tracing the indentations left on the bottom piece by the pen or pencil used to write the message on the top layer. Some of these special techniques include oblique lighting, iodine fuming, and electrostatic detection apparatus (ESDA). Additionally, experts can match torn edges of a document with a portion found at a crime scene, found on a victim, or found in the possession of a suspect. This can also provide important leads.

Reconstructing charred documents. Important documentary evidence is also sometimes found at the scene of a fire or explosion, charred by fire and smoke. Again, using special techniques such as infrared light and infrared photography, examiners can discover what had been written on the document.

Reconstructing shredded documents. Crimes scene investigators may search garbage and recycling bins for evidence of crime. Particularly in the case of white-collar crime, evidence of fraud and other wrongdoing may be found in documents that have been put through a shredding machine. When shredded documents are sent to the forensic laboratory for analysis, document examiners reconstruct the original documents like pieces of a puzzle.

Analyzing photocopies and computer printouts. Forensic document examiners carefully analyze type fonts, imperfections, mechanical oddities, and miscellaneous marks left by photocopiers and printers to identify where the document came from and how it was produced. Photocopiers can leave small marks on copies due to scratches or other damage to the glass or drum. Sometimes perpetrators and others re-photocopy documents several times over, in order to make changes to the document without the appearance of having done so. Examiners also look at the type of toner and how it has been applied to the paper to determine the type of copier or printer that has been used. In addition, computer discs, hard drives, and other devices are also important in criminal investigation.

▲ An erased and altered test sheet. (top)

▲ Alterations of original documents made visible by infrared luminescence. (middle and bottom)

Forensic Science Today

EXAMINING HANDWRITING

Due to the combination of many factors that come together to create a person's handwriting, hand writing is considered to be unique to an individual. Many parts of the handwriting are examined in order to determine if handwriting matches the known source. The slant of the writing, pressure on the surface, connections between letters, style of letters, spelling, and grammar are just a few examples of the parts of the writing that are compared.

▲ Restoration of a portion of the text on a burned document using infrared light and photographic techniques. (top)

▶ Indented handwriting on the envelope (a) can be made visible through the use of oblique lighting techniques (b).

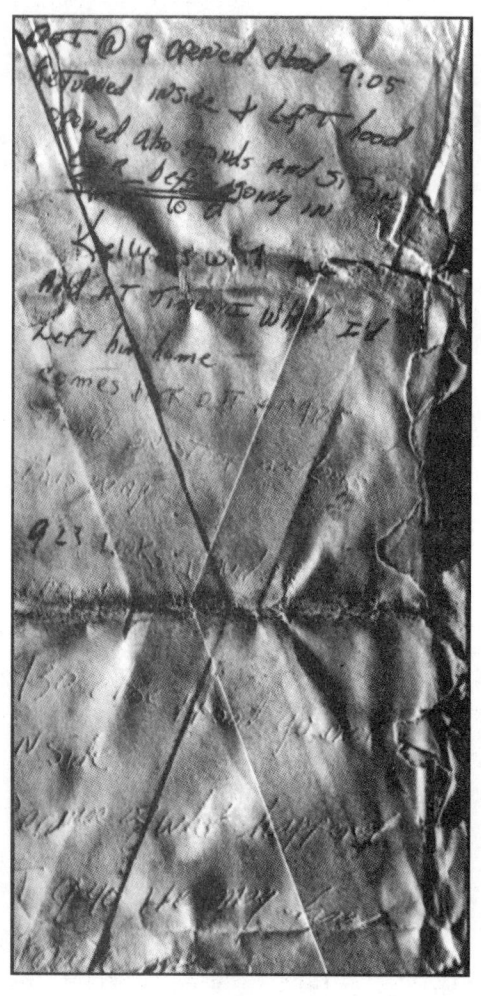

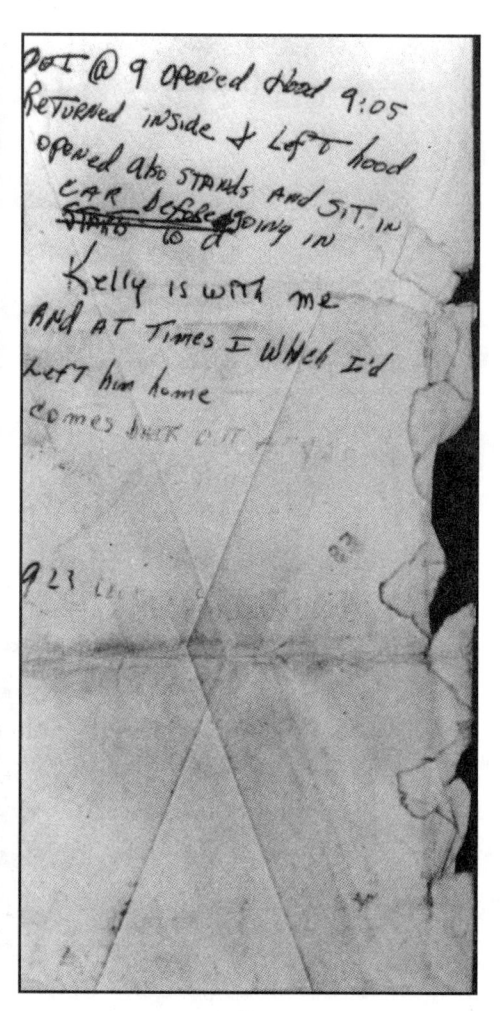

a

b

120

Chapter 10
Questioned Documents

CASE PROFILE
JonBenét Ramsey Murder—
Who Wrote The Ransom Letter?

In Chapter 17 we will learn about the unsolved murder of JonBenét Ramsey, the five-year-old child beauty pageant winner who was found dead in the basement of her house. A key piece of evidence in that case was a three-page ransom note left on the bottom of the stairs. The first page of the letter is reproduced below. What seems odd about it?

> Mr. Ramsey,
>
> Listen carefully! We are a group of individuals that represent a small foreign faction. We do respect your bussiness but not the country that it serves. At this time we have your daughter in our pozession. She is safe and unharmed and if you want her to see 1997, you must follow our instructions to the letter.
>
> You will withdraw $118,000.00 from your account. $100,000 will be in $100 bills and the remaining $18,000 in $20 bills. Make sure that you bring an adequate size attache to the bank. When you get home you will put the money in a brown paper bag. I will call you between 8 and 10 am tomorrow to instruct you on delivery. The delivery will be exhausting so I advise you to be rested. If we monitor you getting the money early, we might call you early to arrange an earlier delivery of the

121

Forensic Science Today

TECHNICAL DEFINITIONS

Document Examiner: A forensic professional who specializes in examining document evidence.

Forgery: A faked or altered document.

Known Sample: A true and genuine sample of the writing of a person (or printing of a computer printer) that is compared to the document that has been seized as evidence to determine authenticity.

Questioned Document: Any document whose origin or authenticity is in question. In other words, is this document real? Does it really come from where the suspect says it came from?

Handwriting Analysis: A specialty within forensic science in which handwriting experts compare samples of writing to determine the document's true author.

Imprints and Impressions

Chapter 11

CHAPTER OBJECTIVES

✔ Distinguish between imprints and impressions.
✔ Discuss what imprints and impressions can tell investigators about a crime.
✔ Understand procedures for documenting imprint and impression evidence.
✔ Identify the methods of collecting imprints, impressions, and known standards.
✔ Understand how forensic scientists analyze imprint and impression evidence in the laboratory.
✔ Define the following vocabulary words and definitions: *impression*, *imprint*, and *electrostatic lifting*.

When a perpetrator leaves the scene of the crime, he or she may leave behind imprints and impressions, such as shoe prints, footprints, or tire marks. Imprints and impressions are useful evidence in identifying suspects and victims as well as reconstructing the sequence of events in a crime.

An *imprint* is a *two-dimensional* pattern found on a hard surface, left behind by residue such as blood, soil, grease, or dust from a shoe, tire, or other object that leaves behind a pattern. An imprint can also be left behind when a shoe or tire actually picks up and removes residue that was already present on the flat surface. The imprint may either be visible to the naked eye, or may require enhancement techniques to be discovered. Two examples of imprints would include an oily shoe imprint in a garage, or a fingerprint left on a window.

An *impression*, on the other hand, is a *three-dimensional* indentation or pattern that is produced when an object such as a shoe, tire, or tool presses into a softer substance like snow, mud, sand, or wet cement. Examples of impressions include tire tracks in mud or shoe prints in the snow.

Forensic Science Today

KEY TERMS

Imprint
Impression
Known Standard
Electrostatic Lifting

WHAT CAN IMPRINTS AND IMPRESSIONS TELL US ABOUT THE CRIME?

Like all pattern evidence, imprints and impressions are extremely valuable in reconstructing crimes. They can be used to prove whether a suspect was at the scene of a crime or whether a witness is telling the truth, to link individuals with crime scenes, or to provide new leads in the investigation.

Imprints and impressions can also provide other vital information about what happened and the sequence of events. For instance, tire and skid mark patterns can tell us how many vehicles may have been involved in a crime, in which direction the vehicles were traveling, whether the drivers applied the brakes, whether any turns were made, and which tire mark was left last. In addition, proper interpretation of footprints and shoe imprint trails can tell us how many persons were present at a crime scene, whether they were moving, how they were moving (i.e., walking, running, or limping), whether they were carrying or dragging along heavy objects, and in what direction they were moving. Sometimes investigators can even get some ideas about a person's weight, sex, shoe size, and stride length just by examining their foot or shoe print.

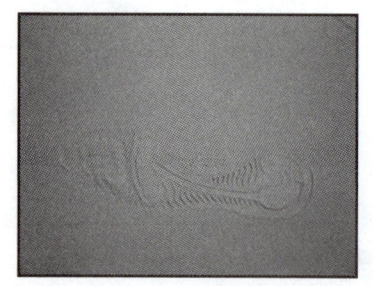

▲ Footprint in volcanic ash in Alaska.

DOCUMENTING IMPRINT AND IMPRESSION EVIDENCE

Before imprint or impression evidence can be collected and transported to the forensic laboratory for analysis, it must be carefully documented through notes, sketches, and photographs. Photos should be taken both with and without a scale in the picture. Proper lighting and angling of the camera are critical to ensure accuracy of the documentation.

COLLECTING IMPRINT EVIDENCE

If the article containing the imprint is small enough to be moved safely from the crime scene, it should be packaged in a manner that preserves the imprint intact. However, if the object with the imprint is too large to be moved (such as a wall), then the imprint will need to be enhanced or "lifted" similar to fingerprints. Depending on the nature of the imprint and the surface upon which it has been deposited, there are a variety of ways in which to collect the evidence:

▼ Dr. Lee using of an electrostatic apparatus to lift a dust print found on the floor.

- **Electrostatic lifting.** This is used for lifting dust-residue prints. Dust prints can be removed from floors, wood, tiles, cloth, pillowcases, bed sheets, newspapers, cardboard, magazines, envelopes, and other similar objects.
- **Gelatin and adhesive tape lifting.** These are similar methods to those used for lifting latent fingerprints.
- **Photographic imprint enhancement.** A forensic photographer can use techniques such as filters and alternate lighting techniques to highlight imprints that may be difficult to see.
- **Powdering.** Imprints can also be enhanced by using fingerprint powders, especially in circumstances such as when a greasy shoe has been tracked across a polished floor.

Chapter 11
Imprints and Impressions

- **Chemical enhancement.** The residues left behind in imprints often contain trace elements, blood, proteins, minerals, and other compounds that may be made more visible with chemical enhancement. Some of these chemicals include iodine, 8-hydroxyquinoline, ninhydrin, amido black, luminol, leucocrystal violet, bloody print enhancement reagent, thiocyanate, and bromcresolgreen.

COLLECTING IMPRESSIONS EVIDENCE

Just as with imprint evidence, if an impression can be moved in its entirety to the forensic lab, it should be packaged very carefully to avoid damage to the evidence. However, impressions are usually contained in mud, snow, or other materials that cannot be transported. In these cases, forensic scientists photograph these types of evidence in 1-to-1 format and make a cast of the impression with dental stone or silicon rubber. In the case of snow prints, a special casting material called Snow Print Wax®, which is a rust-colored aerosol spray wax, is used to preserve the detail of the impression. Many techniques can be used to enhance the impression during photographing. One example is to spray paint up into the air over the impression. When the paint settles onto the impression, this can result in an image with a much higher visualization of details.

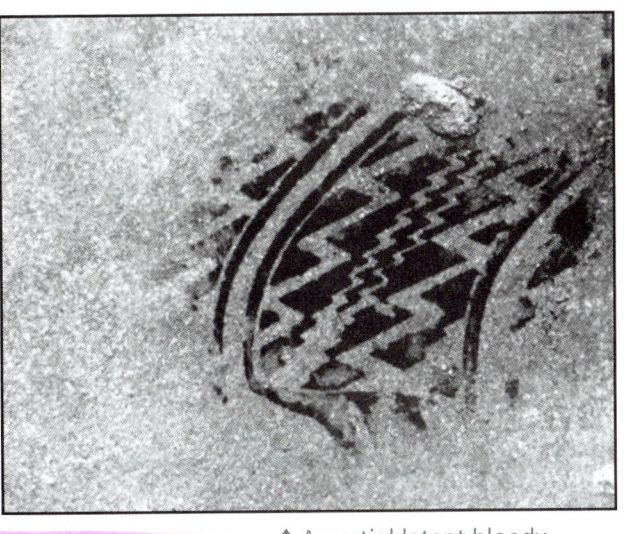

▲ A partial latent bloody shoeprint was discovered at a homicide scene. Chemical reagent Tetrametylbenzendine was used to visualize this shoeprint. (Courtesy of Joe Sudol of the Institute of Forensic Science at UNH.)

Logic Tree: Imprints

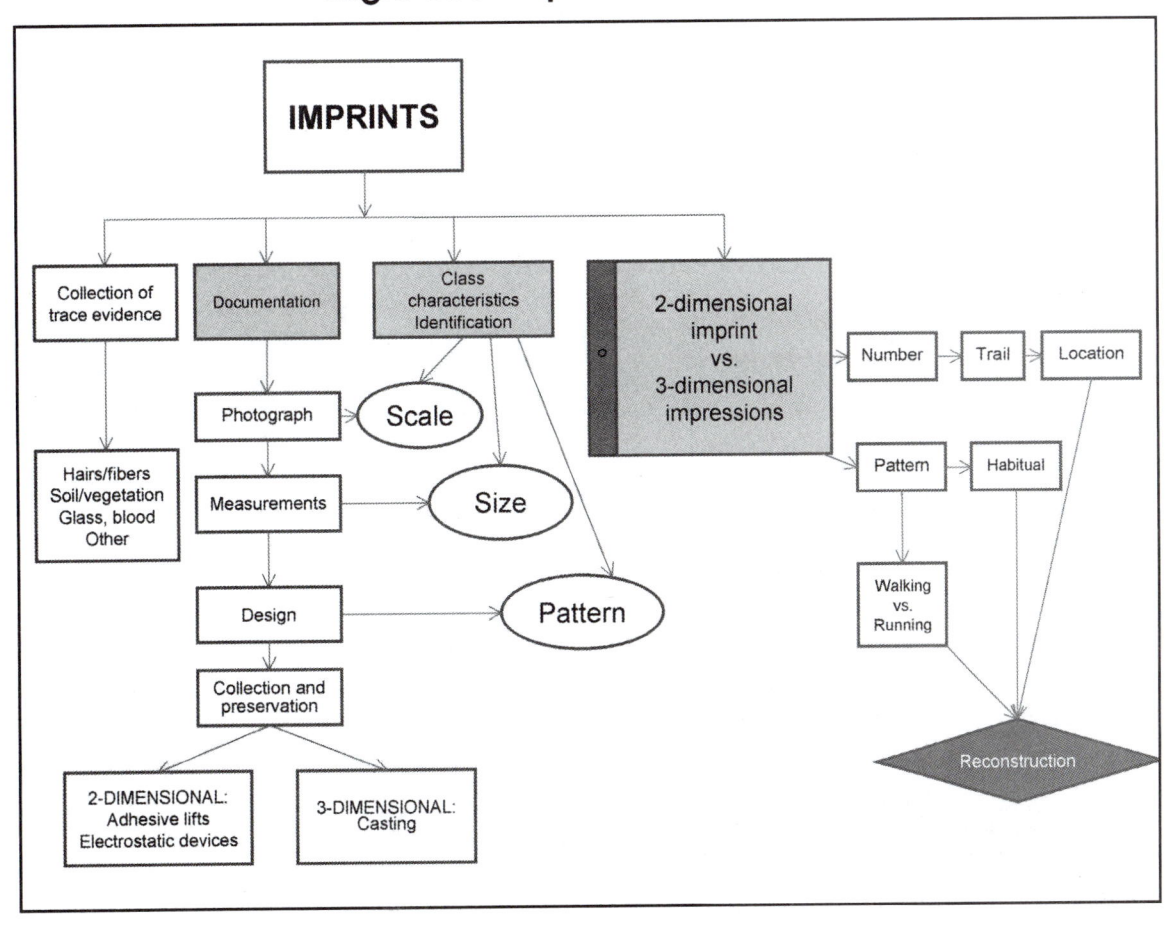

IMPRINTS

- Collection of trace evidence
  - Hairs/fibers
  - Soil/vegetation
  - Glass, blood
  - Other
- Documentation
  - Photograph → Scale
  - Measurements → Size
  - Design → Pattern
  - Collection and preservation
    - 2-DIMENSIONAL: Adhesive lifts, Electrostatic devices
    - 3-DIMENSIONAL: Casting
- Class characteristics Identification
  - Scale
  - Size
  - Pattern
- 2-dimensional imprint vs. 3-dimensional impressions
  - Number → Trail → Location
  - Pattern → Habitual
  - Walking vs. Running
  - → Reconstruction

125

Forensic Science Today

▲ Casting shoe prints in snow.

▲ A shoe impression found in the sand next to a body on a beach. (top)

▶ Suspect's shoe found in a garbage can. It has same class and individual characteristics as the shoe impression found at the scene. (right)

▼ The sole pattern of the suspect's shoe. Compare to the enhanced blood print found on page 125.

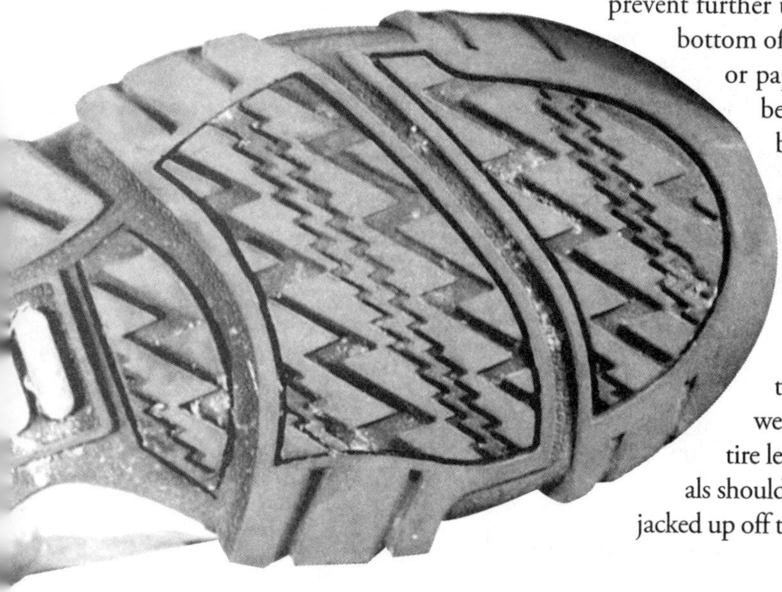

COLLECTING KNOWN STANDARDS FOR COMPARISON

In order to compare the imprint and impression evidence with the object that created it, investigators must first collect the object that created it. The original object is called the *known standard*.

In the case of shoes, the shoe must be sought as soon as possible after the crime to prevent further usage from changing the pattern or material left on the bottom of the shoe. Shoes should be packaged in paper wrapping or paper bags to prevent the loss of trace materials that may be present on the soles of the shoe. Plastic bags should not be used as they can create static that results in important trace evidence sticking to the packaging. A shoe should never be placed directly into the imprint or impression that it made, or else it could destroy the evidence.

For tire mark known standards, investigators must try to duplicate the exact conditions under which tire marks were made, including the same weight of the vehicle that the tire leaving the mark was on. In order to do so, all trace materials should first be removed from the tire, and the vehicle should be jacked up off the ground. Investigators then ink the entire tread circumference of the tire with fingerprint ink, and place approximately ten feet of paper under the tire. Next, the vehicle is lowered onto the paper and slowly driven or pushed along the paper until the tire has completed at least one revolution. Investigators should include identifying data on the paper, such as which tire made the mark, and which side of the tire was facing toward the interior of the vehicle. The process should be repeated on all tires of the vehicle; and then the paper is submitted to the forensic laboratory along with the cast or photograph of the tire imprint or impression from the crime scene for comparison.

ANALYZING IMPRINTS AND IMPRESSIONS IN THE FORENSIC LABORATORY

Forensic scientists analyze imprints and impressions evidence both visually and under a low power stereo microscope to determine not only what made the mark, but also to compare the mark with the known standard. By doing so, the forensic scientist can sometimes even identify the manufacturer of the object, since manufacturers often place unique patterns or logos on their products. For instance, the manufacturer and model of a tire that made a tire tread pattern can be determined by comparing the pattern to those in a publication that compares tire designs. In general, laboratory examiners follow three steps in their analysis of imprints and impressions evidence:

1. **Determining what made the imprint or impression (determining *class* characteristics).** The examiner should note general structural properties such as the size, shape, pattern, and design of the imprint or impression.

2. **Determining *individual* characteristics of the imprint or impression.** If individual characteristics specific to a particular object are present and sufficiently well defined, the examiner may be able to conclude that a particular object made the imprint or impression. Some examples of these types of individual characteristics include damage, wear, cuts, irregularities, and other accidental marks unique to a certain object as a result of its history.

▲ An enhanced image of an imprint upon the truck that hit the police officer (a); and a shoulder patch from the police officer's uniform (b).

◀ A latent imprint was located on a truck that was suspected of being involved in a hit-and-run of a state police officer. (left)

Forensic Science Today

3. **Comparing imprints or impressions with known standards.** When forensic scientists compare an imprint or impression from the crime scene with a known standard, they should reproduce as closely as possible the conditions present when the evidence was created. This includes duplicating the pressure, nature of the surface, and angle of impact upon the surface by the object. After making a comparison of *class* characteristics, the forensic examiner can say whether the evidence and the known standard are similar, or whether a known standard must be eliminated as a possible source of the pattern. After comparing the *individual* characteristics of the evidence and the known standard, the examiner may conclude that the known standard created the imprint or impression if the individual characteristics are the same and there are no unexplained differences. Scientists also have to take shrinkage of casting material into consideration when making comparisons; and they can use charts and formulae to assist in their determinations.

4. **Reconstructing the sequence of events.** From the imprint and impression evidence, an experienced examiner might be able to tell whether a person was walking or running, injured or carrying a heavy object, and the direction of movement. In addition, one can determine the mechanism of transfer; i.e., whether the imprint was caused by a static contact or by a dynamic transfer. The examiner can also determine how many contact events occurred and the possible time lapse.

TECHNICAL DEFINITIONS

Impression: A three-dimensional indentation or pattern that is produced when an object such as a shoe, tire, or tool presses into a softer substance like snow, mud, or sand.

Imprint: A two-dimensional pattern found on a hard surface, left behind by residue from a shoe, tire, or other object leaving behind a pattern.

Electrostatic Lifting: A method used to lift dust residue imprints.

Pictorial: Imprints and Impressions

Chapter 11
Imprints and Impressions

The following photos are contributed by Jim Streeter, senior forensic scientist in charge of imprint and document section, Connecticut State Forensic Laboratory.

▲ Piece of plywood from window displaying partial footwear outsole impressions.

▶ Electrostatic dust lifts of footwear outsole impressions obtained from plywood window.

Forensic Science Today

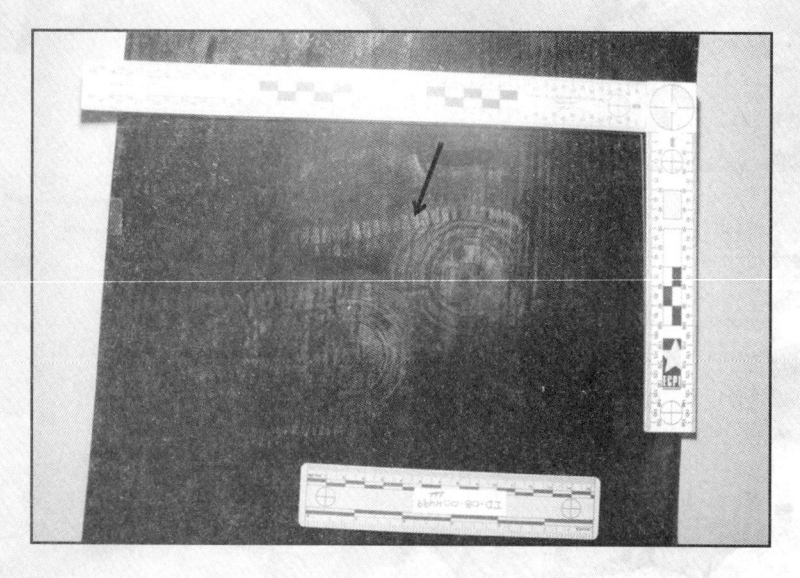

◀ Photograph of outsole impressions present on dust lift. Arrow pointing to an individual identification characteristic (cut).

▶ Photograph of outsoles of NIKE brand "Air" model sneakers.

▼ Photograph of heel area of the outsole of the left article of above footwear. Arrow pointing to an individual identification characteristic (cut).

▼ Enlargement of heel showing identifying characteristic (cut).

Chapter 11
Imprints and Impressions

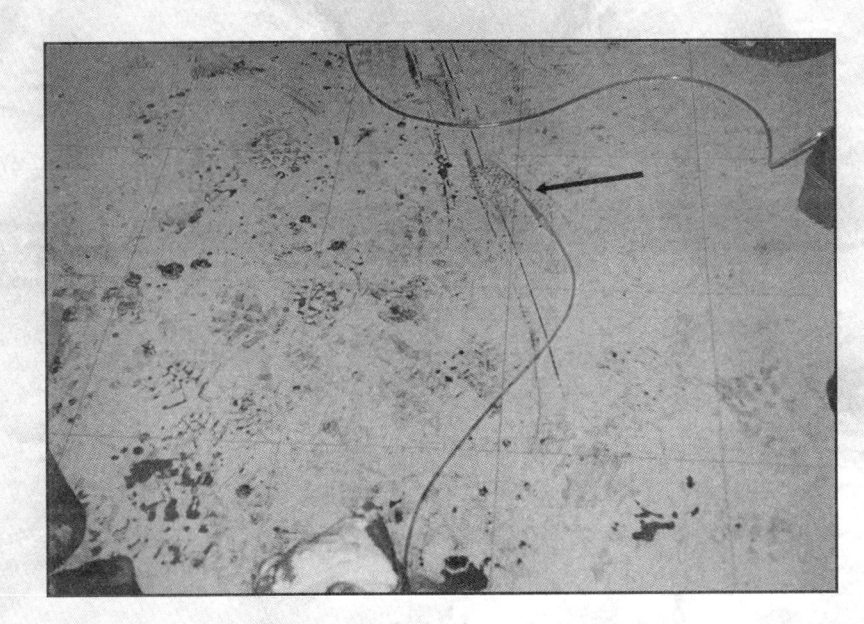

◀ Multiple footwear outsole impressions, in blood-like substance, on floor at crime scene.

▶ Section of floor containing footwear outsole impression, in blood-like substance, similar in design to outsole of suspect's shoe.

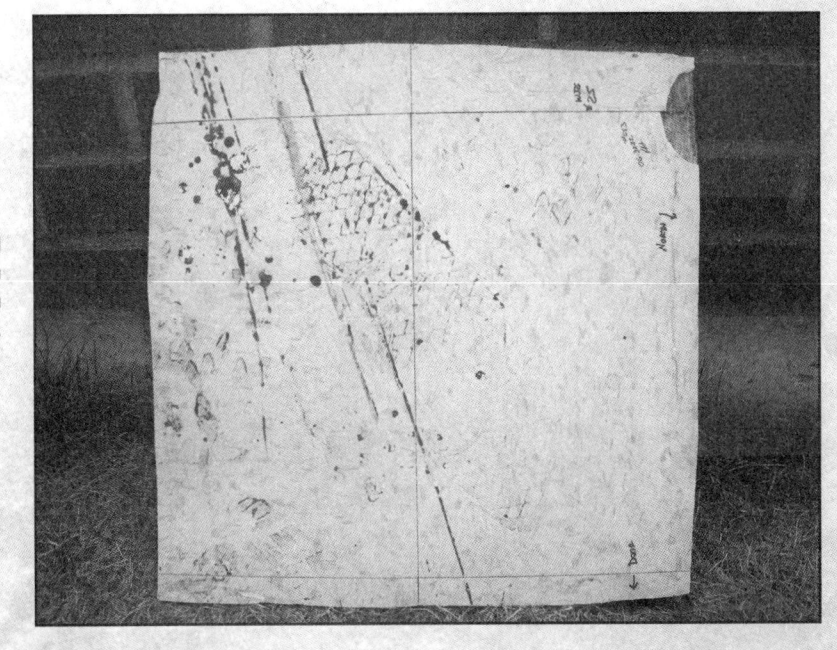

◀ Chemical enhancement of impression in blood-like substance on the section of floor, using Amido Black.

131

Forensic Science Today

◀ Footwear impressions on flooring after chemical enhancement.

▶ Before and after chemical enhancement processing.

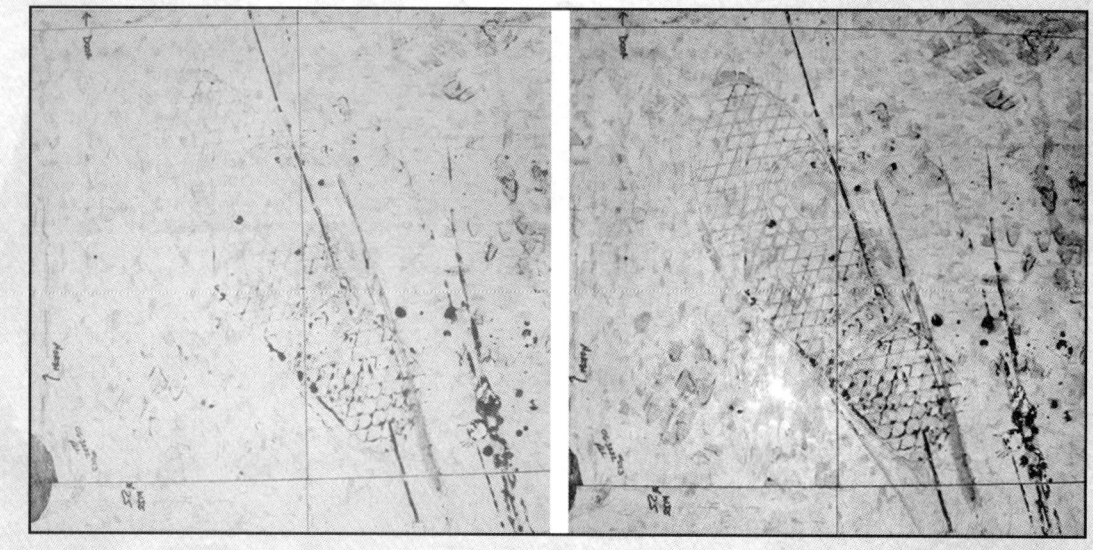

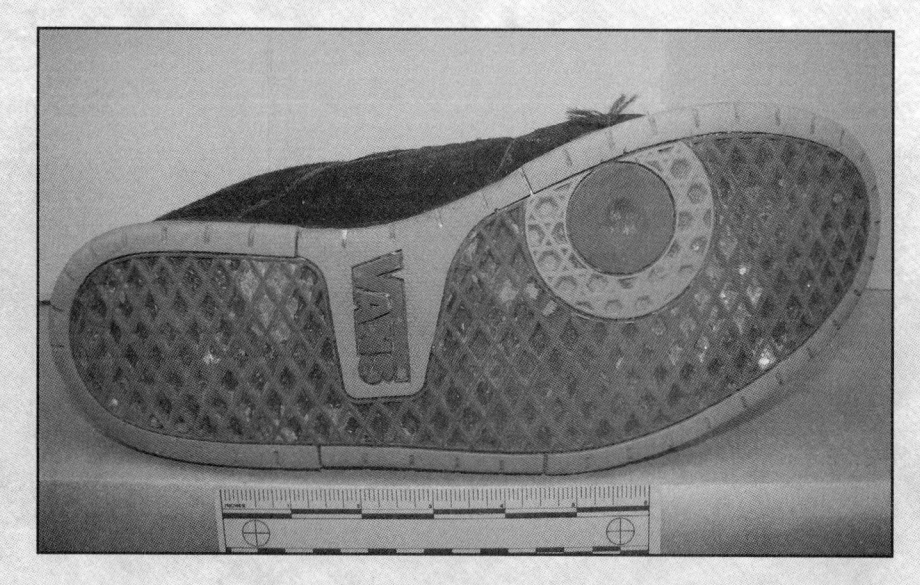

◀ Outsole of the left article of footwear belonging to suspect.

132

Chapter 11
Imprints and Impressions

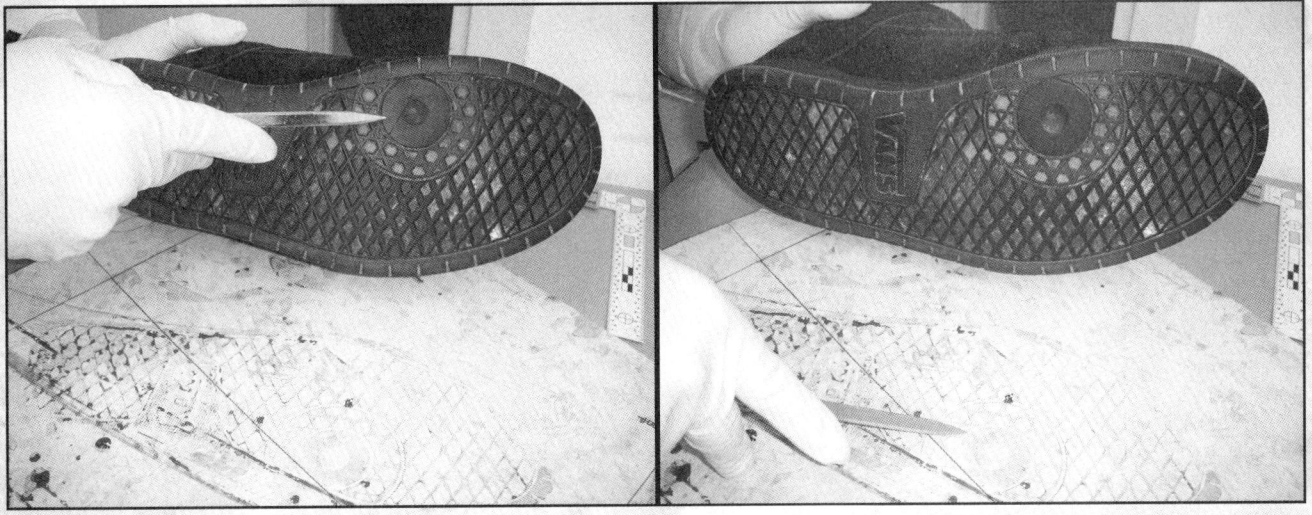

▲ Large cut in bulls-eye pattern on outsole of suspect shoe.

▲ Same cut on impression on flooring.

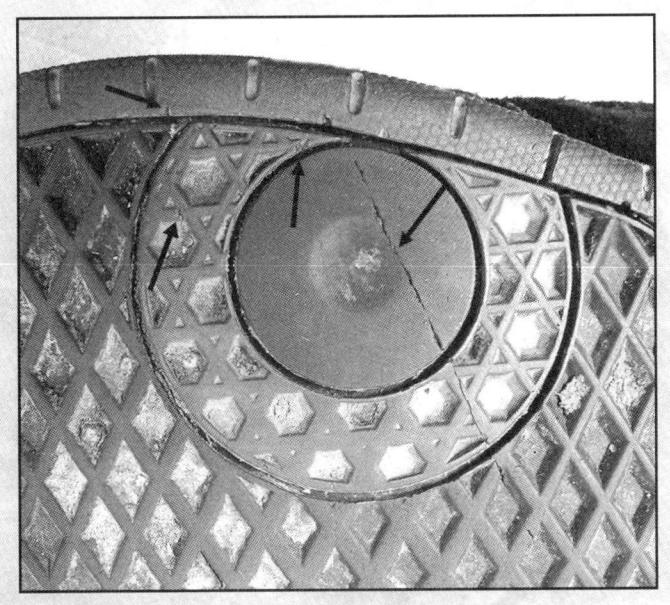

▲ Individual identifying characteristics on outsole of suspect.

▼ Corresponding individual identifying characteristics on impression on flooring.

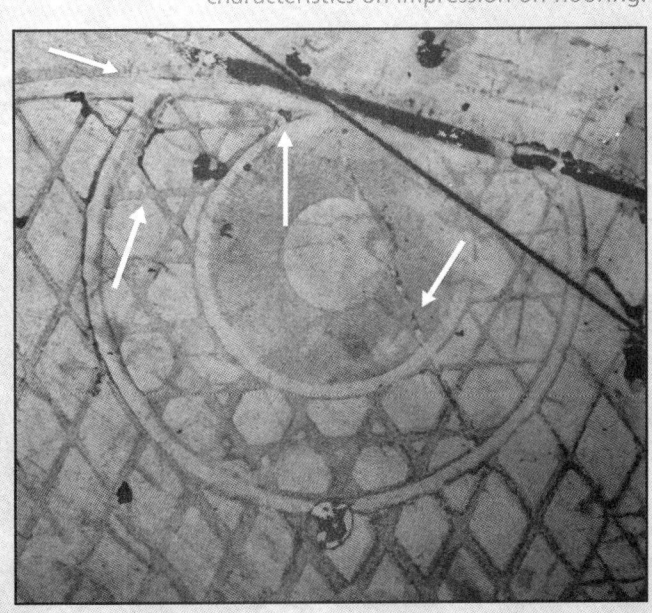

133

Forensic Science Today

▶ Footwear outsole impression on paper surface.

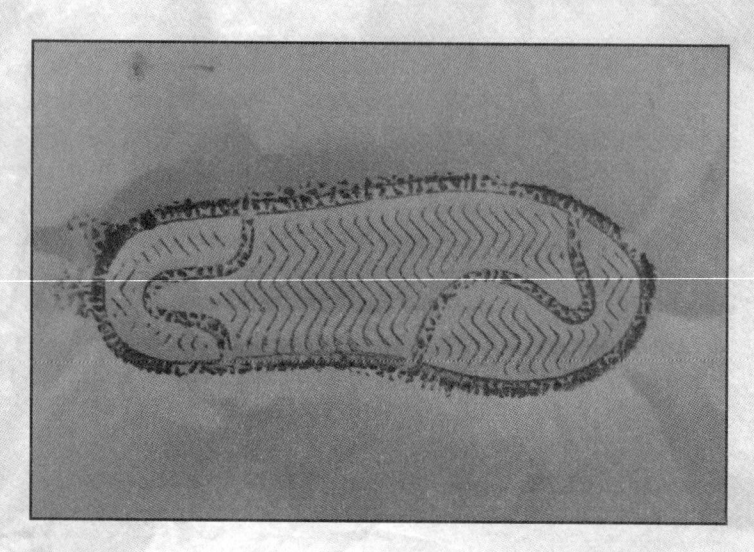

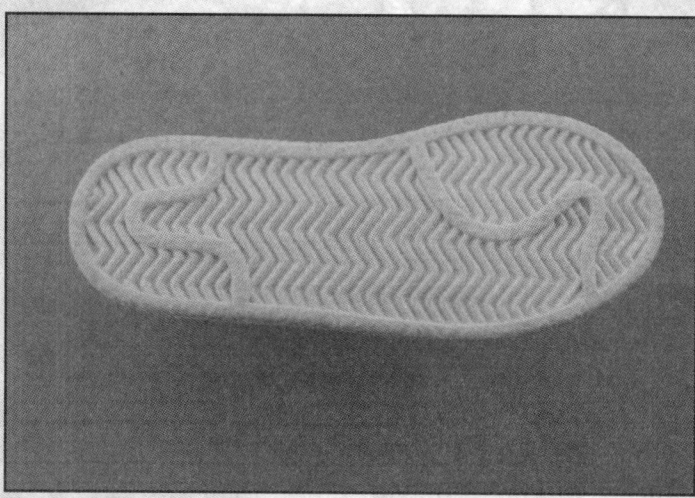

◀ Photograph of outsole of suspect footwear.

▼ Actual footwear that produced impression. ▶

134

Chapter 11
Imprints and Impressions

◀ Piece of wood containing partial footwear outsole impression.

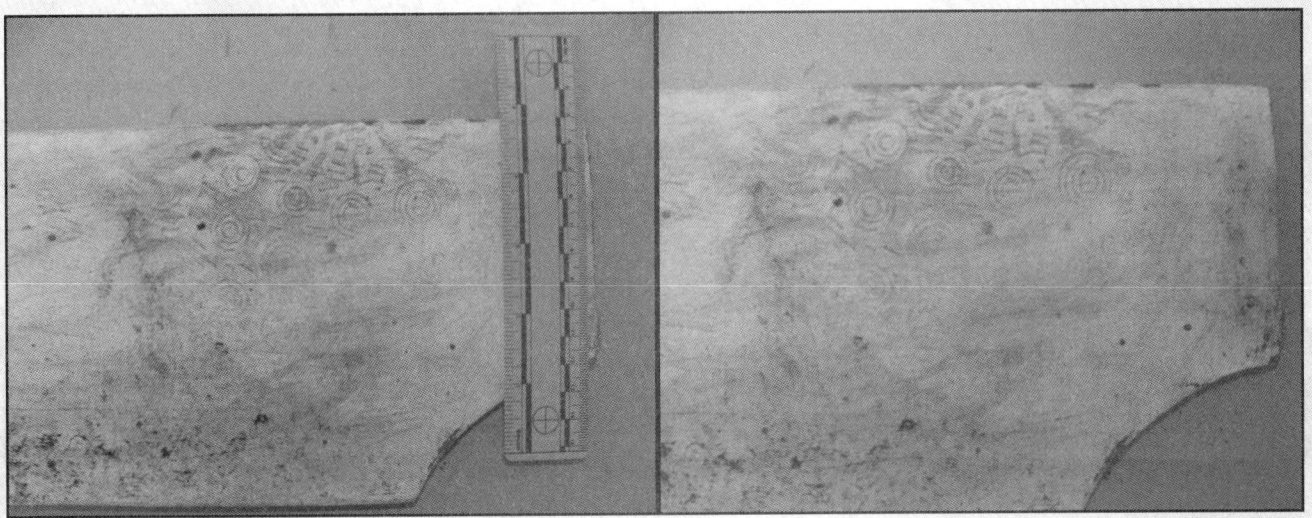

▲ Close-up of partial footwear impression. ▲

▶ SICAR Footwear and Tire [Computer] Data Base System.

135

Forensic Science Today

▲ Footwear Data Base Pattern Coding Guide.

◀ Forensic Examiner imputing information into SICAR System.

136

Chapter 11
Imprints and Impressions

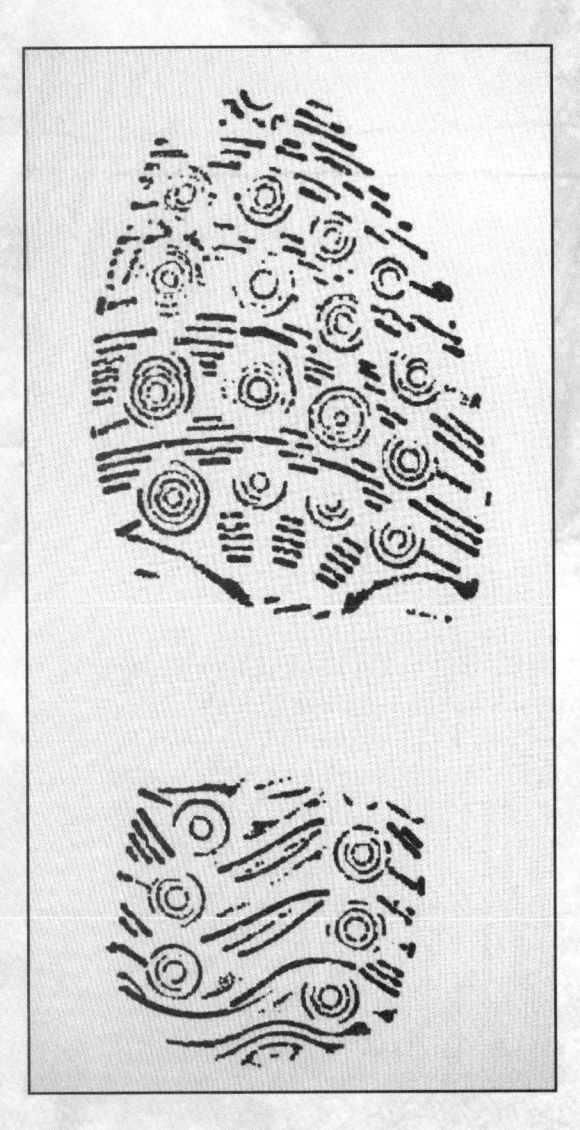

◀ Outsole of article of footwear identified by SICAR data base.

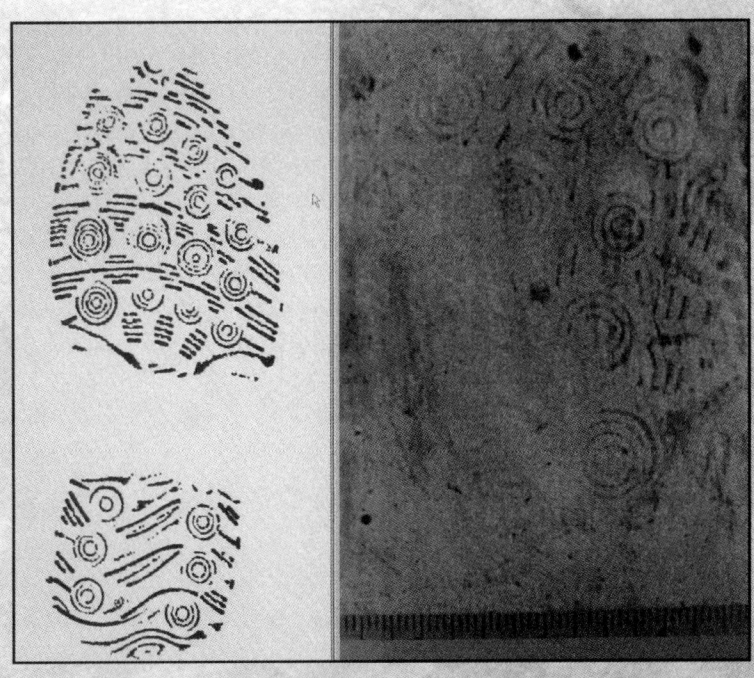

▶ Comparison between suspect impression and outsole of identified footwear.

137

Chapter 12

Toolmark and Firearms Evidence

☞ CHAPTER OBJECTIVES

- ✔ Understand the nature of toolmark evidence.
- ✔ Explain the differences between compression and sliding toolmarks.
- ✔ Follow the steps in proper documentation and collection of toolmark evidence found at a crime scene.
- ✔ Describe how toolmark evidence is analyzed in the forensic laboratory, including comparisons of questioned toolmarks with known standards.
- ✔ Define the following vocabulary words and phrases: *toolmark*, *compression toolmark*, and *sliding toolmark*.
- ✔ Understand the mechanisms and chemical reactions involved in the firing of a gun.
- ✔ Follow steps for proper documentation, collection, and preservation of guns, bullets, and cartridge cases found at a crime scene.
- ✔ Understand the concepts of rifling and striation.
- ✔ Explain how the unique markings on spent bullets and fired cartridge cases are compared to bullets and cartridge cases from guns test-fired in the forensic laboratory.
- ✔ Understand what bullet trajectories can tell us about a crime that involved the shooting of a gun.
- ✔ Explain how bullet trajectories are documented at crime scenes.
- ✔ Understand the nature and properties of gunshot residue (GSR).
- ✔ Discuss what GSR patterns can tell us about a shooting that has occurred.
- ✔ Describe methods for collecting GSR at a crime scene.
- ✔ Understand how GSR is analyzed at the forensic laboratory.
- ✔ Define the following vocabulary words and phrases: *bullet*, *cartridge*, *cartridge case*, *caliber*, *gauge*, *grooves*, *lands*, *rifling*, *GSR*, *point-blank range*, *primer*, *propellant*, *striation*, and *trajectory*.

Forensic Science Today

KEY TERMS

- Bullet
- Cartridge Case
- Ammunition
- Rifling
- Caliber
- GSR
- Point-blank range
- Distance Determination (contact, mid-range, distance)
- Primer
- Propellant
- Striation
- Trajectory
- Toolmark
- Compression Toolmark
- Sliding Toolmark

TOOLMARKS

A toolmark is any cut, scrape, impression, gouge, scratch, indentation or other marking left on an object or surface by another object being forced into or moved across it. Toolmarks are often found at the scene of a burglary or other crime involving breaking and entering when the perpetrator uses, for example, a screwdriver to pry open a window or a door lock. Additionally, toolmarks can often be found on victims who were injured with a weapon or other object. This happened in the famous case of Dr. Sam Sheppard, whose murdered wife was discovered to have a wound pattern on her head from a flashlight, which was later determined to be the murder weapon. (This real-life case later became the inspiration for the television series and subsequent movie, *The Fugitive*.) In another case, toolmarks from a knife used in a stabbing were also evident on the victim's body.

TYPES OF TOOLMARK EVIDENCE

There are two main types of toolmark evidence: (1) *compression toolmarks*, and (2) *sliding toolmarks*. Put simply, a compression toolmark is made by pressing one object into another one. Some compression marks can be dynamic, others can be static. The harder material will sink into the softer material and leave its mark. An example of a compression toolmark is the indentation that a screwdriver can make on a wooden door frame in a breaking and entering situation. A sliding toolmark is made when one object slides across another, leaving behind scrapes and scratches. This can be further classified in two sub groups: forward sliding and backward sliding. Again, the stronger material is forced into and across the softer material.

DOCUMENTING TOOLMARK EVIDENCE AT THE SCENE OF A CRIME

Crime scene investigators must carefully document toolmark evidence at the scene of a crime before taking it away to the lab for further analysis. First, two sets of photographs must be taken of the toolmarks. One set will show the object containing the toolmark, both with and without a ruler beside it to show the nature of the toolmark and its relationship to the crime scene. The other set of photographs will show the

▲ A forensic scientist examines toolmark evidence at a burglary scene.

▲ Forensic investigators check the point of entry and search for toolmark evidence on the screen.

Chapter 12
Toolmark and Firearms Evidence

toolmark itself in closer detail. Secondly, you will need to write down all the toolmarks you see at the crime scene, and include them in sketches of the scene. Also make note of the measurements you have taken of the toolmarks and the objects containing them. Next, remove all trace evidence such as glass, paint, blood, and metal that may also be present on the tool or in the toolmark. Such trace evidence may also need to be further analyzed at the laboratory. Finally, a cast can be made of the toolmark with casting material.

COLLECTING TOOLMARK EVIDENCE FOR FURTHER ANALYSIS

After being documented, toolmarks should be further analyzed in the lab. In order to transport toolmarks to the lab, you should package the actual object with toolmarks on it —if it is small enough—and submit it to the lab along with the suspected tool that was used in the commission of the crime. The tool should be secured in a box to protect all surfaces. The tool and the toolmark should not be packaged together. If the object containing the toolmark is too large to take to the lab, such as a door panel or wall, then the toolmark itself should be cut out of the object, packaged, and secured properly to prevent damage to the toolmark. If it is not possible to remove the toolmark from the large object at the crime scene, the investigator can also make a high-resolution silicone rubber cast of the toolmark, and then bring it in to the lab for analysis.

ANALYZING TOOLMARKS IN THE FORENSIC LABORATORY

Once the toolmark has been transported to the lab, forensic scientists then examine it under the stereo microscope to determine its general characteristics. Using this information, the examiner can determine what type of tool made the marking, such as a screwdriver, tire iron, or gun. In addition, the normal wear and tear of the tool will give it unique qualities similar to the way in which the barrel of a gun also has unique internal markings from being used over time. The friction of two objects rubbing or pressing against one another causes the molecules in contact with one another to be rearranged. This microscopic change of the surface molecules results in an object that has a unique surface.

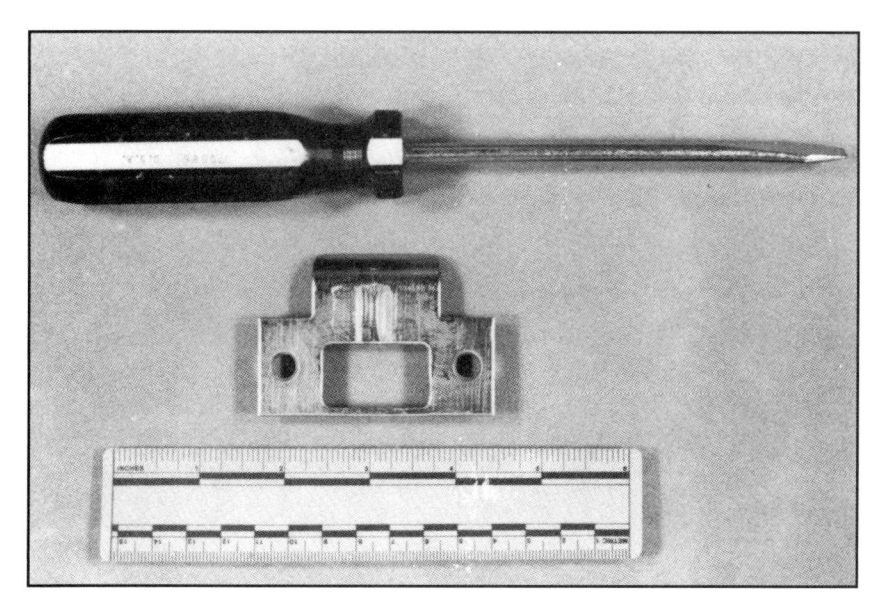

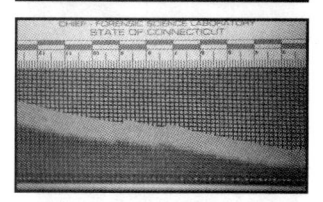

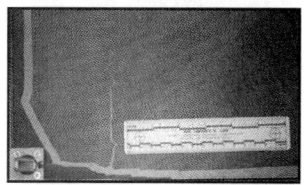

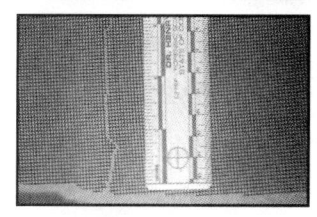

▲ This window screen was determined to have been cut by a sharp instrument.

◀ A toolmark comparison was made between this striker plate from a forced door and this screwdriver found in the possession of the burglary suspect.

141

Forensic Science Today

▲ Toolmark comparison: Left is the test mark made by a screw driver found in a suspect's car; right is the cast of an unknown toolmark found at a burglary scene.

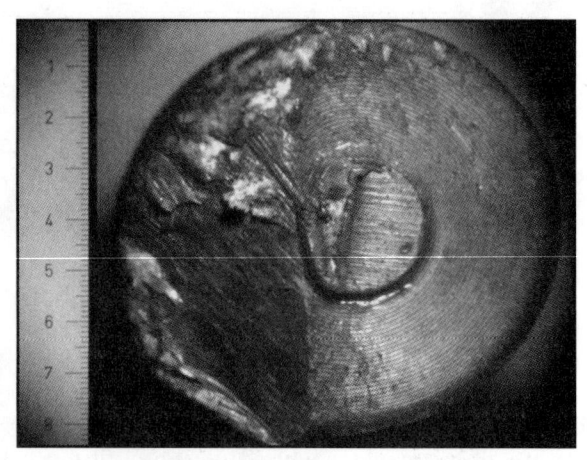

▲ Photomicrograph shows toolmarks on a metal surface.

Forensic tool examiners will produce toolmarks from a tool found at a crime scene, under controlled laboratory conditions. The examiner will look at the toolmarks made by the known tool and the marks from the crime scene, lined up next to one another on the two stages of the comparison microscope. The markings on each are compared. If the toolmarks match, then it can be proven that the tool found at the scene—or in the suspect's home—is the definitive cause of the toolmark made during the crime.

FIREARMS EVIDENCE

Guns are used in a wide variety of crimes, from homicide and sexual assault to robbery. Thus, there is no shortage of work for firearms examiners and criminal investigators who specialize in the examination and analysis of bullet trajectories and gunshot residue.

ANATOMY OF A GUNSHOT

What happens when a gun is fired? A series of chemical reactions happens very quickly—within a split second—resulting in the bullet being propelled out of the gun barrel faster than the eye can see. First, a pull of the trigger sets the process in motion. When the trigger is pulled, the firing pin strikes the base of the primer cup containing *primer* (shock sensitive explosive material and fuel), which is found inside the *cartridge case* (the container holding the bullet, primer, and gunpowder). This causes the primer to explode, which in turn causes the *propellant* (gunpowder) to ignite and rapidly burn. The burning gunpowder produces large volumes of gases, which expand and then propel the bullet forward through the barrel of the gun. This is why gunpowder is called a propellant. As the bullet moves down the length of the barrel, marks and scratches are impressed onto it by the barrel.

According to Newton's Third Law of Motion, for every action there is an equal and opposite reaction. Thus, as the bullet moves forward through an automatic or semi-automatic weapon, the cartridge case that once contained it is forced backward against the rear of the firing chamber. Now that the cartridge case is empty, it must be removed from the firing chamber in order for a new cartridge containing a live round of ammunition to come into the firing chamber. The empty cartridge case is pushed out of the firing chamber by the extractor, and then pushed completely out of the gun itself by the ejector. Like the bullet, marks are impressed onto the cartridge case each time it comes into contact with a metal surface within the gun.

Firearms Examination

Firearms examiners analyze firearms, discharged bullets, cartridge cases, shotgun shells, unusual or homemade weapons, and ammunition components. The firearms examiner tries to answer questions like:

1. What kind of weapon fired the bullet?
2. Was this bullet or cartridge case fired in the weapon?
3. Is the weapon functioning properly?
4. What was the trajectory and distance of the weapon from the impact?

Firearms examiners also compare different types of marks left by tools on evidence, since tools of many kinds are also used as weapons or in the perpetration of crimes.

BULLET AND CARTRIDGE DATABASES

Many violent crimes are solved through the use of national bullet and cartridge case databases. In early days, the FBI developed a database called "DRUGFIRE." Now, a new system called "NIBIN," which stands for National Integrated Ballistic Information Network, has been initiated by the Bureau of Alcohol, Tobacco, and Firearms and Explosives. This system allows law enforcement agencies to view and compare ballistics evidence through its Integrated Ballistic Identification System equipment (IBIS). Close to 200 state and local law enforcement agencies are connected to NIBIN. Police have access to these databases, which store information about and characteristics of particular bullets and cartridge cases used in crimes. Often, the only evidence available in a drive-by shooting or street crime associated with illegal drugs has been a few cartridge cases recovered from the shooting scene. Now that police can compare markings on bullets test fired from guns recovered from crime scenes, they can use the databases to compare the markings to other bullets and cases that were found at other crime scenes as well. This can provide valuable investigative leads in linking one crime to another or one weapon to different crimes.

Forensic Science Today

▶ Spent bullets and casings found at a shooting scene.

DOCUMENTING, COLLECTING, AND PRESERVING FIREARMS EVIDENCE AT A CRIME SCENE

Crime scene investigators should carefully document all firearms evidence found at a crime scene. In general, they should photograph firearms found at the scene and note the location of such weapons in their crime scene sketches. The investigators should also take good notes on when the firearm was found, whether the gun's safety switch was on or off, whether any live ammunition still remained in the gun, the make, model, caliber and serial number of the weapon, and the location of any spent (i.e., fired) bullets or casings found on the scene. Any bullet holes or damaged areas should also be documented.

Guns (Loaded or Unloaded)

Never wipe off guns found at a crime scene as they may contain valuable evidence such as DNA, fingerprints, blood spatters, tissue, or other trace evidence such as hairs or fibers. Loaded guns should be unloaded at the scene prior to transport to lab, but first the chamber location of remaining bullets must be documented. If the gun is a revolver, the chamber position should be marked before removing the bullets. If it is a semi-automatic weapon, any live rounds in the chamber should also be marked. Guns found in water should be removed as soon as possible to prevent rusting.

Bullets and Spent Cartridge Cases

Bullets and cartridge cases should be wrapped in lens tissue and placed in individual envelopes or plastic bags and marked with identifying information. By keeping all bullets and cartridges separated, investigators can ensure that the bullets do not scratch one another or alter the marks that firearms examiners will need to examine for possible identification. Remember that bullets often remain lodged in the human body after being shot. When a doctor is removing a bullet from a victim, he or she must make sure not to use forceps (medical tweezers) to remove the bullet—otherwise

Chapter 12
Toolmark and Firearms Evidence

it may damage the surface of the bullet and destroy microscopic striations. If the bullet is embedded in bone, it must be removed very carefully during surgery if the victim is still alive. If the victim is deceased, however, then the bone itself can be cut out during autopsy and the bullet removed in the forensic laboratory.

Logic Tree: Shooting Scenes

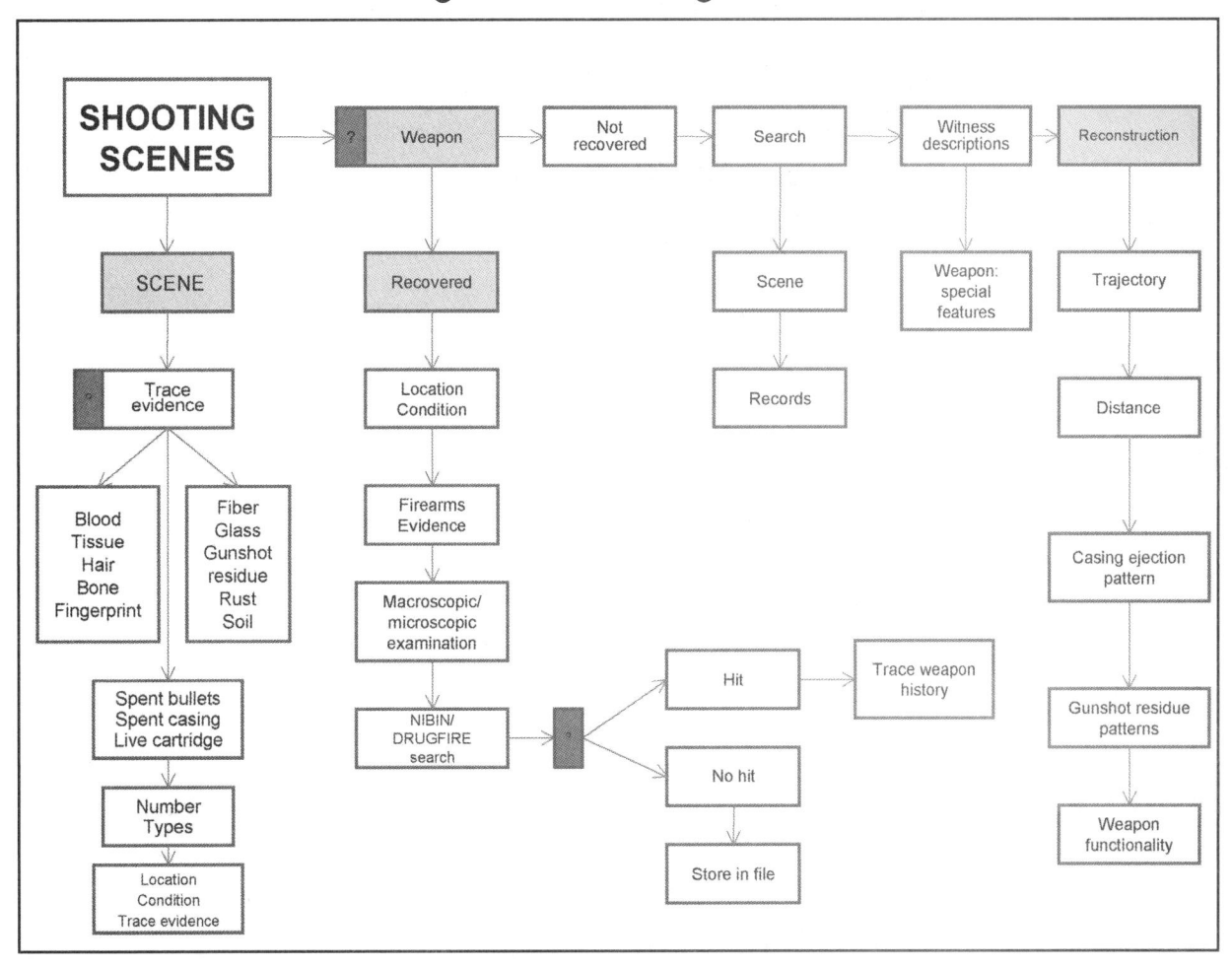

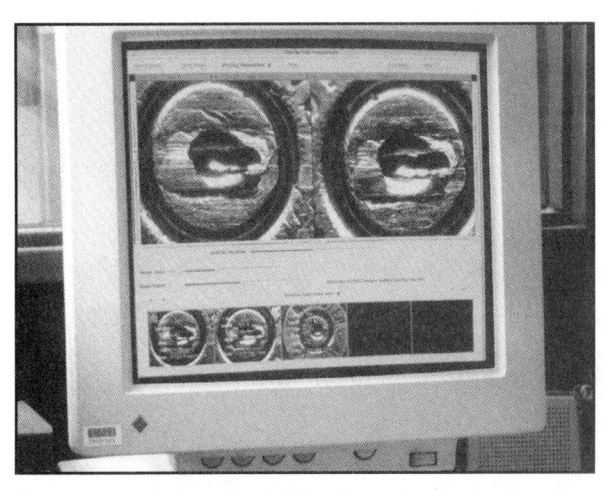

```
SHOOTING SCENES → ? Weapon → Not recovered → Search → Witness descriptions → Reconstruction
    ↓                ↓                          ↓           ↓                      ↓
  SCENE           Recovered                   Scene      Weapon:              Trajectory
    ↓                ↓                          ↓       special features          ↓
  Trace          Location                    Records                          Distance
 evidence        Condition                                                       ↓
    ↓                ↓                                                      Casing ejection
 Blood/Fiber     Firearms                                                      pattern
 Tissue/Glass    Evidence                                                        ↓
 Hair/Gunshot       ↓                                                       Gunshot residue
 Bone/residue    Macroscopic/                                                  patterns
 Fingerprint/    microscopic                    Hit → Trace weapon               ↓
 Rust/Soil       examination                          history                 Weapon
    ↓                ↓                                                      functionality
 Spent bullets   NIBIN/
 Spent casing    DRUGFIRE                      No hit
 Live cartridge  search                          ↓
    ↓                                        Store in file
 Number
 Types
    ↓
 Location
 Condition
 Trace evidence
```

▲ Computer imaging enables forensic firearms examiners to compare firing pin impressions from spent casings. (left)

▲ Striation marks from lands and grooves of fired bullets were used to identify the weapon. (right)

Forensic Science Today

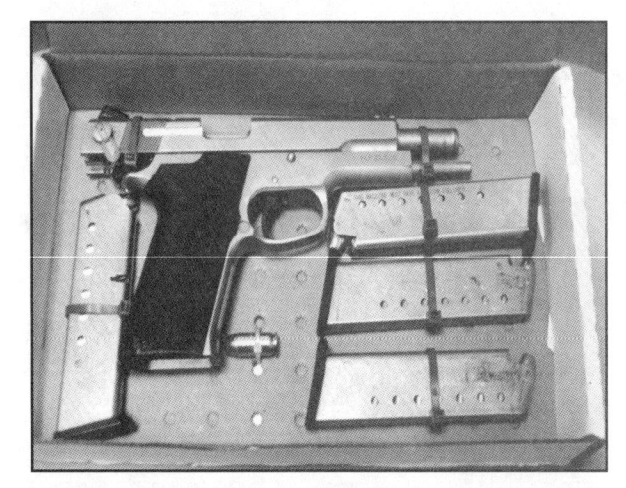

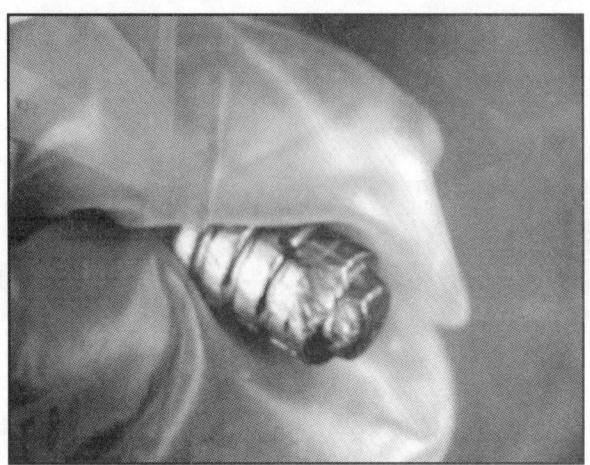

▲ A semi-automatic weapon seized from a suspect involved in a shooting incident. (top)

▲ A spent bullet recovered by a medical examiner from a victim's body. (bottom)

TEST FIRING WEAPONS IN THE FORENSIC LABORATORY

Each rifled gun has a unique "fingerprint"—that is, the individual microscopic markings called *striations* that are made on a bullet as it is fired through the gun's barrel. These markings are unique because they reflect not only the tools used in the manufacturing process of the barrel, but also the normal wear and tear inside the barrel as the weapon is used and fired over time. Even barrels rifled in succession with the same tool are not identical since the tool used for the rifling changes due to its contact with each barrel. A bullet may have markings on it from coming into contact with the gun barrel. The cartridge case may have markings on it from coming into contact with the firing pin, extractor, ejector, and breechface (rear wall of the firing chamber). Each of these gun components will leave markings in a different place on the cartridge case:

- Breechface striations (from contact with the rear wall of the firing chamber) are impressed upon the rear of the cartridge case.
- Firing pin impressions are marked upon the primer cup.
- Extractor markings are visibly traced upon the rim of the cartridge case.
- Ejector marks are impressed upon the side of the cartridge case.

Therefore, by comparing the marks on a bullet with the marks produced by a particular gun barrel, firearms examiners can, for example, identify whether a bullet found in a body or embedded in a floor at a crime scene came from the gun that was found at the suspect's house. By examining a cartridge case for marks from the firing pin, breechface, extractor or ejector, a firearms examiner can determine whether a cartridge case was fired in a particular gun. Firearms examiners do this by test-firing guns in a special forensic laboratory under strictly controlled circumstances in order to obtain a bullet or cartridge case from the known firearm.

After known bullets and cartridge cases are recovered from the crime scene, firearms examiners take measurements to determine the caliber (size) of the bullet, type of bullet, weight and size of the bullet, shape of the bullet, metallic composition, and manufacturers' markings. In addition, the striation marks caused by the passage of the bullet through the individual gun's grooves as it is driven down the barrel are compared microscopically with such markings on bullets found at the crime scene or lodged in the victim's body. These comparisons are done using a comparison microscope. This allows the examiner to view the evidence from the known source (cartridge case test fired in the lab) to the unknown evidence (cartridge case collected at the crime scene). Firearms examiners can also search for a known source of a cartridge case or bullet using a database for firearms evidence—IBIS, Integrated Ballistic Identification System. This database allows examiners to compare firearms evidence collected at a crime scene to firearms, bullets, and cartridge cases collected from other crime scenes. Using this database, bullets can be connected that are found at crime scenes in different cities. It also allows firearms collected at various scenes to be linked to bullets or cartridge cases from different scenes. The markings placed on bullets and cartridge cases by a firearm are examples of toolmarks.

CASE PROFILE
Two Friends, One Wife, and a Gun: Accident or Murder?

Montrose, PA: One summer, attorney Marty Dillon invited his good friend Dr. Stephen Scher to go skeet shooting in the rocky hills of northeastern Pennsylvania. After the two friends spent the day shooting, they headed back to their cabin. According to Scher, Dillon spotted a porcupine running across the trail, so he grabbed Scher's gun and chased the animal a couple hundred feet into the brush. Scher heard a click from the gun and then a loud discharge. Minutes later, said Scher, he found his friend sprawled on the ground and bleeding from his chest. Scher said he tried frantically to revive Dillon, first administering mouth to mouth resuscitation, then sticking his hand on Dillon's chest and massaging his heart. When this didn't work, Scher said that he became distraught and smashed the shotgun against a tree - breaking off the wooden stock - and then went for help.

Hours later, police arrived at Scher's home to tell his wife, Ann, about the shooting. They found her unconscious from a massive overdose of painkillers. It seemed she had attempted suicide. When she awoke in the hospital and learned of Dillon's death, her first response was, "He killed him, didn't he?"

An autopsy on Dillon noted no apparent signs of gunpowder burns on Dillon's chest, which should have been present if the gun had discharged at very close range. The angle of the wound suggested that the shot came from overhead and traveled down. Further, Dillon's body was found within a few feet of the skeet shooting machine, two unbroken clay pigeons under his hand, suggesting that the two friends had not been walking back to the cabin as Scher had said. Dillon's unloaded gun lay nearby.

Police believed that Dillon had been struck by oversized pellets from a 16 gauge high-powered shell that is "normally used for larger game and is not suitable for trap shooting." The four unfired shells still in Scher's gun were typical skeet shooting rounds. No other high powered shells were found at the scene.

Lawrence Dillon, Marty Dillon's father, urged police to investigate further. The elder Dillon told police that Marty Dillon and Stephen Scher had argued that day over Scher's relationship with Patricia Dillon, Marty's wife. However, despite pleas from Lawrence Dillon and others, the shooting was ruled a self-inflicted accident.

Months after the shooting, Marty's widow Patricia Dillon and her two small children moved to Philadelphia. Scher's divorce from his wife Ann was finalized that same year. Two years later Stephen Scher and Patricia Dillon married. No one but close family members reacted to the marriage.

For the next eight years, Marty Dillon's father Lawrence prodded authorities – including three district attorneys and three coroners – to reopen the case. Frustrated, he finally hired his own investigator. Once the investigator concluded that the shooting was a homicide, Lawrence Dillon turned to Dr. Henry Lee to investigate. Lawrence Dillon had read about Lee in the papers.

After taking on the case, investigators began a series of tests. Using pigskin wrapped around a giant sponge to simulate a human body, they concluded that the muzzle of the shotgun would have had to be at least three to five feet away from Marty Dillon to produce the kind of wound Dillon had suffered. It would be impossible for a person to shoot himself with a shotgun at such a distance. Dr. Lee performed additional tests on Scher's boots, which had been stored in a police evidence locker. The tests revealed signs of blood splatter that were consistent with the blood being deposited on Scher's boots by the impact of the shot against the victim's chest while Scher was standing quite close to Dillon. This blood splatter was not evident to the naked eye.

continued on next page...

Forensic Science Today

Based on forensic scientists' tests and findings, state police put two homicide detectives on the case. Within months, the detectives were convinced that Marty Dillon's body needed to be exhumed and a second autopsy performed. At an exhumation hearing, Marty's widow Patricia (now married to Scher) argued that she had much to lose if the new autopsy suggested Dillon's death was a homicide and her new husband, Stephen Scher, had done it. The judge disagreed, and Marty Dillon's remains were exhumed. An autopsy using procedures not available at the time of his death concluded that Dillon could not possibly have shot himself, accidentally or otherwise. Dr. Lee's findings were now confirmed.

Over ten years after the shooting death, police charged Dr. Stephen Scher with the murder of Marty Dillon. Scher pleaded not guilty and the trial commenced.

At the trial, scientists testified about how Marty Dillon was shot. Dr. Lee explained to the jury that Dillon couldn't have shot himself in this way, and that Dillon was shot by someone else. Scientists also testified about the blood spatter found on the shoes. Scher then changed his story, saying that he had argued with Dillon and the gun had accidentally gone off.

The jury deliberated and found Stephen Scher guilty of murder thanks to the persistence of the Dillon family and the evidence examined by the forensic scientist.

▶ Reconstruction of bullet trajectories of multiple shots fired into a vehicle. (Courtesy of Robert O'Brian and Mark Newth.)

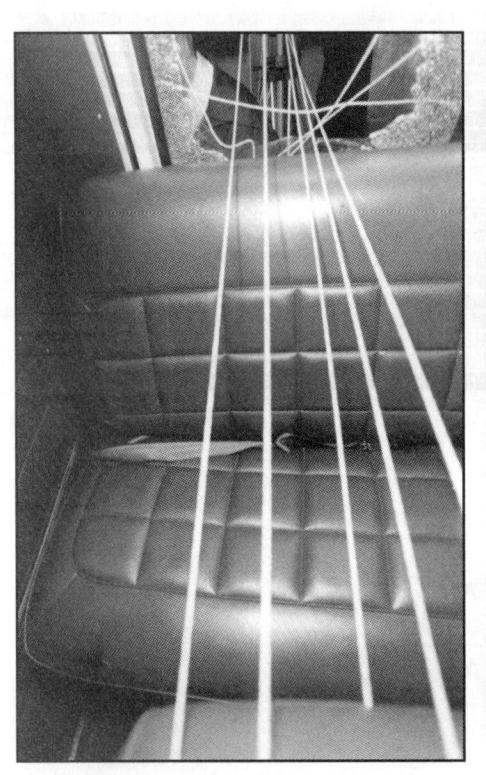

DOCUMENTING BULLET TRAJECTORIES AT THE CRIME SCENE

When a bullet is fired from a gun, the path it takes through the air is called a *trajectory*. Bullets make trajectories both before and after striking a victim or other surfaces. For instance, a bullet can be shot through the ceiling and hit a victim on the floor above. Or, it can go through a victim's arm and then lodge in the wall behind him. Either way, the bullet's trajectory is altered with every object or person it comes into contact with prior to coming to a stop. Therefore, geometry comes into play when measuring bullet trajectories.

Chapter 12
Toolmark and Firearms Evidence

Bullet trajectories can also help forensic scientists reconstruct the crime when combined with statements by witnesses and other physical evidence. Additionally, trajectories can help locate missing bullets to be used in laboratory analysis and crime scene reconstruction.

What can bullet trajectories help us discover? Here are some examples:

- Where the shooter was standing
- Where the victim was standing
- Direction in which the gun was pointed
- Number of shots fired
- How far away the shooter was standing from the victim
- Possible sequence of several shots that were fired
- Where the bullet entry and exit holes are
- Possible projectile deflections or ricochets—that is, objects that the bullet hit en route which then deflected or changed the bullet's flight path.

GUN SHOT RESIDUE (GSR)

Gunshot residue—or GSR as it is commonly known—can be a key indicator of whether a person fired a gun. GSR is a substance that is released in the internal explosion of gunpowder when a shot is fired. After a gun is fired, the detonation and burning of gunpowder results in the creation of gases, soot, and unburned or partially burned gunpowder. These substances are propelled forward with the bullet as it heads toward the target. Simultaneously, some of these materials are blown backwards onto the shooter.

Thus, GSR can often be found on the hands and clothing of the shooter, and also of the victim if the gun is fired at close or point-blank range (contact). GSR residue patterns can also tell us whether the case was a homicide or suicide—for instance, if a gun is found next to the hand of a dead body but there is no GSR on the victim's hand from firing the gun, then somebody else must have fired the gun and the case must be ruled as a homicide. GSR is also useful in determining the distance from the muzzle of the gun to the target, whether a suspect recently fired a gun, the maximum distance between the shooter and victim, and the relative locations of multiple targets.

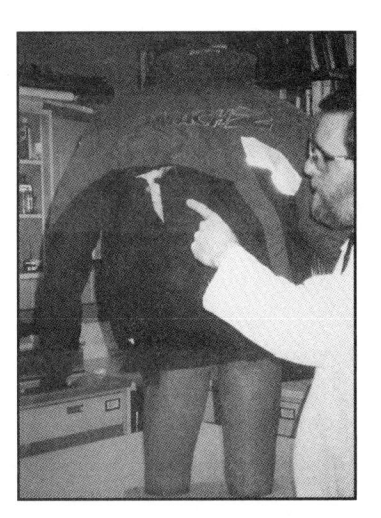

▲ Examining jacket to gunshot residue.

WHAT DO GSR PATTERNS TELL US ABOUT THE SHOOTING?

The analysis of gunshot residue patterns is extremely useful in reconstructing the course of events of a shooting. GSR is especially useful for estimating distances between the gun barrel and target. How GSR patterns are interpreted depends upon the weapon that was used, the type and size of the ammunition, whether an intermediate target was involved, and other factors. Such patterns can usually be seen directly on the shooting victim's clothing or body. GSR patterns can also be transferred to filter paper or treated photographic paper for easier observation.

Gunshot residue patterns will look different depending on whether they came from shots that were fired at contact, close range, medium range, or long distance. In particular:

- Contact/point-blank. When a gun is held in direct contact with the target—i.e., the muzzle of the gun is touching the victim's skin or clothing—soot deposits are visible around the entrance of the bullet hole. Little or no unburned powder will be found.

▶ Proper method for collecting GSR evidence with an adhesive disc for SEM analysis.

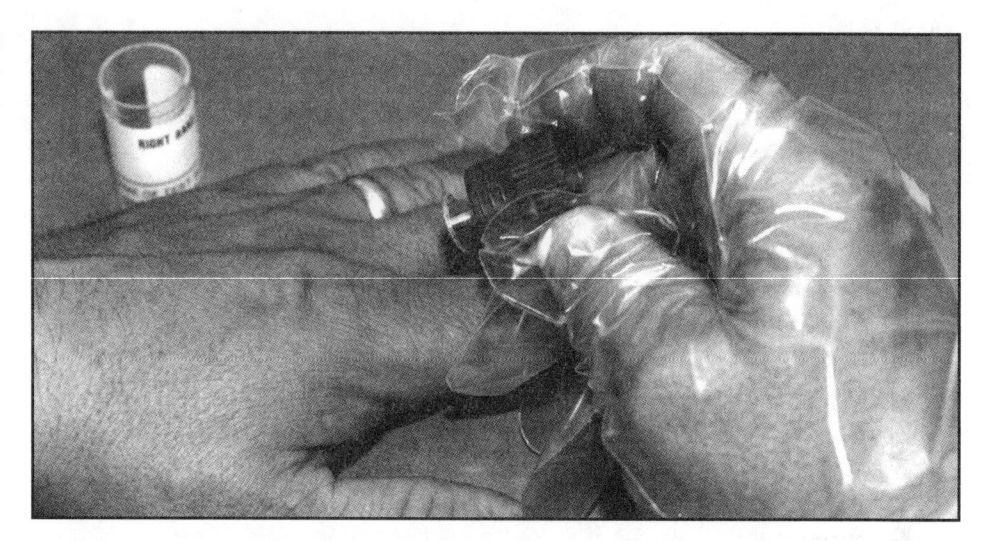

- Close range. When a gun is held very close to (less than six inches), but not touching the target, gunshot residues are deposited around the bullet hole with black soot and smoke, and the pattern is small and dense. As distance increases, the size of the pattern also increases.
- Medium range. When a gun is fired between 1.5 feet and 3 feet from its target, scattered gunshot residue will be seen around the bullet hole but without any black soot deposits. The pattern it creates spreads out further as the distance between the gun and the target increases.
- Long distance range. When a gun is fired from a distance of greater than 5 feet, GSR is usually not found on the target since it does not normally travel that far.

Note that the firearms examiner must use the same weapon for making the above determinations.

COMPONENTS OF GSR
Gunshot residue consists of the following materials:

1. **Primer.** Primer is the material that is detonated in the gun, resulting in the explosion that sends the bullet forward. Primer is made up of four components:
   - *Initiating explosive*—lead azide, lead styphnate, diazodinitrophenol.
   - *Oxidizing agent*—calcium peroxide, barium nitrate, magnesium peroxide, manganese dioxide.
   - *Fuel*—calcium silicide, antimony sulfide, aluminum, titanium, zirconium, lead thiocyanate.
   - *Sensitizing Agent*—powdered glass, tetracene, titanium, calcium silicide.

2. **Propellant gunpowder.** There are two types of gunpowder, which propel the bullet forward upon detonation of the primer:
   - Black powder—75% potassium nitrate, 15% sulfur, 10% charcoal.
   - Smokeless powder—Nitrocellulose (single base); nitrocellulose and nitroglycerin (double base).

3. **Lubricants** consisting of:
   - Substances used to ease the bullet into or out of the cartridge case.
   - Antifriction agents to speed the bullet down the barrel of the gun.
   - Residues of cleaning materials and gun oils.
   - Additives used in shaping and compounding the propellant.

4. **Metallic materials.** Finally, traces of metal may be found in GSR, which may come from the bullet, gun barrel, projectile, jacket, and casing metals. These metallic materials may consist of the following:
   - Lead
   - Copper
   - Zinc
   - Antimony
   - Arsenic
   - Bismuth
   - Chromium
   - Stainless Steel
   - Carbon Steel

COLLECTING GSR AT THE CRIME SCENE

GSR should be collected as soon as possible after the weapon is fired, and always before conducting fingerprinting on the suspect or victim, since many activities remove GSR from the hands. There are three common methods of collecting gunshot residue at the crime scene for further analysis; and the determination of which method of collection is to be used depends upon which type of analysis is going to be conducted in the forensic laboratory.

GSR Adhesive Discs. If the method of analysis to be used in the forensic laboratory will be Scanning Electron Microscopy (SEM), then the method of collecting GSR at the crime scene will be through small adhesive discs that GSR sticks to upon contact. When testing a suspect's or victim's hands for GSR, the investigator will pat both hands and collect GSR from the back of the hands with one of the adhesive discs, and from the palms with another disc. The discs will then be properly sealed in a container and labeled for transport to the forensic laboratory.

Swabs. When the method of forensic analysis is to be an Atomic Absorption (AA) Spectrometer or other elemental analysis, GSR is collected at the crime scene with a swab. Each swabbing kit contains six or seven test tubes, each containing a swab. After labeling each test tube to refer to the various parts of the hand (e.g., RB, RP, LB, LP), the investigator puts on the gloves supplied by the collection kit, removes the swabs from the test tube, moistens the swabs with a 5 percent nitric acid solution, and then swabs the specific area of the hand corresponding to the label on the test tube. The swab is then replaced in its specially labeled test tube, which is properly sealed and transported to the forensic laboratory for further analysis.

Testing for GSR on other items of physical evidence. GSR is not only found on the human body; it is also sometimes found on other items of physical evidence such as a car or clothing. If the item is small, the entire item should be submitted to the forensic laboratory for analysis. Such items should be packaged very carefully to avoid movement or the loss of residue. If the item is too large to send to the lab, it should be sampled at the crime scene using either the adhesive disc method or swab method, and then sent to the lab for testing.

Forensic Science Today

ANALYZING GSR AT THE FORENSIC LABORATORY

Once a garment or material containing gunshot residue reaches the forensic laboratory, it is first examined under a low-power microscope to aid the eye in detecting gunpowder particles, smudges, damage, and burns around a bullet hole. Next, forensic scientists conduct chemical tests to check for the presence of nitrates and nitrites. Several chemical reagents will react with nitrate or nitrite components of GSR to give a color reaction. This is useful for showing the distribution pattern of residue on clothing, or as a screening test to determine other areas that need further testing.

In addition, forensic laboratories use a variety of testing methods to detect the presence of the major characteristic inorganic components of GSR (barium, antimony, and lead). These methods can also determine the distribution and concentration of these components. Such methods include:

- Microchemical tests
- Atomic absorption (AA)
- Anodic Stripping Voltametry (AVS)
- Scanning electron microscope/energy dispersive X-ray analysis (SEM-EDX)
- Neutron activation analysis (NAA)
- Gas chromatography/mass spectrometry (GC/MS)
- High performance liquid chromatography (HPLC)
- Micellar electrokinetic capillary electrophoresis (MECE)

CASE PROFILE

In September 1984, eight-year-old Vicki Lynn Hoskinson was kidnapped in Tucson, Arizona as she rode her pink bicycle to mail a letter at a nearby mailbox. Apart from Vicki's bike left in the middle of the quiet neighborhood street, Vicki's kidnapper left no trace behind. *Or did he?* Following a lead from a school employee who wrote down the license plate number of a suspicious vehicle, police tracked down Frank Atwood. Atwood was a convicted sex offender out on parole, who had violated his parole agreement by moving from California to Arizona.

After arresting Atwood for parole violation, police searched the interior of his car for evidence. However, nothing was found inside the car that could trace Vicki to Atwood. The first real break in the case came when, going back to the area where Vicki's bike had been found, investigators discovered that a nearby mailbox was bent over. Could this mean it had been hit by a car?

An accident reconstructionist was called in to inspect Frank Atwood's car for evidence of a recent collision. He found similar scratches both underneath the car and on the pedal of Vicki's bike. When the bike was laid underneath the car, the bike pedal touched the car in exactly the same spot as where the scratches were located under the car. Could this mean that Atwood knocked Vicki off her bike at a slow speed in order to kidnap her?

Laboratory testing of pink paint found on the car's bumper and material found on Vicki's bike showed a match between the two samples, which verified this hypothesis. The last nail in Atwood's coffin was when Vicki's skull was found in the desert seven months later. Frank Atwood was convicted of the kidnapping and murder of Vicki Lynn Hoskinson, and sentenced to death. A key piece of evidence connecting Atwood to Vicki's disappearance was the existence of toolmarks—the scratches found on both the victim's bike and the suspect's car.

Chapter 12
Toolmark and Firearms Evidence

CASE PROFILE

In a high school scavenger hunt, one item on the list was a balloon. A student looking for a balloon found one on the roof of a nearby house. Unfortunately, the owner of the house came home to find a strange person (the student) on his roof, and, thinking he was being burgled, shot the student with a gun. During the subsequent case investigation, the bullet trajectory was the most important key to establishing what had happened during the course of the events.

TECHNICAL DEFINITIONS

Bullet: A projectile that is pushed forward through the barrel of a gun upon detonation.

Cartridge: The ammunition package consisting of the bullet, gunpowder, and primer.

Cartridge Case: The metal container that holds the cartridge.

Caliber: The size of the interior diameter of the barrel of a handgun or rifle.

Gauge: The size of the interior diameter of the barrel of a shotgun.

Grooves: Indentations in the interior of a gun barrel caused by rifling.

Lands: The portions of a gun barrel interior that have not been cut by rifling, which appear raised relative to grooves.

Rifling: Cut grooves going down the interior surface of the length of the gun barrel, which cause the bullet to spiral as it moves through the air.

GSR (Gunshot Residue): The residue left behind on the fingers, hands, and clothing of someone who has shot a gun. If a gun is fired on contact or near contact, then GSR may also be found on the victim's body or clothing.

Contact (Point-blank) Range: When a gun is discharged directly against the target surface.

Primer: A component of gunshot residue, the material that is detonated in the powder in the casing, resulting in the explosion that propels the bullet forward.

Propellant: Also known as gunpowder, the substance that propels the bullet forward upon detonation of the primer.

Striation: Scratch marks made on a bullet as it is fired from a rifled gun barrel, caused by the shape and markings of the interior of the gun barrel.

Trajectory: The path a bullet takes in the air after being fired from a weapon.

Toolmark: Any cut, scrape, impression, gouge, scratch, indentation or other marking left on an object or surface by another object being forced into or moved across it.

Compression Toolmark: A toolmark that is caused by one object pressing into a surface.

Sliding Toolmark: A toolmark that is caused by one object sliding against a surface.

Fire and Arson Evidence

Chapter 13

CHAPTER OBJECTIVES

- ✔ Identify the two key criteria for investigating an arson scene.
- ✔ Understand the different fire burn patterns found at fire scenes.
- ✔ Describe different types of physical evidence found at fire scenes.
- ✔ Understand the considerations for examining the scene of the fire and searching for evidence.
- ✔ Discuss procedures for collecting and preserving evidence from a fire scene.
- ✔ Understand the basic analysis of solid debris and liquid samples found at fire crime scenes.
- ✔ Define the following vocabulary words and phrases: *arson, accelerant, combustible, charring, trailer, ignition device, point of fire origin, head space,* and *pyrolysis gas chromatography.*

Fire is caused by a combustion reaction - the combination of oxygen with a fuel. Fire investigators define a fire as having the following four requirements: oxygen, fuel source, source of ignition and the ability to be self-sustaining. Fire may be used by criminals to burn down structures, commit crime, or destroy evidence of crime. Forensic scientists study the effects of fire damage in order to determine whether a fire was the result of an accident or arson. When fire is the cause of death, the cause of the fire must be determined. When fire is used to destroy evidence, charred remains are tested in a lab for identification and analysis.

The two key criteria for investigating the arson scene are:

1. **Locating the point of origin**—that is, where did the fire start?
2. **Determining the cause of the fire, such as whether a fire accelerant is present**—that is, did the arsonist (person who intentionally started the fire) use a substance to speed up the fire process?

Forensic Science Today

KEY TERMS

Arson
Accelerant
Combustible
Charring Pattern
Point of Origin
Trailer
Ignition Device
Head Space
Gas Chromatography
Thin Layer Chromatography
Pyrolysis Gas Chromatography

FIRE BURN PATTERNS

There are many different causes of fires, but all fires form burn patterns. These burn patterns are influenced by the environment, how much combustible material is present, how much oxygen is present, and the type of fire that is burning. Fire investigators look carefully at patterns left by fire burning on floors, walls and other objects, in order to discover the point of origin of the fire, how much damage was caused by the fire, and the direction in which the fire traveled. Charring patterns are also used to determine the intensity of the fire and the length of the burn. Finally, not only do investigators look at the patterns that they expect to see, but they also look at deviations—differences—from what they would normally expect from a certain type of fire. Some typical burn patterns found at fire scenes include:

- multiple origin burn patterns,
- "V" shape—or inverted cone—burn pattern,
- low burn pattern configurations,
- direction of charring patterns,
- depth of charring patterns,
- smoke stain patterns,
- trailer patterns,
- material melting patterns, and
- concrete spalling patterns.

What do you think these different patterns might tell investigators about a fire?

▲ A major fire scene located in an industry park. (Courtesy of Dr. Stanley Lee.)

OTHER TYPES OF EVIDENCE FOUND AT THE SCENE OF A FIRE

In addition to fire and smoke patterns, investigators will also search for other types of evidence to determine if a crime has been committed.

Accelerants. The first thing a fire investigator looks for in determining arson is the presence of accelerants. Some examples of accelerants include gasoline, kerosene, and other flammable liquids, chemicals, and gases.

Debris and ashes. Investigators look at the point of origin of the fire and nearby to see if any physical evidence of the cause of the fire is still left. Ashes and debris may still contain traces of accelerant used to start the fire.

Odor evidence. Evidence of an accelerant or a mixture of flammable fluids used to start the fire may be found by a lingering petroleum odor. Arson dogs can be used to detect the residues of accelerants.

Trailers. A trailer is a material designed to spread a fire from one point to another, and can be as simple as a piece of string, or a trail of accelerant poured onto the floor before starting the fire.

Containers. Investigators look for cans, bottles, or other containers that the arsonist may have used to transfer the accelerant to the point of origin.

Tools. Sometimes arsonists leave behind tools used to break into the structure or to crack a safe in the building.

Biological evidence. This refers to evidence of a body or body parts such as bone, tissues, blood, or hair. Such evidence can help determine the identity of victims, or of perpetrators if they were injured during the commission of the crime. It can also help determine whether the fire was started before or after the victim's death.

Documents. Arsonists may also start a fire in order to burn incriminating evidence found in documents, computers, and books; and sometimes evidence of these documents remains unburned after the fire.

Ignition devices. Investigators should look for devices used to start the fire, such as lighters, matches, candles, bomb timers, explosives, or even cigarettes.

Imprints and impressions. Just like with any crime scene, investigators will document and collect fingerprints, toolmarks, tire tracks, footprints, and other types of evidence.

Forensic Science Today

Concepts of Fire Elements

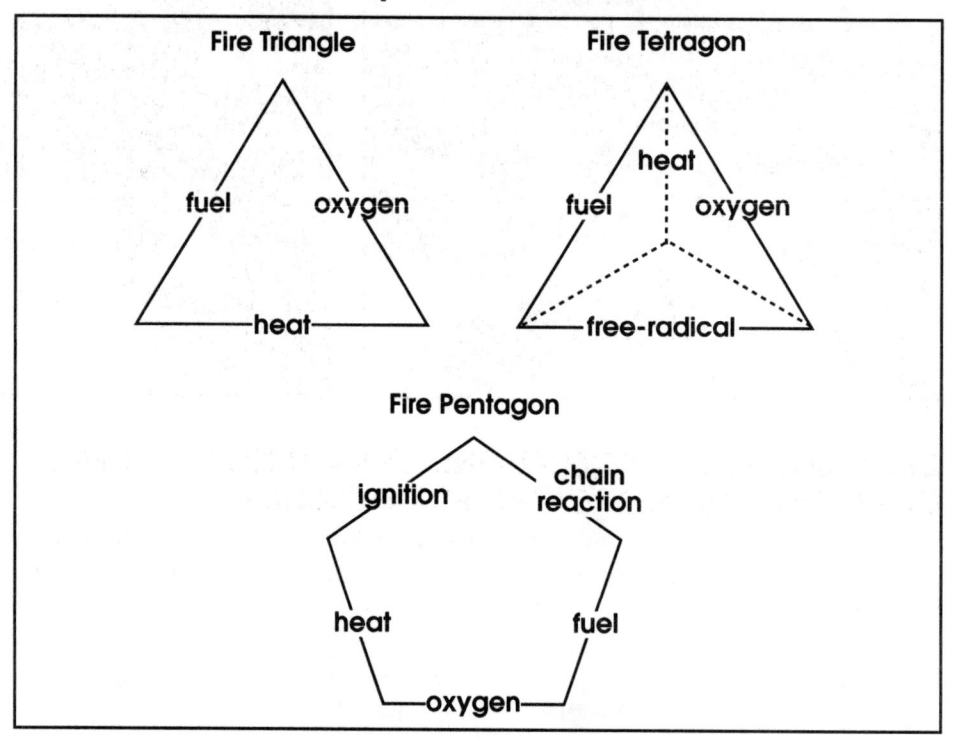

Fire Triangle
- fuel
- oxygen
- heat

Fire Tetragon
- heat
- fuel
- oxygen
- free-radical

Fire Pentagon
- ignition
- chain reaction
- heat
- fuel
- oxygen

Logic Tree: Bombing Explosion Fire/Arson

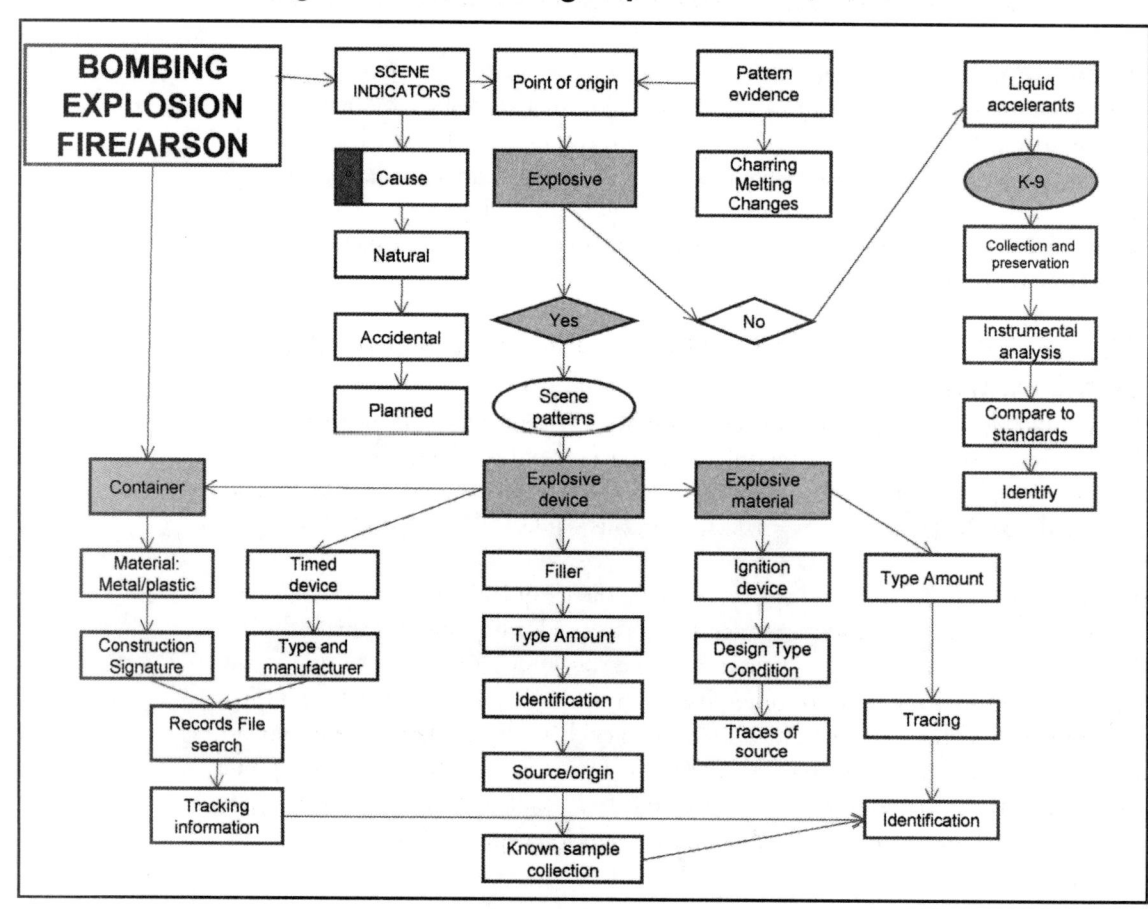

BOMBING EXPLOSION FIRE/ARSON

- SCENE INDICATORS
  - Cause
    - Natural
    - Accidental
    - Planned
- Point of origin
  - Explosive
    - Yes → Scene patterns
    - No
- Pattern evidence
  - Charring Melting Changes
- Liquid accelerants
  - K-9
  - Collection and preservation
  - Instrumental analysis
  - Compare to standards
  - Identify

- Container
  - Material: Metal/plastic
  - Construction Signature
  - Records File search
  - Tracking information
- Timed device
  - Type and manufacturer
- Explosive device
  - Filler
  - Type Amount
  - Identification
  - Source/origin
  - Known sample collection
- Explosive material
  - Ignition device
    - Design Type Condition
    - Traces of source
  - Type Amount
    - Tracing
    - Identification

158

Chapter 13
Fire and Arson Evidence

▲ Fire scene debris collected and sealed in metal cans to preserve the accelerant vapors. (right)

◀ Fire scene debris found near the point of origin. (left)

CRIME SCENE CONSIDERATIONS: EXAMINING THE SCENE OF THE FIRE AND SEARCHING FOR EVIDENCE

The best time to arrive on a fire scene is when the fire is still burning. Remember, when coming to a fire scene, the top priority of forensic investigators is personal safety. If the fire is still burning, the investigator will keep to the side to observe the fire and document its characteristics and conditions. For instance, how high do the flames reach? What color is the flame? What color is the smoke?

When the fire has been completely put out, investigators should take photographs of the fire scene, including destruction of buildings and other objects caused by the fire and smoke. Objects should not be moved unless absolutely necessary in order to ensure safety.

One of the most important things to discover when investigating a fire scene is the point of origin of the fire. Investigators start from the area that has the least damage, and move systematically toward the area with the most damage, making thorough observations and surveying the pattern of charring. By following this path, the investigator will discover the point of origin, which is the place with the most damage caused by the initial explosion or acceleration of flame.

COLLECTING AND PRESERVING EVIDENCE FROM A FIRE CRIME SCENE

Packaging is very important for collecting and preserving evidence from a fire scene. Fragile items should be packed carefully in boxes to avoid breakage, and other types of evidence such as fingerprints, tool marks, and hair samples should be collected and preserved according to the methods detailed in this book.

It is extremely important to package accelerant evidence quickly since accelerants tend to evaporate very rapidly. Evidence of accelerants from a likely point of origin of the fire should be correctly placed into airtight containers as soon as possible. As mentioned above, you can find the point of origin based on fire patterns or places where the accelerant may have pooled. Sometimes there is no evidence left of the accelerant due to the intensity of the fire or the type of accelerant. Therefore, it is critical to collect and preserve absorbent materials that may have absorbed accelerants in them. These types of materials include carpeting, wood floors, soil, fabrics, paper, and other debris.

When collecting fire evidence, you should choose a clean, plain metal container in which to put the evidence. Be sure to leave head space (empty space above the collected evidence) in the container, which will allow the accelerant to evaporate inside the container. Thus, a container should be no more than two-thirds full. Once the container arrives in the forensic laboratory for testing, forensic scientists will often take samples from that head space containing the evaporated accelerant.

ANALYZING FIRE EVIDENCE IN THE LAB

Forensic scientists conduct two main types of analysis on fire evidence: solid debris analysis and liquid sample analysis.

Solid Debris Analysis

Forensic lab scientists are able to collect and test accelerant that has accumulated in solid debris such as ash, wood, or other materials. There are a variety of ways in which this is accomplished.

Cold head space. A sample is taken of accelerant vapors that are trapped within the head space of the container. This is done by making a small hole in the top of the container and withdrawing the sample with a syringe. The sample is then tested using one of the liquid sample analysis methods listed later in this section.

Heated head space. After the cold head space is removed from the container, the container is then heated to 100 degrees Celsius, which causes more vapors to be released into the head space. A sample of this vapor can then be analyzed.

Carbon tube or strip absorption. Another method to release more vapors is to pass a pure dry gas through the container, which forces vapors that are then passed through a charcoal tube that absorbs these vapors. Or, similar to the carbon tube method, an alternative method is to hang a piece of paper coated with a thin layer of activated carbon inside the heated container containing the fire debris, which will then absorb any freed accelerant vapors, allowing the accelerant to be tested. The next step is to extract the accelerant from the carbon tube or strip of paper. Once the traces of accelerant are extracted, they can then be tested.

Solvent extraction. This is an old technique to extract accelerant from debris by washing the debris with a very low boiling solvent. The recovered solvent is then concentrated, and the solvent is selectively removed from the accelerant, which is then tested.

Distillation. Finally, accelerant may be distilled from the debris. This method is no longer used by forensic laboratories since it is very time-consuming and susceptible to contamination. However, it is good to know how accelerants were extracted prior to new technology. First, the debris is placed in a flask and either water or ethylene glycol is added. The mixture is heated to boiling point and the vapors are then passed into a condenser. When the vapors are cooled, a liquid layer of accelerant will float on the condensed water or ethylene glycol. The accelerant can then be tested.

Chapter 13
Fire and Arson Evidence

Liquid Sample Analysis

Once the liquid accelerant is obtained either directly from the site or after extracting it from debris as described above, there are several different methods of testing liquid samples of fire evidence. These include:

- Gas Chromatography
- Infrared Spectroscopy
- Combination Gas Chromatography and Infrared Spectroscopy
- Combination Gas Chromatography and Mass Spectrophotometry
- Refractive index
- Boiling point
- Flash point
- Thin Layer Chromatography

See Chapter 18 (Forensic Toxicology) for more information about these high-tech methods of testing liquid samples.

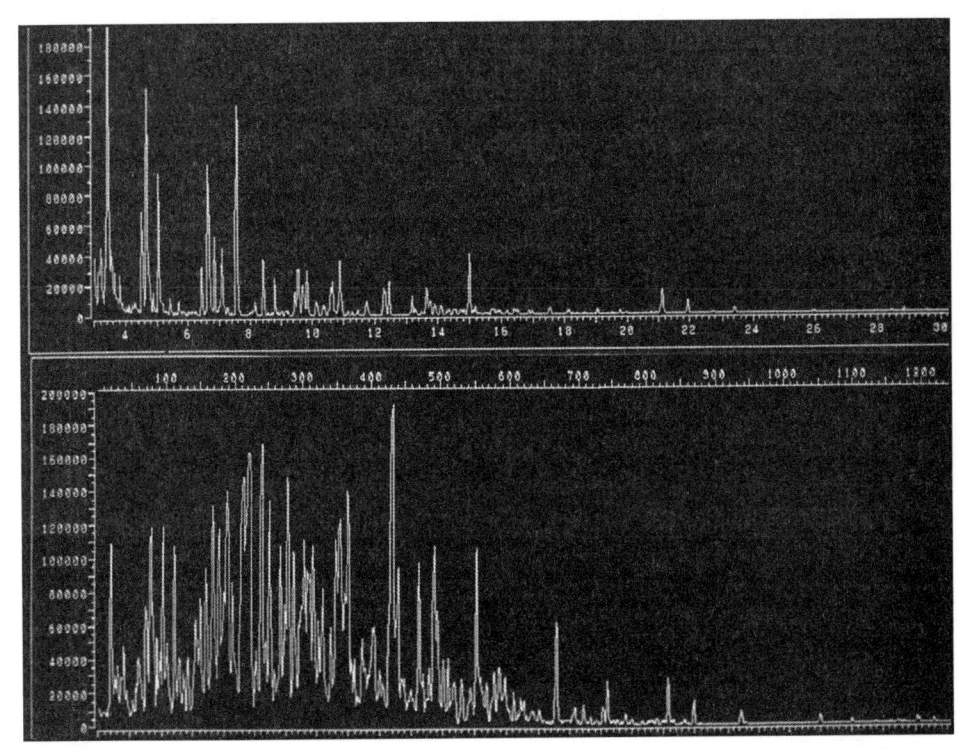

▲ Two gas chromatograph (GC) charts show the presence of accelerants in the fire scene debris. Top: sample from fire scene. Bottom: known gasoline standard.

TECHNICAL DEFINITIONS

Arson: A fire that has been set intentionally.

Accelerant: Any substance that aids in the development or spread of a fire, such as gasoline.

Combustible: A material that easily catches fire.

Charring: Patterns left behind by a fire burning walls and other objects.

Trailer: Material designed to spread a fire from one point to another.

Ignition Device: A device used to start a fire or explosion.

Point of Fire Origin: The point at which a fire is first ignited.

Head Space: Space left at the top of a container used for storing and transporting chemical evidence, to allow for evaporation without leading to explosion under pressure.

Pyrolysis Gas Chromatography: A form of chromatography that analyzes solids whereby the sample is heated to a temperature at which the material decomposes into its gaseous components.

Chapter 14

Chemical Evidence

CHAPTER OBJECTIVES

✔ Understand the nature of chemical evidence.
✔ Give examples of solid, liquid, and gaseous chemical evidence that may be found at a crime scene.
✔ Explain the proper techniques for collecting and packaging chemical evidence.
✔ Discuss the different types of laboratory analysis of chemical evidence, including preliminary examination, chemical testing, and instrumental techniques.
✔ Define the following vocabulary words and phrases: *chemist, organic, inorganic, solubility, melting point, boiling point, vapor, vapor point, vapor pressure, pH, index of refraction, color, chromatography, gas chromatography, thin-layer chromatography, spectroscopy,* and *mass spectrometry.*

Unknown chemical substances found at a crime scene can provide important information about the crime that was committed. Crime scene technicians can conduct field tests that detect many chemical residues found at crime scenes, vehicles, or on clothing, such as fire accelerants, blood, explosives, gunshot residue, drugs, cosmetic material, construction material, and trace metals. These materials should be collected and packaged correctly for further analysis at the forensic laboratory.

TYPES OF CHEMICAL EVIDENCE FOUND AT CRIME SCENES

Chemicals found at a crime scene can be present as a form of solid, liquid, or gas:

Solid chemicals encountered at a crime scene can be any *organic* or *inorganic* chemical or mixture thereof, and might include poisons, drugs, soil, cement, minerals, explosive residues, gunshot residues, cosmetics, and trace materials.

Liquid chemical evidence can come as either pure liquid or as a mixture, in water or other solvents. Examples of liquid chemical evidence include acids, gasoline and other accelerants, body fluids, beverages, chemicals for drug labs, cleaning fluids, medications, and poisons. Some common liquids change to gases at various temperatures (known as their *vapor point*), such as Toluene, a chemical found in glues that are sniffed (a lethal activity leading to serious brain damage and, in some cases, death).

Forensic Science Today

KEY TERMS

Chemist
Organic
Inorganic
Solubility
Spectroscopy
Mass Spectrometry
Chromatography

▼ This milk carton containing unknown chemical substances was collected from an explosion scene.

Gaseous chemical evidence might include natural gases (such as those found at fire scenes), refrigerants (such as Freon), butane from cigarette lighters, propane, tear gas, and propellants from a variety of household canned sprays.

COLLECTING AND PACKAGING CHEMICAL EVIDENCE FOR FURTHER ANALYSIS AT THE LAB

If field tests conducted by crime scene technicians indicate that a substance is chemical in nature, then the substance should be carefully handled and packaged to prevent loss or contamination of the evidence, as well as to prevent any possible injury to the person handling the chemicals. If an unknown chemical is still contained in its original packaging, then the package should remain unopened and be submitted to the lab along with any box, label, instructions, or forms associated with the chemical. However, if a chemical is found outside its original package, then it should be collected in a non-reactive, leak-proof container.

When transferring liquid chemicals to a container, technicians must be careful not to use a container that will dissolve or react on contact with the liquid. If the chemical evidence is found in a solid powdery form, it should be placed into a piece of paper folded in what is known as the "druggist's fold" (see Figure 1) and then placed into a paper envelope. The druggist's fold is a way of folding trace evidence or particles into a piece of paper forming a type of small envelope from which the substance cannot escape. Unknown solids should never be placed in metal containers due to the possibility of corrosion.

Sometimes chemicals are found as transfer evidence on clothing, vehicles, or parts of the body, particularly at a burglary, homicide, or sexual assault crime scene. Such chemicals might include lipsticks, cosmetics, lubricants, plaster, and other types of powder or smeared chemical substances. In these cases, the clothing should be packaged in paper and submitted to the forensic laboratory for complete examination and analysis.

ANALYZING UNKNOWN CHEMICALS IN THE FORENSIC LABORATORY

When an unknown chemical arrives in the laboratory for analysis, forensic scientists face an array of testing options. One thing to keep in mind is that chemicals can often be mixed or contaminated, which can complicate matters. Depending on the chemical or mixture thereof that has been submitted, its laboratory analysis could be as simple as a quick microscopic identification or very complicated, requiring many days and several instrumental techniques to discover its properties.

Preliminary Examination in the Lab

During the preliminary examination of chemical evidence, the forensic scientist can determine a number of physical characteristics of the substance, such as melting point, boiling point, odor, color, pH, and index of refraction. (These terms are defined at the end of this chapter.) If the chemical is found on an item of clothing, it should be carefully examined under the microscope to locate cosmetics, lipstick, or other trace evidence. Once the chemical is located and its pattern is recorded, a sample can be carefully removed for further analysis.

Chapter 14
Chemical Evidence

Figure 1. THE DRUGGIST'S FOLD

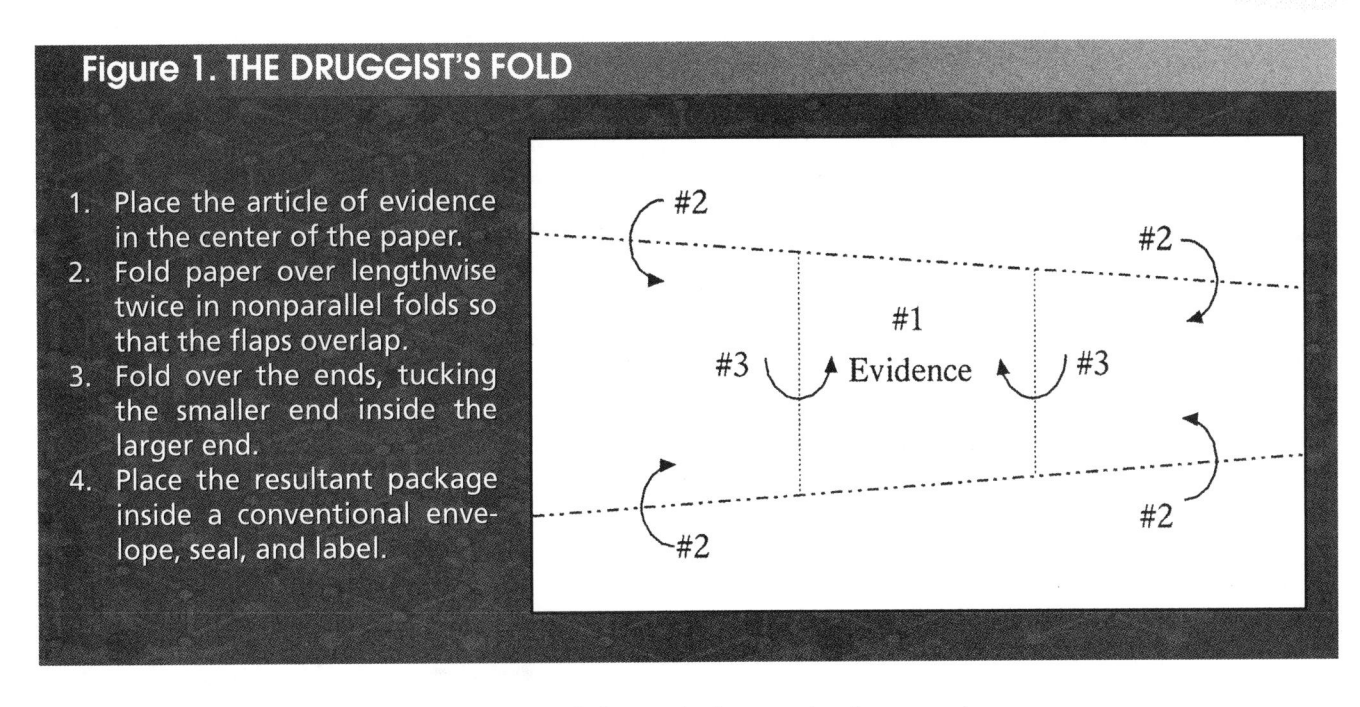

1. Place the article of evidence in the center of the paper.
2. Fold paper over lengthwise twice in nonparallel folds so that the flaps overlap.
3. Fold over the ends, tucking the smaller end inside the larger end.
4. Place the resultant package inside a conventional envelope, seal, and label.

Vapor point is another physical characteristic of chemicals that can be determined in the lab. A vapor is a substance that, though present in the gaseous phase, generally exists as a solid or a liquid at room temperature. Vapor point is the temperature at which a particular liquid chemical is transformed into a gas, and vapor pressure is the pressure exerted when a solid or a liquid is in equilibrium with its own vapor. In other words, a substance does not have to boil to have vapors, although boiling will completely convert it to the gas form, which is then equivalent to vapors. For many substances, a small increase in temperature and/or a small decrease in pressure will cause vapors to occur. Many substances have vapors at room temperature (for example, iodine solid has purple vapors), but many require higher temperatures or pulling vacuum to lower the pressure to create the vapors. The converse is that lowering the temperature or increasing the pressure will suppress the vapors—that's why you smell gasoline at the pump more strongly in the hot summer, but much less so in the cold winter. Each substance has its own definitive vapor pressure.

Conducting Chemical Tests in the Lab

One of the tests that forensic scientists conduct on chemical evidence is solubility. Solubility refers to the "dissolve-ability" of a substance, as affected by variables such as what type of liquid it will dissolve in, and the percentage of material dissolved in different solvents. For example, asphalt—which is commonly seen as a stain on bodies in hit and run cases—is much more soluble in chloroform than in acetone. Moreover, the solubility of organic materials differs from that of inorganic materials. Solubility of inorganic materials is affected by the nature of the solvent, the molecular structure and intrinsic nature of the compound, and the pH of the solution if it is a water solution. On the other hand, the solubility of organic compounds such as paint or grease is different depending on whether it is soluble in water, dilute base, dilute acid, chloroform, acetone, or other organic solvents. While many organic materials show little water solubility, most inorganic or biological materials have some solubility in water.

Forensic Science Today

Concepts in Chemical Evidence Analysis

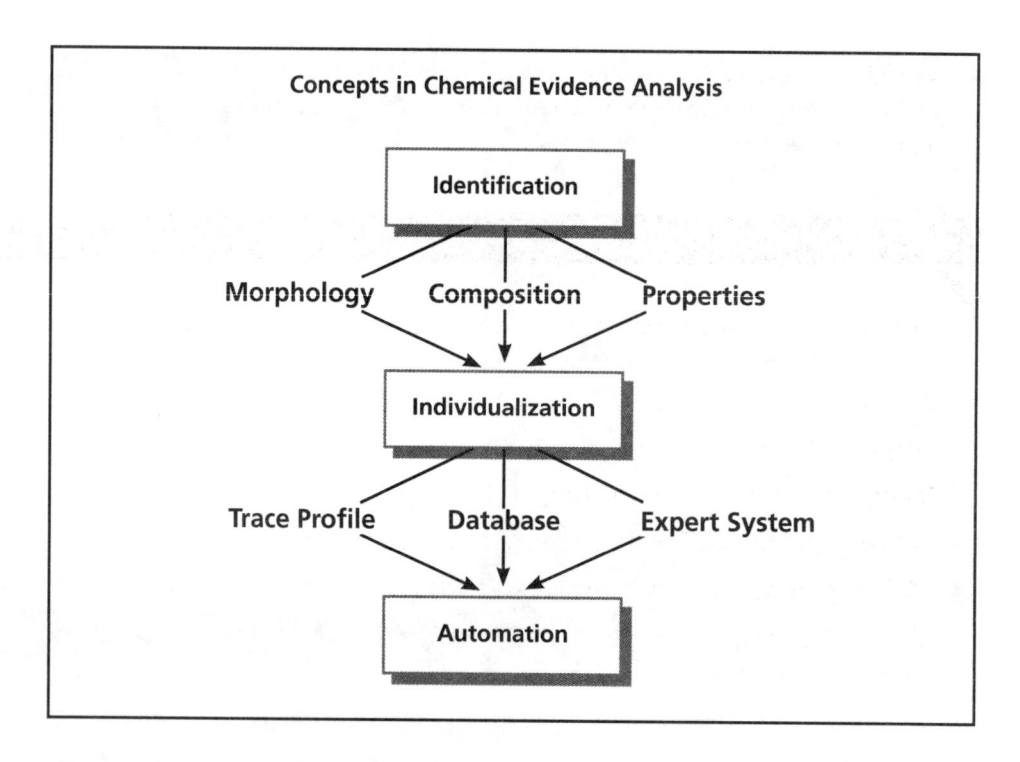

Identification → (Morphology, Composition, Properties) → Individualization → (Trace Profile, Database, Expert System) → Automation

Using Instrumental Techniques to Analyze Chemicals in the Lab

Finally, forensic scientists use a wide variety of instrumental techniques to determine the properties of chemical evidence. In Chapter 18, Forensic Toxicology, we will learn in more detail about chromatography, gas chromatography, and thin-layer chromatography. Chromatography is useful for analyzing drugs, inks, dyes, explosives, and accelerants. Thin-layer chromatography, liquid chromatography, and gas chromatography can separate mixtures to help identify unknown samples. The gas chromatograph can also be combined with mass spectrometry to identify the chemical structure of unknown substances. It is interesting to note that iodine vapors are used in thin layer chromatography to "develop" the chromatogram by staining the organic compounds after they have eluted to their different positions on the layer.

In addition, forensic chemists also use the following technologies in their analysis of chemical evidence:

▼ An SEM photomicrograph shows the enlargement of unknown chemical substances.

- *Infrared spectroscopy* can identify the major organic components of the sample, and is very useful for qualitative identification of organic substances such as paints, fibers, explosives, and drugs.
- *Ultraviolet and visible spectroscopy* are used to examine colored materials, such as dyes or lipstick; can be used to screen substances; and can gather quantitative information about drug and biological samples.
- *Mass spectrometry* can rapidly analyze arson residue, drugs, and unknown chemical samples.
- The *scanning electron microscope*, with an energy dispersive X-ray analyzer, is useful to determine the elemental composition of many types of materials, ranging from gunshot residue to metal filings.

Chapter 14
Chemical Evidence

■ *High Performance Liquid Chromatography (HPLC)* separates, identifies, purifies, and quantifies various chemical compounds, and is useful for identifying drugs and other toxins in blood and urine.

TECHNICAL DEFINITIONS

Chemist: A scientist who specializes in chemistry. Qualified forensic chemists often work as technical advisors to criminal investigation teams, especially in the case of clandestine laboratories.

Organic: Derived from living matter.

Inorganic: Not derived from living matter.

Solubility: The ability of one substance to dissolve into a solvent.

Melting Point: The temperature at which a solid is transformed into a liquid.

Boiling Point: The temperature at which a liquid is transformed into a gas (at standard atmospheric pressure).

Vapor: A substance which, though present in the gaseous phase, generally exists as a solid or a liquid at room temperature.

Vapor Point: The temperature at which a particular liquid chemical is transformed into a gas.

Vapor Pressure: The pressure exerted when a solid or a liquid is in equilibrium with its own vapor.

pH: A measure of the acidity or basicity of a substance. On a scale of 0 to 14, 7.0 is neutral; less than 7.0 is acidic; and greater than 7.0 is basic.

Index of Refraction: For a substance, it is the ratio of the velocity of light in a vacuum to its velocity in the substance. (For substances with suitable indexes, like a glass prism, shining white light through it will break—refract—the white light into its various color components. This happens because the speed of light slows perceptibly as it passes through the substance. A rainbow is sunlight broken/refracted by the mist of water particles in the air after raining).

Color: The perception by the eye of various wavelengths of refracted light components.

Chromatography: A method of laboratory analysis to identify and quantify individual chemical compounds within a substance.

Gas Chromatography: A type of chromatography that separates liquids and gases.

Thin-Layer Chromatography (TLC): A type of chromatography that deposits the sample and solvent on a plate of glass thinly coated with adsorbent.

Spectroscopy: The study of spectra in order to ascertain the chemical composition of substances and the physical properties of atoms, ions, and molecules.

Mass Spectrometry: A laboratory method of determining the chemical makeup of a substance by passing it through a high-vacuum chamber where electrons collide with the molecules, and

Chapter 15

Digital Evidence

by Michael Buratowski and Dr. Jeff Salyards

☞ CHAPTER OBJECTIVES

✔ Understand how and where digital information is stored
✔ Identify two different types of exams and the roles a computer can play in an investigation
✔ Identify the key steps in a computer forensic examination
✔ Define the following vocabulary words and phrases: *operating system, file system, hash value, hexidecimal, file signature, unallocated space, graphical user interface (GUI), bit, bytes, digital evidence,* and *malware.*

Computers and the digital information they use and store can be found just about anywhere. They can be embedded in a tiny hearing device in the ear or can be used to guide the space shuttle. Not only are they used in our homes, our phones, in schools, hospitals and crime labs, but they are also used in countless forms of criminal activity. Hackers use them to break into business and banking networks to commit fraud and theft. Sexual predators use them to contact victims through the Internet or to download child exploitative materials. They are even used by organized crime and clandestine drug labs to track money and contacts. In short, computers are an intrinsic part of life today and as such can play a key role in criminal investigations.

HOW DIGITAL INFORMATION IS STORED

A computer does not understand language or even letters and numbers like humans do. In fact, computers don't even understand anything more than 1's and 0's. It is amazing to think that for all the functions that a computer can perform, it is really only conducting mathematical calculations using these Binary Digits, or **BITs**. The data that we enter in as numbers and words are translated from what we see on the computer screen into the 1's and 0's the computer actually uses.

The computer uses an **Operating System** and a **File system** to store, retrieve and use these 1's and 0's. The Operating System is a collection of programs that control memory, handle system requests, control the devices used to input and output data, and provide a Graphical User Interface (GUI). Examples of Operating Systems are Microsoft Windows Vista, Mac OS X, Ubuntu Linux, and Unix. The second system that is needed is the File System. The File System is a set of rules that govern how data is to be saved, tracked, retrieved, and deleted from a digital storage device. Most people never have to deal with the File System as it is set up very early in the preparation of a digital storage device. The File System is created during the formatting process. Examples of File Systems are NTFS, EXT3, FAT32, and HFS.

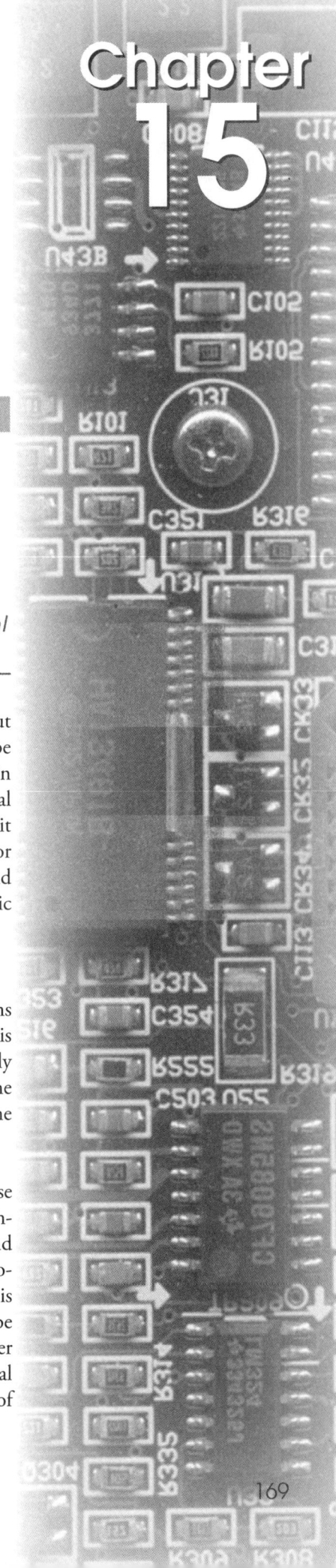

Forensic Science Today

KEY TERMS

Operating System
File System
Hash Value
Hexidecimal
File Signature
Unallocated Space
Graphical User Interface (GUI)
Bit
Bytes
Digital Evidence
Malware

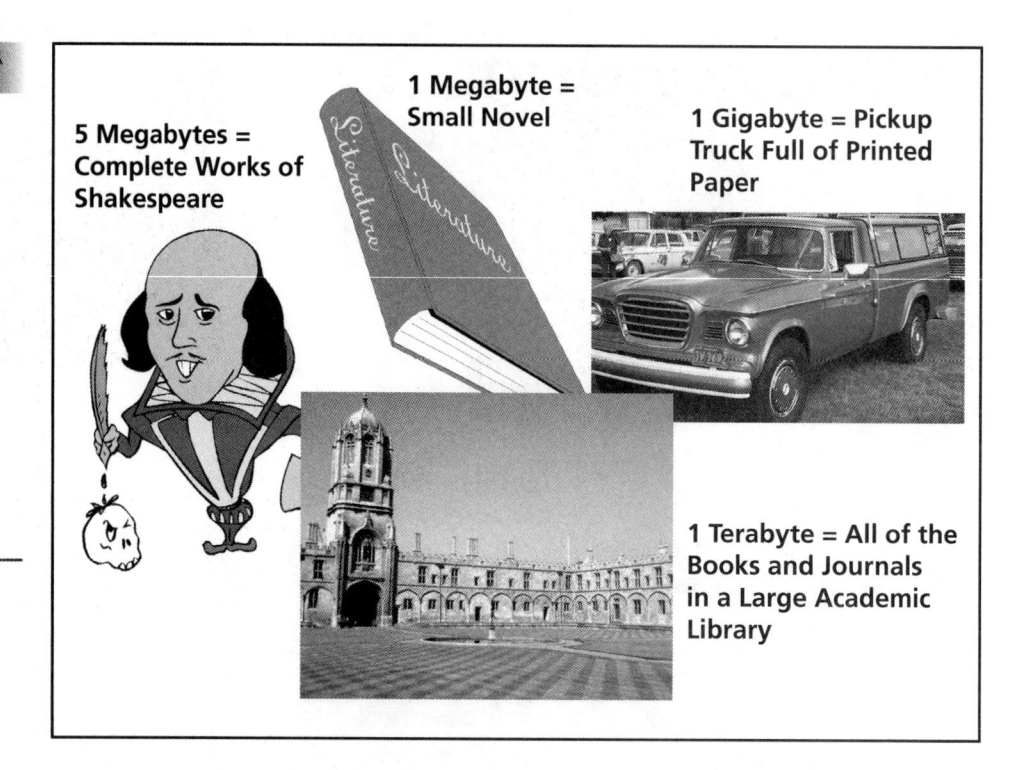

5 Megabytes = Complete Works of Shakespeare

1 Megabyte = Small Novel

1 Gigabyte = Pickup Truck Full of Printed Paper

1 Terabyte = All of the Books and Journals in a Large Academic Library

Data can be stored in many different types of media (storage devices). Over the years the amount of data that can be stored has grown exponentially, while the physical size of the storage media has shrunk. It is now possible to store 8 gigabytes (GB) worth of data on a card that is about the size of your fingernail and no thicker than a small stack of papers.

One of the most common storage devices is a hard disk drive. Hard disk drives are found in nearly every office computer, laptop, MP3 player, newer game consoles, and even some thumb drives. A hard disk drive uses platters made of glass or metal. The platters are coated with a magnetic film that allows the disk to store polar positive and negative charges. These polar positive and negative charges equate to the 1's and 0's that the computer can understand. The platter spins at a very high speed while the read/write heads float just above on the air generated by the spinning platter. Another common form of media is a flash drive. This type of media is contained in thumb drives, MP3 players, cell phones, digital cameras, and even some newer computers as the main storage device. These devices store memory using a technological concept that has been around for many years. The data is stored in much the same way as old EEPROM chips are written to.

Name of original file:

Forensic.doc

Name after being deleted:

_orensic.doc

Remember, the 1's and 0's are stored by the Operating and File Systems working together. Files that are not deleted and that are in the logical file system can be seen by the average computer user. When you delete a file, it does not get removed from the logical file system right away. Instead the file is renamed often using an underscore.

The renamed file is no longer presented to the normal computer user. But all of the information about these files remains in the directory entry (or folder as they are often called). A Computer Forensic Examiner can recover these files completely intact with all relevant information including Modified, Accessed, and Created dates and times (commonly referred to as MAC times), file names, full path (location the file is stored) and potentially even which user account the file was created by.

Chapter 15
Digital Evidence

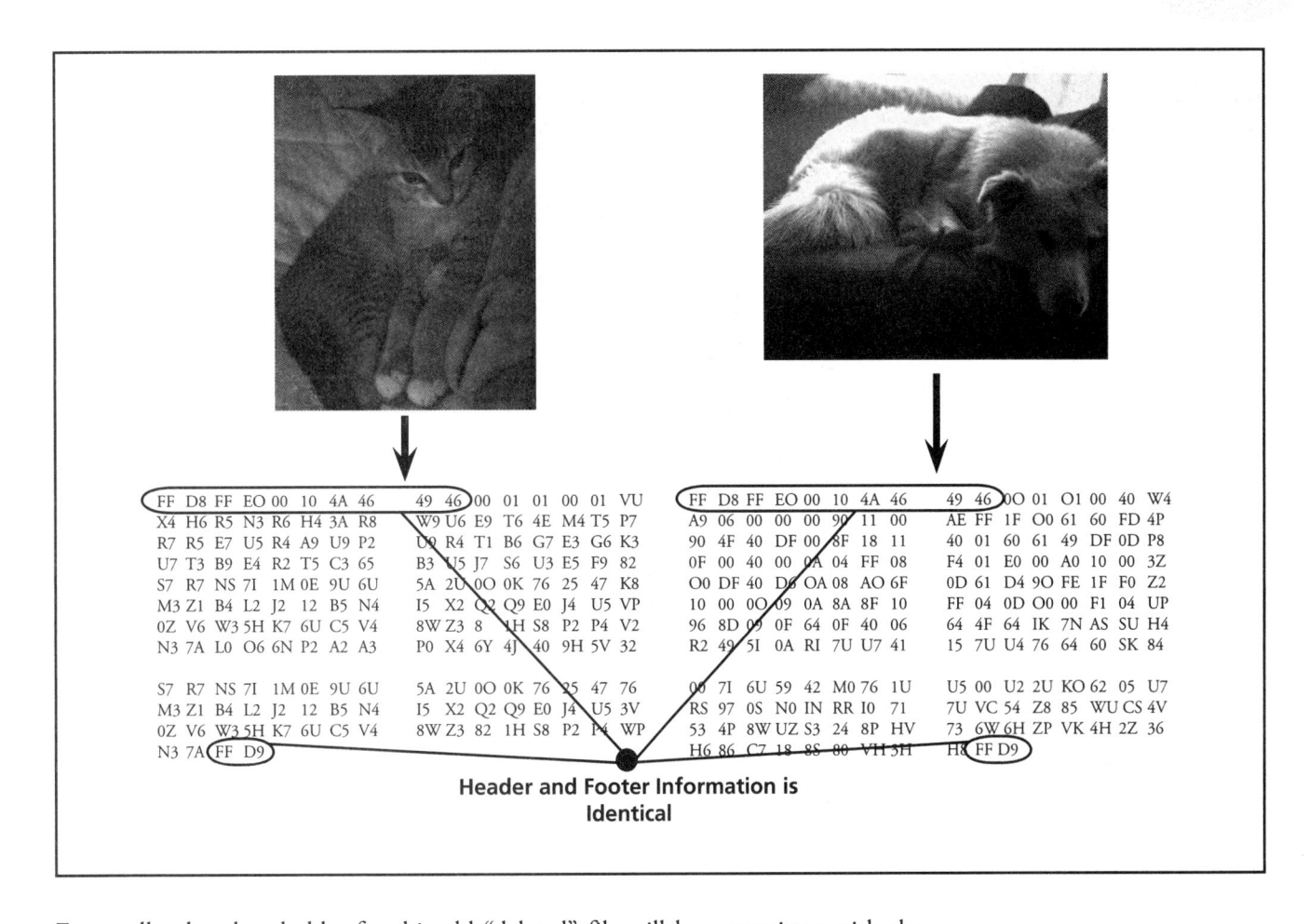

Header and Footer Information is Identical

Eventually, the place holder for this old "deleted" file will be overwritten with the name of a new file. Luckily for Computer Forensic Examiners, most Operating Systems and File Systems do not immediately overwrite data that is deleted. It may seem strange that this not automatically built in, however think of how many times you have deleted a file only to realize you still needed it. The actual 1's and 0's for these "deleted and overwritten" often remain undisturbed for months and even years on a typical hard drive in an area called **unallocated space**. In order to carve unallocated space, the forensic examiner uses special tools that are able to locate a string of 1's and 0's known as the file header or file signature, which tells the computer what type of file to expect.

Ultimately, the 1's and 0's may be truly replaced with new 1's and 0's for a new file. Once this happens, the information has been wiped away. While some scientists are experimenting with using scanning electron microscopy to examine the hard drive one bit at a time to look for "shadows" of 1's and 0's that have been wiped away, this technique is not practical.

There is one last hiding place for the 1's and 0's from an old deleted file. This is a concept known as slack space.

171

Forensic Science Today

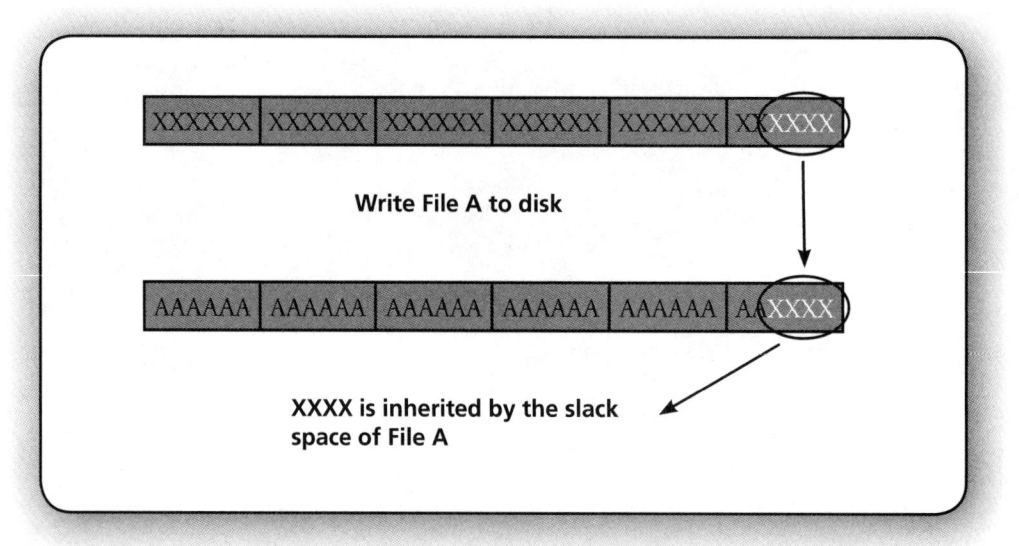

```
XXXXXX | XXXXXX | XXXXXX | XXXXXX | XXXXXX | X(XXXX)
```

Write File A to disk

```
AAAAAA | AAAAAA | AAAAAA | AAAAAA | AAAAAA | A(XXXX)
```

XXXX is inherited by the slack space of File A

On digital media there are containers called sectors. These sectors each hold 512 bytes of data. Consecutive sectors are put together in equal groups of one or more sectors that are known as clusters. A cluster is the smallest addressable area to which the Operating System can store data. So when new data is written to that cluster but does not fill the entire area, the data from the previous file may still be partially there.

WHAT ARE THE DIFFERENT TYPES OF EXAMS AND THE ROLES A COMPUTER CAN PLAY IN AN INVESTIGATION?

Computer forensic examinations will have a different focus based on how the computer was part of the crime. Let's look at a few examples of how a computer can be:

- the weapon
- the victim
- a storage container
- the witness

The Computer as a Weapon

Many of us have received annoying spam emails (like the one in Figure 1) that attempt to trick us into giving money or financial information to a fraudulent source.

On the other end of the spectrum, some criminals use computers in much more sophisticated ways to break into a financial institution or a multinational cooperation. Once inside, the criminal can steal, vandalize, and spy. Remember, it's not just information that is at stake; a hacker could theoretically disable the electronic doors in a hospital intensive care unit. Forensic Examinations in these cases focus on identifying who and where the criminal is and what type of software is being used and exploited to conduct the attacks.

The Computer as a Victim

Almost all companies have business secrets, intellectual capital (e.g., inventions, recipes, marketing materials, copyrights, etc.) and sometimes even national secrets like blueprints to weapon systems. Because computers play such a big part of our personal and professional lives, this protected information is stored on numerous desktop computers and networked storage (e.g., NAS, DAS, SAN, etc.). Business professionals also freely communicate about these projects using email. If a competitor or foreign government can break into a company's secure network, they can gain access to all types of information. They can download important documents and spreadsheets. They can monitor email communications. They can even delete or modify information to create confusion, sabotage a project, and gain a competitive advantage. Forensic examinations in these cases have three goals:

1. Discover the path of attack used by the criminal. This information will help the system administrator patch the hole and make the network more secure in the future.
2. Detect the type of **"malware"** (malicious software) used to give the criminal remote access to the network.
3. Identify if any information was stolen or exfiltrated. Learning what information was lost will help the company do a damage assessment.

The Computer as a Storage Container

Up until the 1970s and even into the 1980s, many businesses still used paper and file cabinets to store all of their business information. Now, almost all of these data are stored electronically. As a result, the key evidence in fraud and embezzlement investigations is most often found in various computer files. Even some organized crime operations (e.g., large drug dealers, prostitution, illegal gambling) have gone to a paperless computer operation. In these cases, the forensic examiner tries to discover the relatively few evidentiary files mixed in with the hundreds of thousands of computer files that are not pertinent to the investigation. The examiner may also play a key role in piecing together a timeline of when transactions and criminal activity occurred. This type of analysis helps the investigator in the field by identifying potential witnesses to interview, places to search, and new potential suspects to surveil. Sadly, digital cameras and computers have greatly increased the production, storage, and transmission of child pornography, and these forensic examinations focus on detecting the illegal images, determining where they came from, and identifying to whom they were sent.

Dear Customer

XXX.com service would like to inform you, that you received money transfer from Andreas at andreas666@earthlink.net. Amount is $217. In order to receive that amount from XXX.com you have to register your ATM card to prove you are our customer.

Your e-mail is not registred with us, you need to setup an account and verify your identity. Please complete the form below to be enrolled to XXX.com service. Once you register, the money will appear in your account. Check the balance in your overview page. You can withdraw the outstanding balance to your credit or debt card's bank account...

▲Figure 1. Sample Fraudulent Email

Forensic Science Today

INTERNET IDENTITY THEFT

Did you know that your computer stores all kinds of information about you? Every time you browse the Web and enter information it is stored in cache files, browser history and other temporary Internet files. This is hidden deep in your hard drive and can be used to reconstruct your Internet habits. This includes personal information such as names, addresses, credit card numbers, usernames and passwords. This type of stored personal information can make you vulnerable to identity theft. Identity theft is a criminal offense that involves stealing someone's personal information – such as a credit card number – for personal gain (like using the stolen credit card number to make unauthorized purchases).

Identity thieves use a variety of ways to gain access to this personal information stored on your computer:

- Phishing - sending fake emails that appear to come from legitimate institutions such as banks, Ebay, etc. These emails attempt to get you to reveal personal information.
- Browsing social network websites such as Facebook, MySpace, etc. to glean personal information.
- Retrieving information sent over an unsecured transmission (always make sure that you send personal information only to websites beginning with "https").
- Sending a virus that will install malicious software on your computer (like spyware) that will collect everything they need and automatically send it back to them.

It is important to clear your internet cache every time you enter any sort of personal information onto a website, and keep up-to-date with your anti-virus software in order to keep your identity safe.

—K. Taylor

The Computer as a Witness

Computers and related electronic equipment like GPS, cell phones, and portable media players are playing an increasing role in our society. Many of our conversations happen on cell phones. Most of our correspondence is done with email. Most of our research is done on the Internet. Many of our purchases are done on-line. Much of our travel is guided by a GPS machine. And digital cameras take most of our photographs. This increased reliance on computer technology in our everyday lives has revolutionary consequences for forensic science.

Digital evidence can be any and all of the binary (1's and 0's) information stored on these electronic devices. In the late 1980s, DNA testing provided a radical improvement in our ability to link a person to a biologic sample. Likewise, digital evidence has the potential to tell us not only where people have been and what they have said, but also what they are thinking, where they are planning to go, and what they are planning to do.

Chapter 15
Digital Evidence

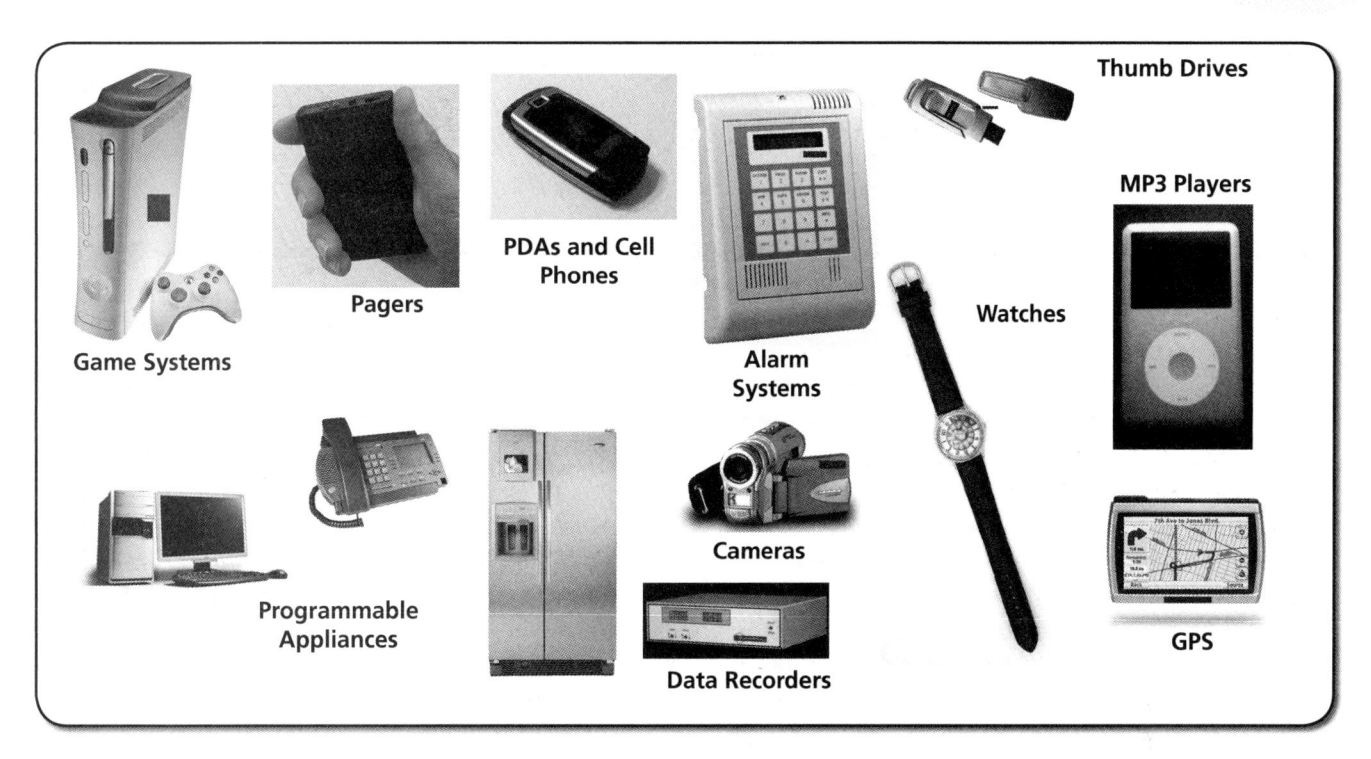

▲ Different types of electronic evidence could become evidence.

You are under surveillance!
Just think about all of the digital information you might produce by lunch:

a) send a text message to two friends during breakfast.
b) listen to a song on an MP3 player.
c) set the house alarm
d) turn on the GPS on the way to work
e) buy coffee and doughnuts with an ATM card
f) use a magnetic swipe card to open the door
g) surf the internet for news stories and an update of an on-line dating service
h) send and receive several dozen emails

If you were missing, the victim of a homicide, or the suspect in sexual assault case, all of this information could be critical in resolving the investigation!

Crimes, like sexual assault, would have had nothing to do with computers 10 years ago. Now digital evidence can provide the "smoking gun" like a digital photograph of the victim on the suspect's cell phone, an email where the suspect makes a confession to a friend, or GPS coordinates that place the suspect's car at the scene of the crime.

Internet Predators and Victim Grooming— How to Stay Safe

Violent sexual predators may use the Internet to meet potential victims and build a relationship of trust. They do this by posing as a teenager or someone similarly close to the intended victim's age. When the victim is under the age of 18, this is known as "child grooming." The aim of victim grooming is to establish a relationship and get the victim to lower his or her inhibitions about meeting the predator in person. The sole purpose of such relationship-building is to enable the predator to commit a sexual offense against the victim. The use of computer and cellular technologies to groom a child for eventual sexual offenses is illegal and many countries have implemented strict legislation against it. Computer software that monitors suspicious chat room dialogues and other Internet activity is available from several different companies. It is vitally important to remember that anyone you meet online could be a predator posing as a "regular" person. Do not ever use your real name or photograph in chat rooms or online forums. Use a fun username and avatar instead. It is best to keep online relationships where they belong – online. Absolutely never meet an online friend in a private place away from the public eye.

—K. Taylor

WHAT ARE THE KEY STEPS IN A COMPUTER FORENSIC EXAMINATION?

Almost all computer forensic examinations consist of six steps:

Step 1 – Make an exact bit for bit copy or image of the evidence

Most computer examiners do not work directly on the hard drive from a computer that has been seized as evidence. Instead, all of the digital information is copied onto a forensically sterile hard drive that belongs to the laboratory. The image is verified using a mathematical algorithm called a Message Digest 5 Hash (or just MD5 Hash for short) to ensure that the copy contains exactly the same digital information as the original. The MD5 value of a data stream will change dramatically if even one bit is different.

Step 2 – Look for known good and bad files

Every computer has thousands of files that are a routine part of the operating system and commonly used programs like word processors, internet browsers, and email programs. The National Software Reference Library (NSRL) is a database of these "known good" files. There are other data bases that keep track of "known bad" files. For example, the National Center for Missing and Exploited Children has a database of known child pornography, and several private companies have databases of malware and computer viruses. Using these types of data bases, a forensic computer examiner can significantly reduce the number of files that have to be manually inspected.

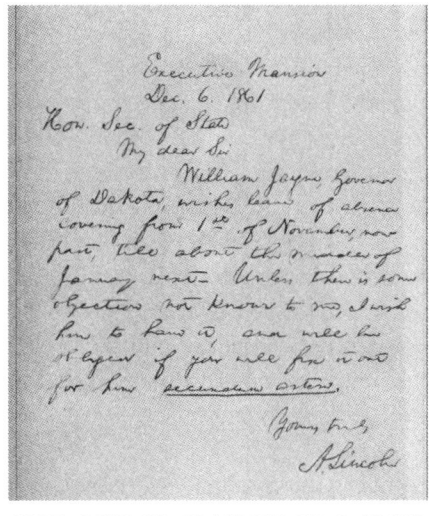

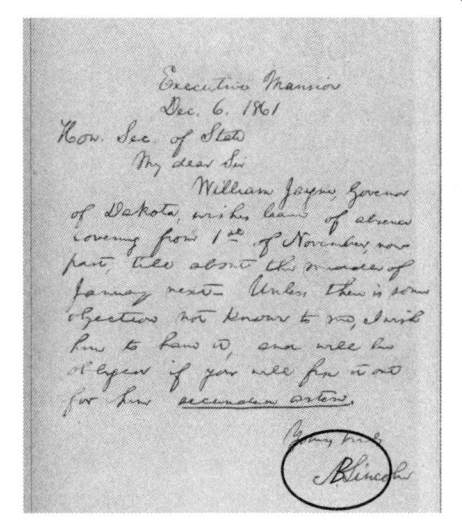

MD5: 14f3b37a90dd0265a78c8c14d02 **MD5: c46d9e24qc59aa0b401a422bc0a**

**Small Change in Document =
Big Change in MD5 Hash Value**

Step 3 – Search for evidentiary files that have been saved by the user

The computer forensic examiner will conduct searches of the existing files (patent evidence) for key words, dates and times, and binary information that is connected to the investigation. For example, if a husband is accused of poisoning his wife, the examiner may search for all files that have the words, "poison," "cyanide," "arsenic," and "ethylene glycol."

Step 4 – Search for files, deleted files, meta-data, and other latent digital evidence

As explained earlier in this chapter, digital evidence does not necessarily go away because a file is deleted, and the 1's and 0's may still be hiding on the hard drive. Many computer files also have information hiding in them that is not readily apparent to the user. For example, the photograph in Figure 2 (page 178) is what you would see in a typical photo viewer. The same photograph in Figure 3 (page 179) shows you what a computer forensic examiner would be able to see.

Step 5 – Interrogate the system, log, registry, and activity files

All computers create files and store information about what has happened to them as they are being used. They keep track of programs that have been installed and executed. They monitor updates to software. They keep a history of communication with the Internet and other computers. These types of files can be helpful in any case, but they are especially important in intrusion investigations, where the examiner is trying to determine how and when a certain computer was compromised.

Step 6 – Publish a report

Like all forensic disciplines, this step is critical. The challenge is writing about very technical work in a way that is understandable to people who may not know a lot about computers.

▲ Figure 2. Picture

HOW TO BECOME A COMPUTER FORENSIC EXAMINER

Historically, people have joined the ranks of computer forensic examiners by taking one of two paths. Some started out working with computers and got interested in forensics; others started in law enforcement or another forensic discipline and then got interested in computers. Most of the people in the second group had some sort of personal interest in computers like gaming, building, or programming. All successful computer forensic examiners have a genuine interest and curiosity about how computers work as well as an inquisitive mind, a willingness to continually research and learn new technologies, the ability to stay organized, and the ability to be methodical when dealing with large tasks.

While any knowledge of computer science is advantageous to a Computer Forensic Examiner, it is not mandatory to be a programming master, network architect, or have a bachelor's degree in computer science or electrical engineering. However, this type of education would be a plus. A good starting point to get a wide breadth of knowledge in computer hardware and operating systems is to pass the CompTIA A+ Technician certification. This certification was developed by many of the largest computer software and hardware companies. In today's computer environment, nearly all computers are connected to the Internet. So, another good step would be to complete the CompTIA Net+ certification. This certification provides a good understanding of network concepts and principles.

With the attention that computer forensics has gained over the past few years, many colleges and universities now offer degree programs for computer forensics. These degrees can be very valuable and provide a fair foundation to start a career in computer forensics. While formal education is available now, there is also specialized technical training available. Organizations like the High Technology Crime Investigators Association (HTCIA), the International Society of Forensic Computer Examiners (IS-FCE), and International Association of Computer Investigative Specialists (IACIS) offer training that will teach you the underlying principles of computer forensics. This type of training will teach you how the computer works and stores information.

Additionally, some software companies also produce computer forensic software suites to use in the examination of digital evidence. These companies also offer training that is exceptional, but the curriculum is generally focused on the use of their particular tool/software in conducting analysis. The two major forensic software vendors are Guidance Software that created EnCase and AccessData that created the Forensic Tool Kit (FTK). Additionally, there are other smaller vendors such as X-Ways Forensics who are also quite capable as well as task specific tools such as those created by Paraben that focus on cell phones and other portable devices.

Finally, someone trying to learn about computer forensics should sign up for one or more of the available list servers and websites focused strictly for computer forensics. Reading the postings and seeing what practicing examiners are facing will give you an idea of the challenges and solutions in computer forensics.

Chapter 15
Digital Evidence

Tag	Value
----------EXIF----------	
Manufacturer	CASIO
Model	QV-4000
Orientation	top - left
Software	Ver1.01
Date and Time	2003:08:11 16:45:32
YCbCr Positioning	centered
Compression	JPEG compression
x-Resolution	72.00
y-Resolution	72.00
Resolution Unit	Inch
Exposure Time	1/659 sec.
FNumber	f/4.0
ExposureProgram	Normal program
Exif Version	Exif Version 2.1
Date and Time (original)	2003:08:11 16:45:32
Date and Time (digitized)	2003:08:11 16:45:32
ComponentsConfiguration	Y Cb Cr -
Compressed Bits per Pixel	4.01
Exposure Bias	0.0
MaxApertureValue	2.00
Metering Mode	Pattern
Flash	Flash did not fire.
Focal Length	20.1 mm
MakerNote	432 bytes unknown data
FlashPixVersion	FlashPix Version 1.0
Color Space	sRGB
PixelXDimension	2240
PixelYDimension	1680
File Source	DSC
InteroperabilityIndex	R98
InteroperabilityVersion	(null)
Sensing Method	One-chip color area
File Source	Digital Camera
Scene Type	Directly photographed
Custom Rendered	Normal
Exposure Mode	Auto
White Balance	Auto
Digital Zoom Ratio	0
Focal Length in 35 mm Format	112
Scene Capture Type	Standard
Gain Control	None
Contrast	Normal
Saturation	Normal
Sharpness	Normal
Compression	JPEG (old-style)
Orientation	Horizontal (normal)
X resolution	180
Y Resolution	180
Resolution unit	inches
Thumbnail Offset	10144
Thumbnail Length	9776
Y Cb Cr Positioning	Co-sited
----------MakerNotes----------	
Image Quality	Normal
Firmware Version	0.1.1.5
White Balance	Auto
Focus Mode	Auto, Continuous
Spot Mode	Unknown (320)
Image Stabilizer	On, Mode 1
Macro Mode	Off
Shooting Mode	Scenery
Audio	No
Data Dump	Binary data 8200 bytes
White Balance Bias	0
Flash Bias	0
Internal Serial Number	(F60)2008:06:17 no. 0416
Color Effect	Off
Burst Mode	Off
Sequence Number	0
Contrast	High
Noise Reduction	Standard
Self Timer	Off
Rotation	Horizonatal (normal)
Color Mode	Normal

▲ Figure 3. Picture with Meta-data

CASE EXAMPLE

An Internet café is a place where you can use a computer to surf the web, play games, email family, or chat with friends. Some cafés offer not only Internet connectivity but also a computer for use. These Internet cafés sometimes lack user logging and accountability measures due to the high volume of computer users. The lack of accountability combined with an element of privacy can promote an environment where an individual can misuse the Internet café's resources.

A case was submitted to a forensics laboratory where child pornography images were discovered on an Internet café computer. The suspect of the case was unknown. There were no eyewitnesses. The computer contained only one generic Internet café user account, and a manual sign-in log was not maintained. A preliminary examination of the case at the laboratory revealed that a total of four child pornography images were found on the computer. The case agent resubmitted the case to see if the individual responsible for the child pornography could be determined. Similar Internet café cases submitted in the past could not be solved.

The digital forensic analysis was aimed at searching for computer activity that occurred during the creation times of the discovered child pornography images located on the desktop and within the recycle bin. Temporary Internet cache usually provides quick clues for the forensic examiner, but the cache during the timeframe of the images was either intentionally deleted or overwritten. However, examination of several data remnants recovered from a chat program, Internet email and web pages provided necessary clues.

By examining a timeline of events and by piecing together the recovered data remnants, a computer forensic examiner was able to learn the suspect's online email account name and password, his current employer and even which high school he graduated from. This detailed information led the investigating case agent to the successful arrest and prosecution of the suspect.

About the Authors of this Chapter

Michael Buratowski is the program manager for the largest digital evidence laboratory in the nation, the Defense Computer Forensics Laboratory, where he is responsible for a staff of over 140 computer forensic examiners, forensic imaging specialists, software developers, and administrative personnel. He started his law enforcement career in 1995 with the Montgomery County Police in Maryland. During his tenure there he worked various patrol assignments, served as the department web developer, and finally as a detective in the Computer Crimes Unit. In 2002, he took a position with General Dynamics as a contractor computer forensic examiner at the Defense Computer Forensics Laboratory. His computer forensic certifications include the CFCE, EnCE, DoD Basic Computer Forensic Examiner, and Paraben Certified Examiner for PDA's and cell phones.

Dr. Jeff Salyards is a Principal Analyst with Analytic Services (ANSER). He is nationally recognized as a leader in the forensic sciences community. He was the director of the Defense Computer Forensic Laboratory from 2006-2008. He holds a PhD in Chemistry from Montana State University and a Masters of Forensic Sciences from George Washington University. In 21 years, Dr. Salyards' experience ranges from conducting hundreds of investigations to developing and teaching courses. He was elected to the Board of Directors for the American Society of Crime Laboratory Directors/Laboratory Accreditation Board, and appointed to the Department of Justice National Steering Committee for Regional Computer Forensic Laboratories and the Council of Federal Forensic Laboratory Directors.

Chapter 15
Digital Evidence

TECHNICAL DEFINITIONS

Operating System: The Operating System is a collection of programs that control memory, handle system requests, control the devices used to input and output data, and provide a Graphical User Interface (GUI).

File System: The File System is a set of rules that govern how data is to be saved, tracked, retrieved, and deleted from a digital storage device.

Hash Value: A fixed length hexidecimal value that is calculated by putting a stream of data through a mathematical algorithm. The length of the Hash Value is dependant upon which Hash algorithm that is used.

Hexidecimal: A numeral system that uses 16 (instead of 10) as a base. It uses letters in addition to numbers to represent these base value. Counting to 15 in hexadecimal looks like, 0, 1, 2, 3, 4, 5, 6, 7, 8, 9, A, B, C, D, E, F. This number system is especially useful in computer science because four bits of information represents 16 combinations of 1's and 0's. Each combination can be assigned a unique value from the zero to F range of choices.

File Signature: A common series of bytes at the beginning of files of the same type. There is often, but not necessarily, a common series of bytes at the end of files of the same type.

Unallocated Space: Area on a digital storage device that is currently not assigned to a computer file.

Graphical User Interface (GUI): an interface that uses icons, pictures and symbols to allow people to interact with a computer. Older and more primitive computers and electronic devices us a text interface. GUI lets you "drop and drag," and "point and click." Text interface requires learning operating system and programming languages like MS DOS.

Bit: Comes from the phrase binary digit and refers to the 1's and 0's that make up the foundation for storage of information in a computer.

Bytes: A collection of 8 bits of information. Some examples are: a) 0 0 0 0 0 0 0 0, b) 1 1 1 1 1 1 1 1, c) 0 1 0 1 0 1 0 1, and d) 1 0 0 1 0 1 1 0.

Digital Evidence: A broad term meant to include all electronic devices that contain binary information that may be useful in an investigation. Examples include: a) computers, b) cell phones, c) gaming stations, d) digital cameras, e) GPS Systems, f) MP3 Players, etc.

Malware: Any software designed to infiltrate a computer system and make changes without the owner's knowledge and consent. These programs are malicious software. Over time these two words were combined into Malware. Sometimes "computer virus" is used as synonym for Malware. In fact, a virus is a specific type of malware. Other types include worms, trojan horses, rootkits, and spyware.

Forensic Art

Chapter 16

CHAPTER OBJECTIVES

- ✔ Identify, differentiate, and explain the different types of forensic art used to solve crimes today.
- ✔ Give at least two real-life examples of when forensic art was used to solve a crime.
- ✔ Identify the steps needed to pursue a career in forensic art, if desired.
- ✔ Define the following vocabulary words and phrases: *composite*, *age progression*, *facial reconstruction*, and *image modification*.

Forensic art refers to any type of artistic services that aid in the pursuit of justice. Forensic artists are often associated with police sketches of criminal suspects based on a description given by a victim or witness. Forensic artists also show age progression or other physical changes in perpetrators who have been on the run from the law for many years. Finally, forensic art is used to identify missing persons and crime victims, whether it be through age-progression of a missing child, or facial reconstruction of a deceased individual based on an unidentified skull. Traditionally, most forensic art is hand-done artwork. Certain procedures may be done using computer software, but the artist must still have the fundamental knowledge of art and anatomy.

TYPES OF FORENSIC ART

There are several types of forensic art; each relying on a solid knowledge of human anatomy.

Composite Imagery refers to drawings of people (sometimes the face only; sometimes the entire body) or drawings of evidence, made up of a combination of parts that have been individually described. You will often see composite sketches of wanted criminals shown on television or at the post office. These sketches are based on descriptions of the criminal by the victim and/or witnesses.

Sketches courtesy of Karen T. Taylor.

Forensic Science Today

KEY TERMS
Forensic Art
Composite
Age Progression
Facial Reconstruction
Image Modification

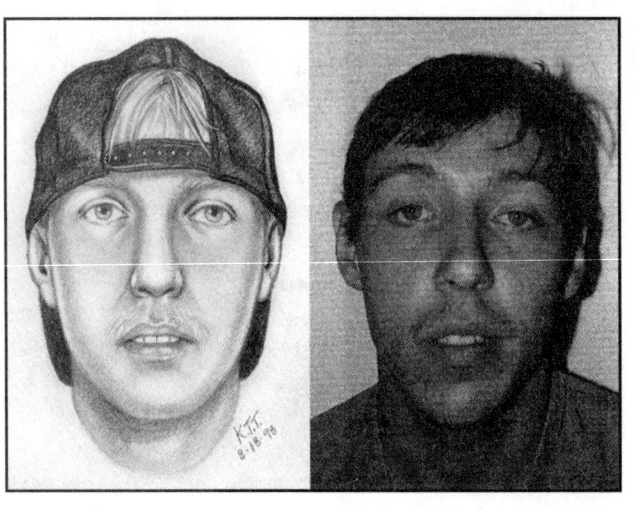

◀ Example of a composite sketch based on a description given to artist Karen T. Taylor.

▼ Example of a child age progression sketch rendered by forensic artist Karen T. Taylor.

Age Progression is where forensic artists use their art skills together with their knowledge of human anatomy to show what someone might look like today, based on photographs of the person taken in the past. In some cases the artist predicts growth of a face and in others they predict aging. This is typically used in the cases of child abduction, to show what missing children might look like years after being abducted. Age progression is also used to track fugitives who have been on the run from the law for many years. Age progression must be done by an artist with knowledge of the complex patterns of craniofacial growth and aging.

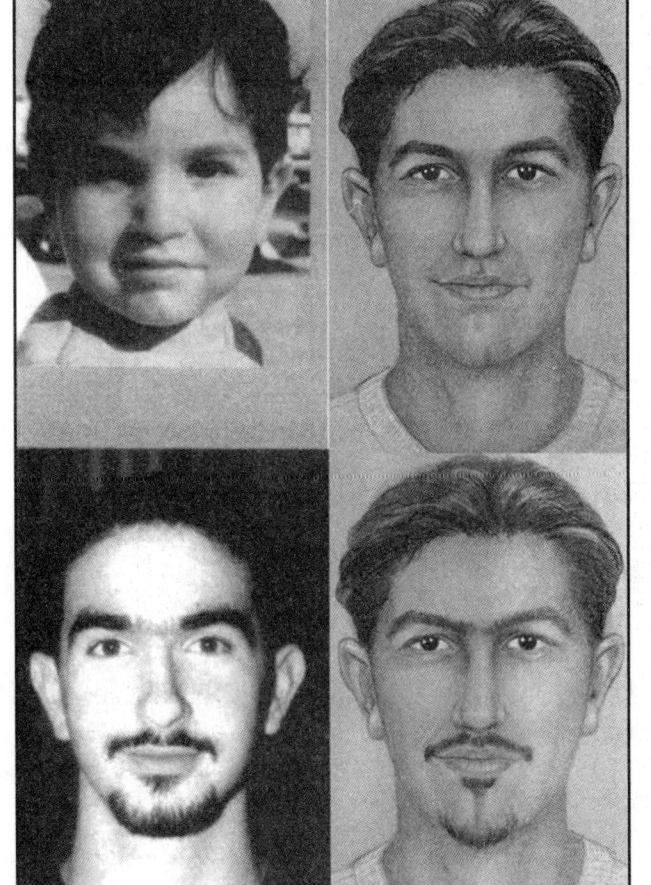

Image Modification and Image Identification refers to methods of digital enhancement of existing photographs to make them more useful to investigators. It also refers to comparisons of one facial image to another to aid with identification. Since forensic artists are facial identification specialists, they can help make alterations to outdated photos. This can involve adding facial hair, changing hairstyles, showing weight gain or loss, and so on. Artists can also help investigators with comparisons of faces, such as comparing a face from a video surveillance photo to a mug shot.

Postmortem Imagery is the use of drawings or computer images of faces to help identify dead bodies. The forensic artist uses morgue photos to produce images of how the deceased person might have looked when alive. These are not as unpleasant as morgue photos and therefore can be distributed in the news media. If someone recognizes the face, then records like fingerprints or dental x-rays can be located and used to legally identify the person.

Three-dimensional facial reconstruction is the use of a victim's skull to make a sculpture of what the victim's face and head might have looked like while the victim was still alive. Then, someone who knew the person might see the reconstruction and help identify the victim. Here, the forensic artist collaborates with an anthropologist and a forensic odontologist to construct the facial features of the unknown individual based on the underlying cranial structure. Using knowledge of anatomy and tissue depth, the forensic artist attaches small pegs to different parts of the skull to indicate how deep the tissues go below the skin. Clay or other molding material is then used to fill in the gaps between the pegs, and to create the layers of tissue surrounding the skull. Next, the forensic artist uses other facial measurements—such as the eye socket and nasal aperture—to construct the intricate details of the individual's facial characteristics.

Two-dimensional facial reconstruction is similar to three-dimensional facial reconstruction, except that here the forensic artist uses a photograph of a victim's skull rather than the actual skull, to create a drawing—as opposed to a sculpture—of what the victim might have looked like in life. This also involves collaboration with an anthropologist and forensic odontologist, as well as knowledge of anatomy and the underlying cranial structure. The artist places transparent paper over front and side view photos of the skull and draws the face. The same tissue depths, formulae, and measurements are used for drawing as for sculpture.

ANCIENT BO MAN: FORENSIC TECHNIQUES IN A HISTORICAL CONTEXT

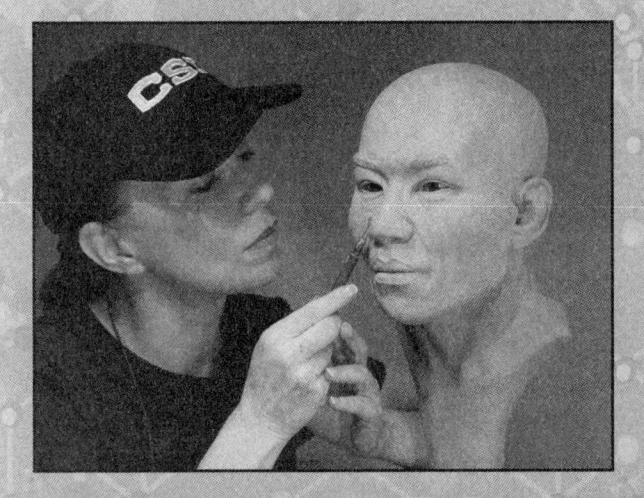

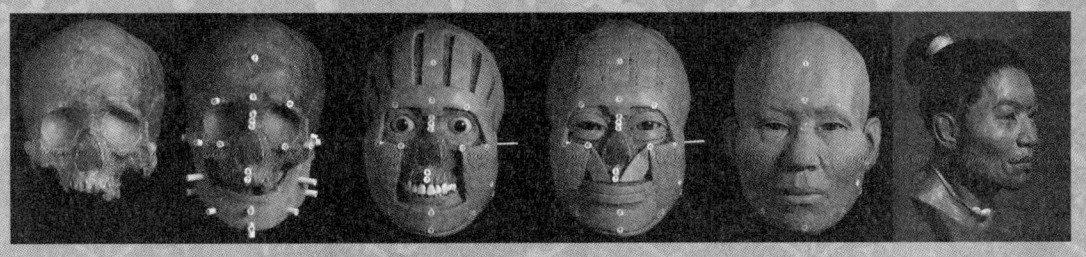

▲ Stages in skull reconstruction of the Bo Man.

Using a resin copy of a fragile 800 year-old skull from southern China, forensic artist Karen Taylor reconstructed what the ancient Bo Man might have looked like while alive. There was no mandible with the skull, so she sculpted the missing part as well as the teeth to get the most accurate reconstruction possible. This reconstruction was done for a documentary by the Discovery Channel. Taylor is shown putting the finishing touches on the reconstruction, after adding hair to the sculpture.

"AS SEEN ON TV"—REAL-LIFE FORENSIC ARTIST KAREN T. TAYLOR

You may be familiar with a character named Teri Miller who is a forensic artist on the popular television crime series *C.S.I.* This character is roughly based on real-life forensic artist Karen T. Taylor, who has spent over twenty years helping to solve crimes with her skills. Writers and researchers for C.S.I. sought Karen Taylor's input while they were creating the program, to ensure factual accuracy. Karen Taylor's artwork has also appeared on the program; and you may have even seen her hands at work reconstructing the face of a victim based on a skull.

▲ Karen T. Taylor consulting on the set of the television series C.S.I.

Karen Taylor studied art at the School of Fine Arts at the University of Texas and the Chelsea School of Fine Art in London, where she was also a freelance portrait sculptor at Madame Tussaud's Wax Museum. She decided to put her talent to work solving crimes and spent eighteen years at the Texas Department of Public Safety. She also taught art at the FBI Academy, and her art has been featured on most television channels, including the program, America's Most Wanted.

In her own words, this is what Karen Taylor says about what it is like to be a forensic artist:

"For the forensic artist, one image can literally be responsible for the recovery of a precious stolen child, stopping a serial rapist or murderer, or providing closure for the family who has lost a loved one to homicide. It is an awesome responsibility….As an artist, I have known the greatest possible satisfaction, seeing the tangible real-life effects my forensic art has had. On many occasions there have been phone calls to say, 'you know that drawing you did'…of that pedophile or murderer…'well, we got him.' What could be better than that?"

Karen Taylor has worked to professionalize the field of forensic art, and has also written a comprehensive textbook on the subject, entitled, *Forensic Art and Illustration*. (See "Resources" at the end of this book for more information.) She believes that everyone involved in criminal investigations and prosecutions should seek to understand more about forensic art, since it is such an important tool in identifying both perpetrators and victims.

For more information on Karen Taylor and her work, please see her website at **www.karentaylor.com**.

Chapter 16
Forensic Art

CASE PROFILE
AMERICA'S MOST WANTED: THE CHICAGO JANE DOE CASE

On January 21, 2007 the remains of a murder victim were discovered in a large box in a snowy alley. The entire box was transported to the Cook County Medical Examiner's office. It was determined that death had occurred 3-4 days previously, and due to decomposition and the position of the body in the box, the facial features were unrecognizable. The only thing detectives could tell was that the victim was a white woman with long reddish-brown hair in a ponytail. The Medical Examiner advised that she had most likely been between the ages of 18 to 25 years old. After a police sketch released to the public failed to yield any calls, the Medical Examiner contacted forensic artist Karen Taylor to create a 3D reconstruction of the victim's face. They sent her the skull, a hair sample and several autopsy photos. Taylor first created a 2D reconstruction as an aid in doing a more accurate 3D sculpted version. After taking photographs of the skull with tissue depth markers on it, she created a sketch of what the face might have looked like (See image below). She then layered strips of clay over the skull according to the tissue depth markers, and used prosthetic eyes in the same color as the victim's eyes. In December 2007, the television show, *America's Most Wanted* aired a segment on Chicago's Jane Doe featuring Taylor's 2D and 3D reconstructions of the victim. The reconstructions were also published that month in a dental publication where they caught the eye of a local dentist and his receptionist. Using the reconstructions and dental information, they were able to identify the victim as being 17 year old Marlaina Reed. As yet, the murderer of this young woman has not been apprehended. For more information, go to the website of America's Most Wanted at **http://www.amw.com/fugitives/case.cfm?id=49245**.

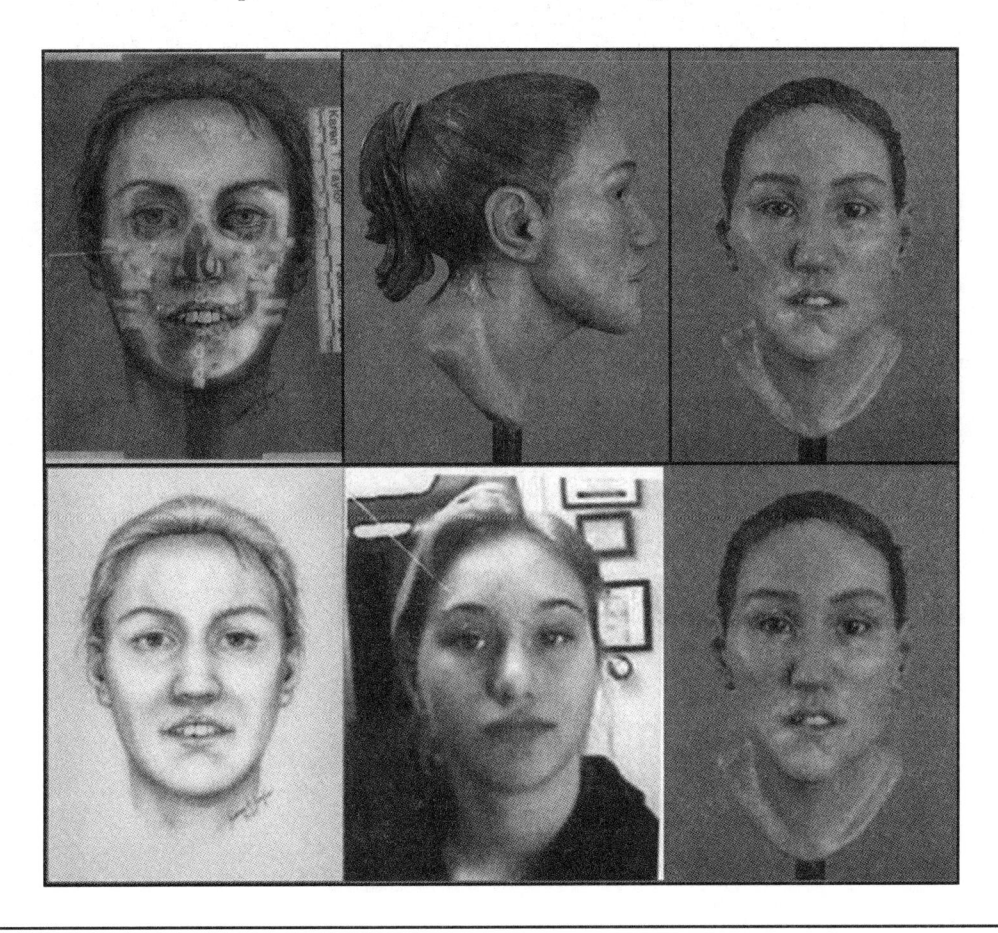

Forensic Science Today

CASE PROFILE
Reconstructing a Lost Face

In 1994, a human skull and a few scattered bones were found by hunters in Lubbock, Texas. Based on eleven knife marks on the ribs, authorities determined that the deceased person had been murdered. The skeleton was judged to be of female origin, in the age range of eighteen to twenty-four. Apart from one high-heeled shoe found near the bones, there were no clues as to the identity of the victim: A search of dental records from local missing women had failed to bring a match.

Finally, forensic artist Karen Taylor was asked to create a likeness of the dead woman's face based on the skull. Using her knowledge of anatomy and relative tissue thicknesses, Taylor attached rubber pegs to the skull at various locations to indicate the thickness of tissue. She then photographed the skull with the pegs on it, and placed transparent paper over the photograph. Taylor then sketched the facial contours and drew in a nose and eyes. Because no hair was found with the skeletal remains, she deliberately drew the hair in an ambiguous manner in order to encourage interpretation.

The finished two-dimensional facial reconstruction was released to the media and public; and two weeks later a call came in from a woman whose missing daughter, Belynda Kay Tillery, resembled the sketch. Indeed, Belynda's dental records matched the x-rays taken of the teeth and skull that had been found by the hunters, and Taylor's sketch even matched the photo on Belynda's driver's license! Eventually, Belynda's boyfriend, Troy Armstrong, was tracked down and convicted of Belynda's murder, for which he is serving a life sentence. Troy had physically abused Belynda while she was alive; and as happens in many domestic violence situations, the abuse had escalated until it finally ended in murder.

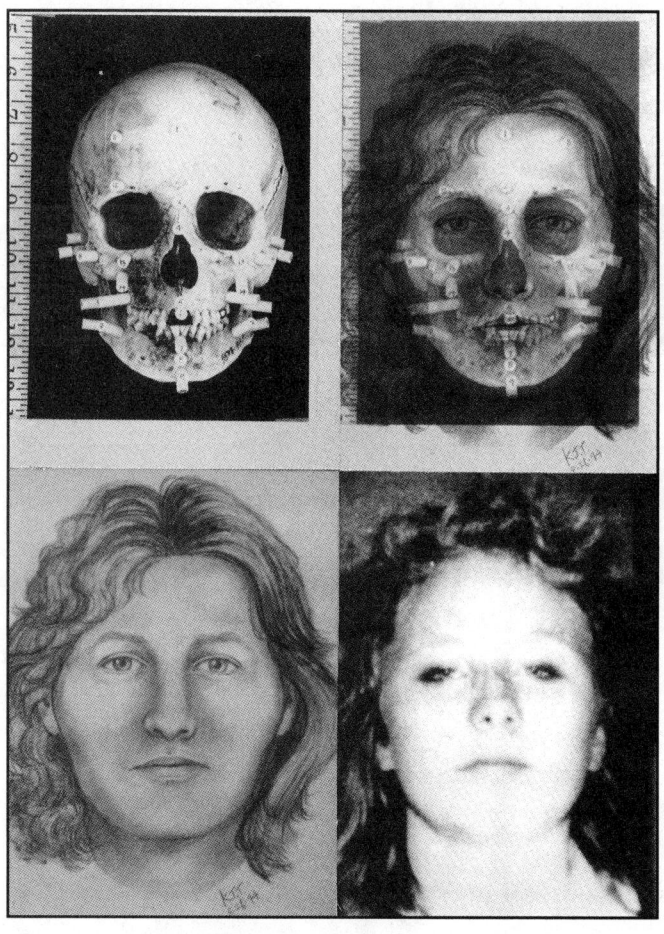

Using Forensic Art as Evidence at Trial. Forensic art can also be introduced as evidence in a court trial. This is sometimes called "demonstrative evidence." Examples of forensic art that can be admitted into evidence include drawings, sculptures, models, computer graphics, and animation.

HOW CAN I BECOME A FORENSIC ARTIST?

As yet, there is no one particular path to becoming a forensic artist. There are many different types of forensic art being used in the world of criminal justice, and just as many ways to learn the skill. Although the first requirement is obviously an aptitude for visual art, most forensic artists supplement their artistic skills with courses or workshops on the varieties of forensic art. Forensic artists also benefit from studying human anatomy; and a familiarity with forensic anthropology and forensic odontology is also extremely useful. In addition, it is highly recommended that forensic artists who work with victims or their families take psychology courses in sensitive memory-enhancing techniques. Finally, like all careers in forensic science, forensic art requires diligence and a strong commitment to your work.

AGE PROGRESSION USED TO SOLVE AN INTERNATIONAL CRIME

Here's an example of how America's foremost forensic artist, Karen Taylor (highlighted earlier in this chapter), used age progression to solve an international crime: In a case publicized worldwide, Karen Taylor's fugitive update helped locate and capture Cuban political assassin, Virgilio Paz Romero, the murderer of Chilean ambassador Orlando Letelier. After fifteen years at large, Paz was apprehended while wearing a red shirt, a detail Karen had actually predicted!

TECHNICAL DEFINITIONS

Composite: Drawing of individuals or evidence that is made up of a combination of parts that have been individually described.

Age Progression: The use of photographs of missing victims or suspects to show what a person who has not been seen in many years might look like today.

Facial Reconstruction: The use of the human skull and knowledge of cranial anatomy to either make a sculpture or draw a sketch of what a deceased victim may have looked like while alive.

Image Modification: Methods of digital enhancement of existing photographs to make them more useful to investigators.

Forensic Pathology

Chapter 17

☞ CHAPTER OBJECTIVES

✔ Understand the roles of medical examiners, forensic pathologists, and coroners.
✔ Differentiate between the cause of death and manner of death.
✔ Discuss the changes that occur in the human body after death.
✔ Identify five different categories of physical evidence that may be found on the human body.
✔ Discuss what happens during the autopsy procedure.
✔ Identify the steps necessary for a career in forensic medicine.
✔ Define the following vocabulary words and phrases: *autopsy*, *pathology*, *forensic pathologist*, *medical examiner*, *coroner*, *rigor mortis*, *algor mortis* and *livor mortis*.
✔ Define role of death investigator and medical examiner's assistant.

Forensic medicine is the application of various medical specialties to legal issues. Pathology refers to the study of disease; and a forensic *pathologist* is a medical doctor who studies injuries and disease to determine the cause and manner of death. A *medical examiner* is a forensic pathologist who conducts autopsies to determine the cause, manner, and mode of death by examining the internal organs of the corpse. In some instances, medical examiners also examine people who are injured but still alive.

The most common situation in which forensic medicine is involved is when a person dies unexpectedly. In such a situation, a forensic pathologist or medical examiner is called in to investigate. Each state has a law requiring that either a coroner or a medical examiner oversee the investigation of unexpected deaths. Is a coroner the same as a medical examiner? No—coroners are usually officials elected by the people in their community; and they don't even need to have a medical degree! So, if a coroner is not actually a pathologist, he or she will usually direct the investigation and rely upon someone who is trained in pathology to conduct the more technical medical examination. Medical examiners, on the other hand, are physicians who have earned a doctorate degree at medical school and have then specialized in pathology so that they can perform autopsies as part of their job.

Forensic Science Today

KEY TERMS
Pathology
Forensic Pathologist
Medical Examiner
Autopsy
Coroner
Rigor Mortis
Livor Mortis
Algor Mortis

Forensic pathologists and medical examiners investigate the circumstances surrounding a death, which allows them to determine both the *cause* and *manner* of death. The cause of death refers to the medical or forensic cause, whereas the manner refers to the way in which the death was accomplished. There are five "types" of manner of death:

1. Natural Causes
2. Accidental
3. Suicidal
4. Homicidal
5. Undetermined

For example, a person who was strangled to death would have the cause of death listed as asphyxiation and the manner of death listed as homicide. In this way, a forensic pathologist can classify whether a death was by a natural manner, or if it was an accident, suicide, homicide, or undetermined. Frequently, a medical examiner is assisted in their search for information about the manner of death by a medicolegal death investigator. A death investigator collects evidence at the death scene and investigates the decedent's social and medical history to assist in the determination of cause and manner of death. Sometimes forensic pathologists also investigate non-violent deaths, such as those of otherwise healthy individuals, those who have never seen a doctor, deaths that could have been caused by medical malpractice such as in surgery, or deaths occurring in public institutions. Sometimes, a forensic pathologist may even be consulted when a victim survives a violent attack or an accident, in order to examine and interpret pattern injuries.

Depending on the country or state in which you live, the law dictates whether a death must be reported to a medical examiner or to a coroner. Approximately one-third of all counties in the U.S. use coroners rather than medical examiners. After a death is reported, it is up to the medical examiner or coroner to decide whether an autopsy is necessary in order to determine the cause and manner of death.

WHAT HAPPENS TO THE BODY AT DEATH?

Upon the moment of death, the human body undergoes a natural progression of changes as described below.

Algor Mortis refers to the change of temperature that occurs after death due to the lack of energy being produced and expended. The normal body temperature for living humans is 98.6 degrees Fahrenheit. Under normal indoor conditions, the body temperature of a corpse drops at the rate of 1.5 degrees Fahrenheit per hour. The body may still feel somewhat warm to the touch up to three hours after death. However, around four to six hours after death, the body becomes cool to the touch; and after twenty-four hours the body cools down to the temperature of the external environment. In the days following death, if a body lays undiscovered it will reach the "bloated" stage, where internal bacteria release gases during a process known as "putrefaction." During this stage, the body becomes very hot (up to 127 degrees Fahrenheit) due to the activities of carrion insects. (See Chapter 20 for more information on insect activity after death.) Eventually the body will crack open and allow the gases to escape, which brings the temperature back down to that of the surrounding environment. Body temperature can also increase after death if the surrounding temperature is higher than that of the body: for example, the body is in the trunk of a car on a 100°F day.

During an autopsy, the medical examiner removes tissues and organs from the body, examines specimens under the microscope, conducts tests, and requests analyses of body fluids such as blood or urine, which provide information about diseases the victim may have had or poisons the victim may have ingested.

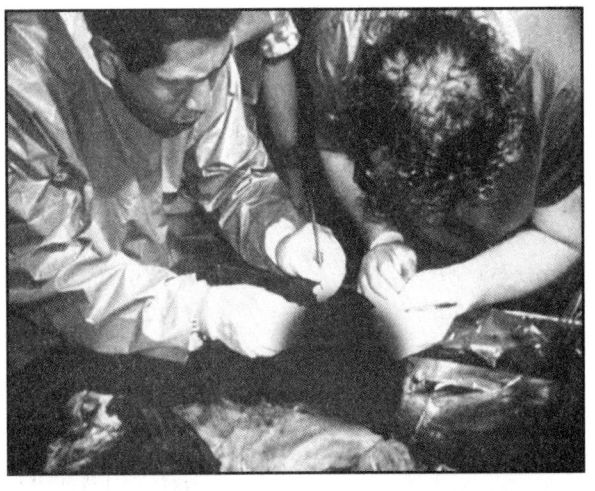

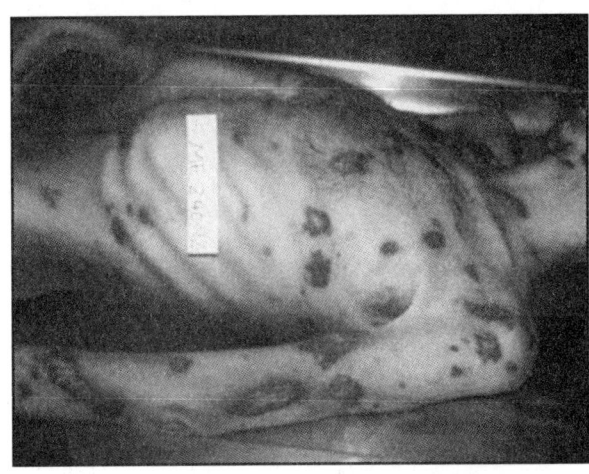

Chapter 17
Forensic Pathology

Rigor mortis begins approximately four hours after death. This refers to a stiffening of the body resulting from the lack of oxygen pumped to the muscle cells after the heart stops beating. When blood is no longer being pumped through the body, ATP (adenosine triphosphate)—the source of energy required for muscle contraction—stops being produced as well. Rigor mortis begins in the eyelids and small muscles of the lower jaw, and then spreads down through the neck and the rest of the body. The timing of rigor mortis is affected by several variables. If there has been a violent struggle just prior to death, then rigor mortis will set in more quickly since the decedent's ATP has already been used up. Warmer air temperatures will also speed up rigor mortis, while cold air slows it down. In addition, rigor mortis sets in more slowly with overweight individuals and more quickly with very thin people. In most cases, rigor mortis will have completely set in within twelve hours of death, and lasts approximately thirty to thirty-six hours, after which the body goes back to a non-rigid state. The loosening of rigid muscles follows the same progression as when rigor mortis sets in, beginning with the face and jaw, and spreading down through the rest of the body.

Livor mortis refers to the discoloration of the skin after death caused by the gravity-induced stagnation and settling of red blood cells in the small blood vessels. Commonly referred to as "lividity" (reddish coloration), livor mortis becomes fixed approximately six to eight hours after death. This means that the discoloration of the skin can no longer be changed by shifting the position of the body. Because lividity is caused by gravity settling the blood, it occurs in the part of the body that is lying closest to the ground. Thus, a homicide victim lying face-down will develop lividity in the face and chest. In some murder cases, one of the types of mistakes that investigators may make is to turn the victim's body face-up in order to prevent lividity from forming in the face.

▲ Autopsy is an important aspect for death investigation. Here, a medical examiner and a forensic scientist work together to determine the manner, cause, and facts related to the death. (top)

▲ This man had used dirty needles to inject drugs and developed AIDS. Toxicological analysis of autopsy specimen is important for determining the cause of death. (bottom)

PHYSICAL EVIDENCE ON THE HUMAN BODY

The human body can be a crime scene in itself. Therefore, when a forensic pathologist performs an autopsy, this is considered a unique type of crime scene investigation. You might imagine that as soon as a dead body is wheeled into the morgue, the medical examiner cuts open the cadaver with the body saw right away. However, this is not the case. Prior to dissecting the body, it is first necessary to examine it for physical evidence.

Some examples of evidence found on the human body are as follows:

- **Conditional evidence.** This refers to evidence of certain conditions; and examples would include lividity, rigor mortis, odors from poisons or perfumes, insects, body location, body position, and clothing.
- **Pattern evidence.** This refers to patterns made on the human body, such as bite marks, imprints, impressions, weapon marks, insertions, injuries, and clothing patterns. It might include marks made by weapons like gunshot

wounds, or wounds by blunt or sharp instruments. It might include marks made by tools, such as ligature (strangulation) marks, handcuff marks, or tire marks. Other examples of pattern evidence are marks made by clothing like shoe prints, logos, or jewelry impressions; and marks made by animals such as dog bites, claw marks, and insect bites.

- **Biological evidence.** Examples would include grass, leaves, wood, plants, pollen, flowers, fruits, seeds, insects, maggots, worms, and so on.
- **Chemical evidence.** This includes cosmetics such as makeup and perfume, firearms evidence such as gunshot residue or lubricant, or vehicle-related chemicals like rubber, paint, grease, or bits of metal (such as from a hit-and-run incident). Other chemicals found on the human body may include drugs, alcohol, poisons, ink, dye, oil, wax, cleaning fluids, acids, caustic powders, soil, glass, or minerals.
- **Physiological evidence.** Examples include blood, semen, urine, fecal materials, tears, milk, bile, tissues, bone, or hair.

WHAT HAPPENS DURING THE AUTOPSY

After documenting, collecting, and preserving any trace evidence from the deceased body, the medical examiner then prepares the body for autopsy. In some jurisdictions a medical examiner's assistant helps the medical examiner (ME) with the autopsy by aiding in the collection of organ samples and performing the physical examination of the exterior of the body. First, the weight, height, and other physical characteristics of the body are documented. Next, the body is photographed, x-rayed, and examined with an alternate light source for stains. This can lead to important information about bullets and other objects lodged within the body. The medical examiner makes note of distinguishing characteristics such as tattoos, scars, and birthmarks; scrapes under nails for evidence of foreign DNA, and clips the fingernails and toenails for further analysis. If the case is a suspected rape, the examiner swabs the oral, anal, and vaginal cavities for semen and other evidence; and if the case is a suspected suicide by shooting, the hands are tested for gunshot residue (see Chapter 12, Toolmarks and Firearms Evidence, for more information on gunshot residue).

STAGES IN THE POSTMORTEM INTERVAL

- **Perimortem Stages ~ 4 hours**
  Pupils dilated, sphincter relaxed, body flaccid, lip moist, livor mortis nonfixed.

- **Postmortem Stages ~ 4–48 hours**
  Rigor mortis begins, livor mortis fixed, rigor lyses (tissue softening), decomposition begins.

- **Decomposition Stage**
  Autolysis, fatty acid, gases, ammonia production, marbling, liquefaction, putrefaction, desiccation, skeleton.

Chapter 17
Forensic Pathology

Next, the decedent is fingerprinted. To do this, the medical examiner, assistant, or criminalist from the identification section of the crime lab dries the fingers and palms of the cadaver. The fingers are inked directly with an ink roller or ink pad, and each finger is pressed into the fingerprint card. The palms of the hand are also printed by inking in the same manner and then rolling a large cylindrical object such as a bottle or beaker, wrapped in paper secured by rubber bands, over the palms of the hand. What if the body has begun to decompose? There are many interesting ways to print fingers in this situation, such as inserting a solution beneath the top layer of skin so that the ridge patterns become more visible. Or, the outer layer of skin may be slipped off and placed over the gloved fingers of a live person for easy fingerprinting.

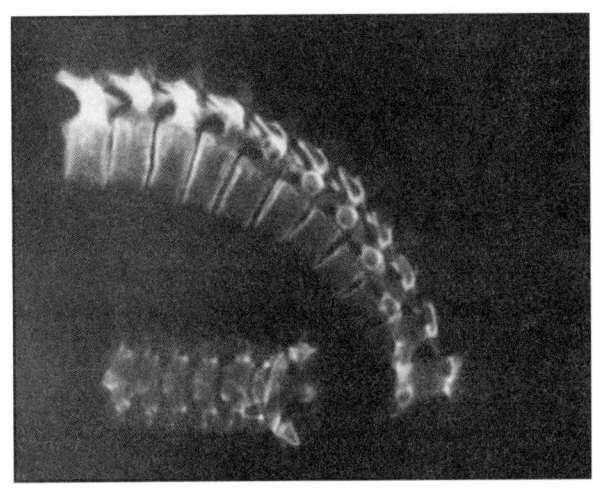

▲ Taking x-rays of the body is an important step for autopsy.

After the external information is noted, the medical examiner then conducts the autopsy according to procedure. After the body is rinsed and the head placed on a stabilizing block, the examiner makes a "Y-incision" which involves cutting down from each shoulder and down the middle of the abdomen into the pelvis. The completed cut looks like the shape of the letter "Y" (hence the name Y-incision). A saw will be used to cut through the collarbone and ribs, and the ribs are spread open in order to reveal the internal organs. The internal organs are then removed, individually weighed, and measured; and samples of blood and tissue are taken and sent to the toxicology lab for further analysis. The final analysis also includes a toxicology report to see if drugs or other poisons are present in the body.

Once the organs of the body are examined, the head is examined externally. If necessary, the medical examiner then cuts into the scalp at the back of the head and peels the skin forward up and over the face. Using a power saw, the examiner opens up the skull and examines and weighs the brain. The brain will be dissected and preserved for further examination.

After taking note of all the factors both external and internal to the body, the medical examiner then comes to a final conclusion as to the cause and manner of death.

ESTIMATING TIME OF DEATH

External Conditions
- insects
- plants
- maggots
- animals
- environment
- corroborating factors

Body Condition
- color
- temperature
- eyes
- rigor mortis
- livor mortis

Body Internal Condition
- gastrointestinal content
- degree of putrefaction
- vitreous fluids electrolyte
- biochemical changes
- decomposition

Forensic Science Today

CASE PROFILE
Who Killed JonBenet Ramsey?

On Christmas Day 1996, the lifeless body of five-year old JonBenet Ramsey, a child beauty pageant winner, was discovered in the basement of her sprawling Boulder, Colorado home. She had rope wound tightly around her neck and one of her wrists. The case gained immediately national attention and all eyes were drawn to her parents, Patsy and John Ramsey. Despite the best subsequent investigation and forensic efforts, there was just not enough solid evidence to link to any definitive suspect and to make an arrest. To this day, the case has never been solved. Listed below are some key items of evidence. What do you think happened? Which forensic techniques would be appropriate for recognizing and analyzing the evidence? What seems strange about this case? What went wrong?

The Ransom Note

At the foot of the stairs leading to the bedrooms was a three-page letter written in a black felt-tipped pen, purporting to be written by kidnappers demanding $118,000 ransom for the return of JonBenet. Facts about the note are as follows:

- The ransom sought was the exact amount of John Ramsey's annual bonus at work.
- The language used in the note was not typical of ransom notes used by kidnappers.
- The ransom note was written in disguised hand printing.
- The paper and ink used for the note came from the Ramsey household.
- Police found two practice pages with similar handwriting on them in a garbage can in the house.

The House

- A small window in the basement had been broken.
- Black scuff marks were found on the inside surface of the wall near the broken window.
- A shoe print was found on the floor of the wine cellar in the basement.

JonBenet's Body

- Upon discovery in the basement, JonBenet's body was cool to the touch and stiff.
- Some of her blonde hairs were twisted in the rope embedded in her neck.
- Cause of death was determined upon autopsy to be ligature strangulation and head trauma.
- Other injuries included contusions (bruises) on the scalp, right brain hemisphere, and the tips of the temporal lobe; dotted injury patterns on the cheek and body; and abrasions on the right cheek, posterior right shoulder, left lower back, posterior left lower leg; and small abrasion and vascular congestion of vaginal mucosa.

Chapter 17
Forensic Pathology

HOW CAN I BECOME A MEDICAL EXAMINER?

If you think you have an interest in examining dead bodies, read on. All medical examiners are medical doctors who have generally taken pre-med coursework during college or university before going on to earn their doctorate in medicine. During medical school, you will need to specialize in pathology. After completing medical school, you should complete at least four years of training in postgraduate pathology, and then enter a one-year forensic pathology residency program. Most of these residency programs are with the medical examiner's office affiliated with a medical school in a large city. Currently, there are only about forty such programs in the U.S., so the competition can be tough. However, the satisfaction of meeting the challenges of working as a medical examiner make the long years of study and competition well worth it, as many medical examiners will tell you.

TECHNICAL DEFINITIONS

Autopsy: A postmortem surgical procedure conducted by a medical examiner to determine the cause, manner, and mode of death by examining the internal organs and the corpse.

Pathology: The study of disease and pathogens.

Forensic Pathologist: A medical doctor who studies disease and injury to determine the cause and manner of death. A forensic pathologist sometimes also performs autopsies.

Medical Examiner: A forensic pathologist who is a local government official responsible for performing autopsies to determine the cause and manner of death.

Coroner: A county official, not always a medical doctor, who investigates and declares the official cause of death with the assistance of a forensic pathologist.

Rigor Mortis: The temporary stiffening of the body after death after the heart stops pumping oxygen to the muscle cells.

Livor Mortis: The discoloration of the skin after death caused by gravity-induced pooling of red blood cells in the small blood vessels. (Also known as lividity).

Algor Mortis: The change of temperature that occurs after death due to the lack of energy being produced and expended

Chapter 18

Forensic Toxicology

CHAPTER OBJECTIVES

✔ Understand the role of the forensic toxicologist in solving crimes.
✔ Identify situations in which a forensic toxicologist would likely be called in.
✔ Explain how absorption, distribution, and elimination of ethyl alcohol works in the human body.
✔ Understand the principles of Blood Alcohol Concentration (BAC) testing and the cost to society of drunk and drug driving.
✔ Describe the different environments in which a forensic toxicologist might work.
✔ Identify several high-tech laboratory instruments and methods used in forensic toxicology to identify and analyze toxic chemicals found in evidence.
✔ Identify the steps needed to pursue a career in forensic toxicology, if desired.
✔ Define the following vocabulary words and phrases: *toxicology, forensic toxicologist, Blood Alcohol Concentration (BAC), postmortem toxicology, chromatography, thin-layer chromatography, gas chromatography, mass spectrometry,* and *spectrophotometry.*

"All things are poison and nothing is without poison, only the dose permits something not to be poisonous." (Or, "The dose makes the poison.")
—Paracelsus, 16th Century

Forensic toxicology is the application of toxicology to the purposes of the law. A forensic toxicologist identifies and analyzes poisons, drugs, and chemicals, and then studies the effect of such substances on the human or animal body. Toxicologists often identify the presence of unexpected chemical substances in body tissues or body fluids. They also determine the quantity of that substance present in the sample taken from the body. In addition, not only do toxicologists determine the effect a substance has on humans, they also look at the effects of different substances on animals such as horses or dogs that are involved in the racing industry.

KEY TERMS

Toxicology
Toxicologist
Blood Alcohol Concentration
Postmortem Toxicology
Chromatography
Thin-Layer Chromatography
Gas Chromatography
Mass Spectrometry
Spectrophotometry

In order to find out whether chemicals or drugs contributed to a person's death, forensic toxicologists must often work together with law enforcement officers, forensic pathologists, crime scene investigators, and other forensic scientists. Forensic toxicologists are especially important to help determine whether toxic or illegal substances have been ingested by a person in many different types of criminal and civil cases. The most common situation is the case involving driving while drunk or under the influence of substances that impair one's ability to drive, such as illegal drugs or certain prescription medications. Other scenarios involve testing for Drug-Facilitated Rape (DFR) drugs in the system of a rape or sexual assault victim. (See Chapter 6, Drug Evidence, for more information on drug-facilitated rape).

Forensic toxicology is also very important for helping the medical examiner determine the cause of death during an autopsy. The medical examiner will analyze postmortem (post-death) body fluids or tissues to help determine whether alcohol or other drugs, chemicals, or poisons were involved in a questioned death. Forensic toxicology is used to determine the presence or absence of drugs and their metabolites, chemicals such as ethanol and other volatile substances, carbon monoxide and other gases, metals, and other toxic chemicals in human fluids and tissues. The medical examiner can then evaluate the role of these substances as a determinant or contributory factor in the cause and manner of death. A variety of factors can influence the concentration and effect of drugs found in the human body after death, such as the body storage temperature, postmortem gastric diffusion, time to autopsy, body position, age, and gender.

"The Dose Makes the Poison"

This famous quote by 16th century physician Paracelsus refers to the fact that any substance, administered in a high enough dose, can be poisonous. The toxic effect of a substance will increase in relation to the increase in dose. All chemicals have a dose response curve. This is a range of doses that result in a graded effect between the extremes of no effect and a toxic effect. All chemical substances, given a large enough dose, will exhibit a toxic effect. Even drinking water, if taken in a large enough quantity, will cause death. Thus, the toxic potency of a chemical is determined by the amount of the chemical that produces a specific response in a specific biological system.

THE TOXICOLOGY OF ALCOHOL (ETHYL)

The alcohol we know as spirits in drinks like wine and beer is technically known as ethyl alcohol or ethanol. Like any substance, alcohol can be deadly if taken in the wrong dose. In fact, alcohol poisoning is a very serious problem on college campuses or wherever "binge" (excessive) drinking takes place and has been the cause of death in many cases.

Chapter 18
Forensic Toxicology

The oldest recorded archaeological evidence of wine may have originated in the mountains between the Black and Caspian seas and has been found as a residue in jars in Northern Iran circa 5000 B.C. Alcohol has had a varied history since then, and in 1920 the 18th Amendment (Prohibition) took effect in the U.S. prohibiting the manufacture, sale, transportation, import or export of beverages containing ethyl alcohol. Prohibition lasted until 1933. There are various opinions as to the effectiveness of Prohibition but there is no dispute that the amount of alcohol consumption in the U.S. decreased considerably during that time and remained lower until long after 1933. Many feel that if alcohol had first come on the market in recent times it would have been illegal. The cost to all of us from alcohol use and abuse is considerable in health, property and loss of life.

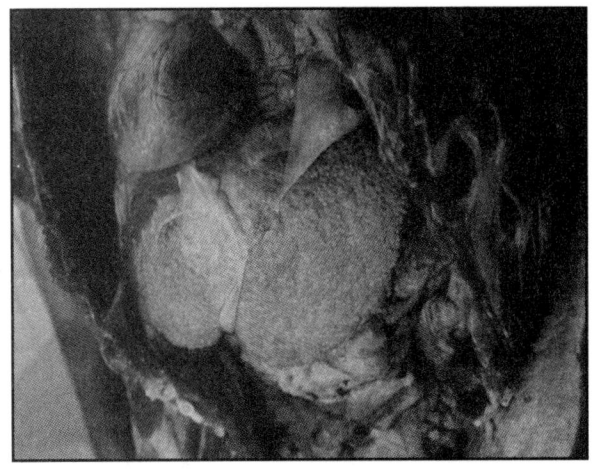

▲ An autopsy shows the condition of the liver from a victim of chronic alcoholism.

Production

There are other alcohols in addition to ethyl alcohol such as methyl alcohol and ethylene glycol (used in antifreeze) to name two. All alcohols are toxic and when consumed in quantity will cause death. Alcohol can be produced naturally through fermentation or synthetically through a chemical process. Alcohol (ethyl) for human consumption must be naturally produced, that is, by fermentation.

Approximate Percentages of Alcohol

Beer	4-6%
Wine	12-15%
Distilled Spirits	40-50%
Grain alcohol ("Everclear")	98-100%

Proof = Divide the proof by 2 to get % of alcohol.

Absorption

Ethanol can enter the body through ingestion, injection and inhalation. Injection of alcohol can be extremely dangerous. The most common method is ingestion by mouth. Alcohol is one of the only substances that does not have to be digested but is absorbed unchanged by mucous membranes directly into the blood stream. The rate of absorption can be slowed by food in the stomach and is much faster on an empty stomach.

The type of beverage consumed and the concentration of alcohol in that beverage affect the absorption rate. Alcohol consumed at higher altitudes is absorbed more rapidly. All of these factors combine with others to determine the specific ethanol absorption rate in a particular individual. As a general rule, the ethanol from a single alcoholic beverage is usually absorbed in forty-five minutes to an hour on an empty stomach.

Forensic Science Today

▶ Absorption of an equal dose of alcohol (full stomach vs. empty stomach).

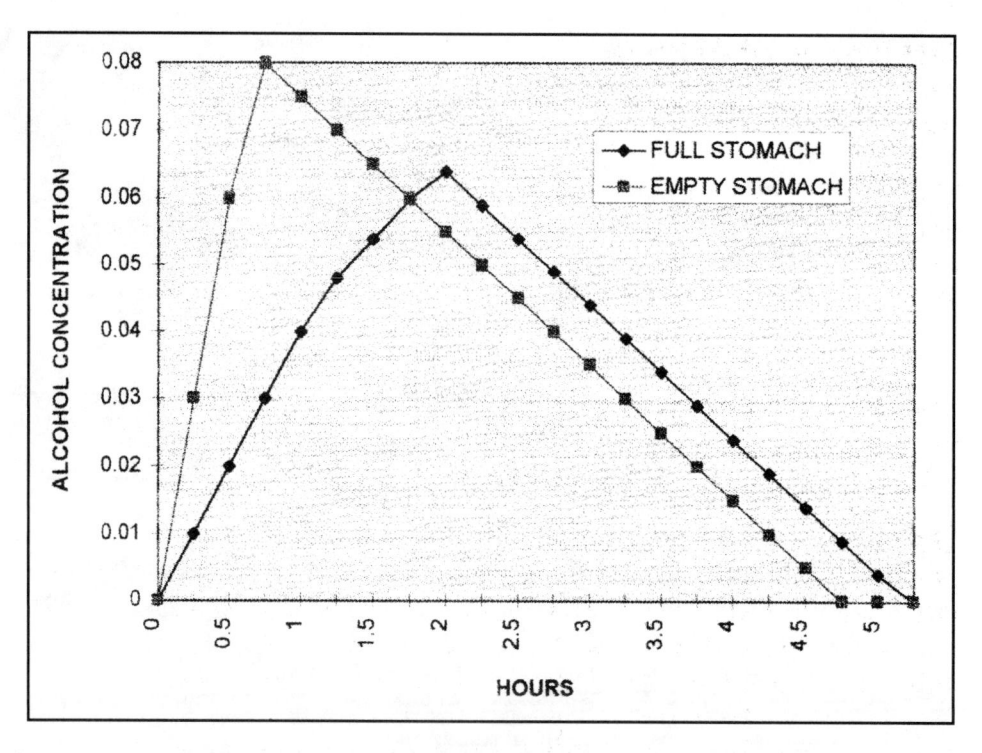

▶ A representation of ethyl alcohol distribution pathways (going through body systems). Courtesy of the Texas Breath Alcohol Manual.

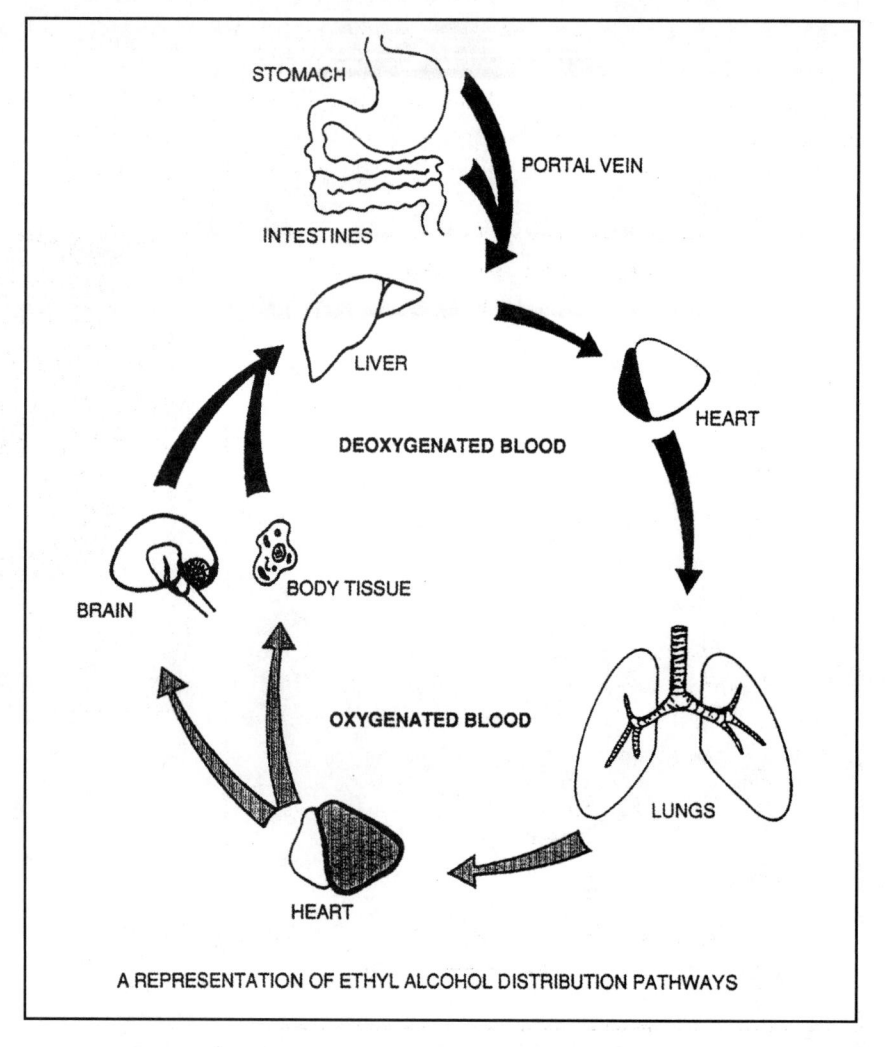

A REPRESENTATION OF ETHYL ALCOHOL DISTRIBUTION PATHWAYS

Chapter 18
Forensic Toxicology

Distribution of Alcohol
Once absorbed, alcohol is transported throughout the entire body and absorbed into the bloodstream.

The concentration of ethanol in body tissue is directly dependent on the water content of that tissue. For example, water content of muscle tissue is greater than that of bone but less than that of fat tissue. An overweight person has less water per pound of body weight than a normal weight person as fat tissue has very low water content. Within these constructs alcohol concentration will vary due to body weight.

Females have less water per pound of body weight due to additional adipose tissue in breasts, buttocks and thighs. The heavier person must consume a greater amount of alcohol to reach a specific alcohol concentration.

Elimination of Alcohol
Alcohol is eliminated in several ways: metabolism, excretion and evaporation. Metabolism accounts for most of the alcohol eliminated and is oxidized by the Alcohol Dehydrogenase enzyme (ADH) in the liver to acetaldehyde and acetic acid, and then to carbon dioxide and water. Reported rates for alcohol elimination in individuals are from 0.010% to 0.025% per hour. This represents the amount of alcohol in grams in 100 milliliters of blood or 210 liters of breath. Studies have shown this correlation between the amount of alcohol in the above stated amounts of blood and breath to be the same. Elimination of alcohol is a physiological process and is not significantly affected by exercise or stimulants such as caffeine. So, the only way to sober up is a factor of time elapsed. An alcohol drink for comparison is one 12 ounce bottle of beer or one ounce of 100 proof distilled spirits, which would both contain a similar amount of alcohol.

◀ Relationship of body water content to alcohol concentration.

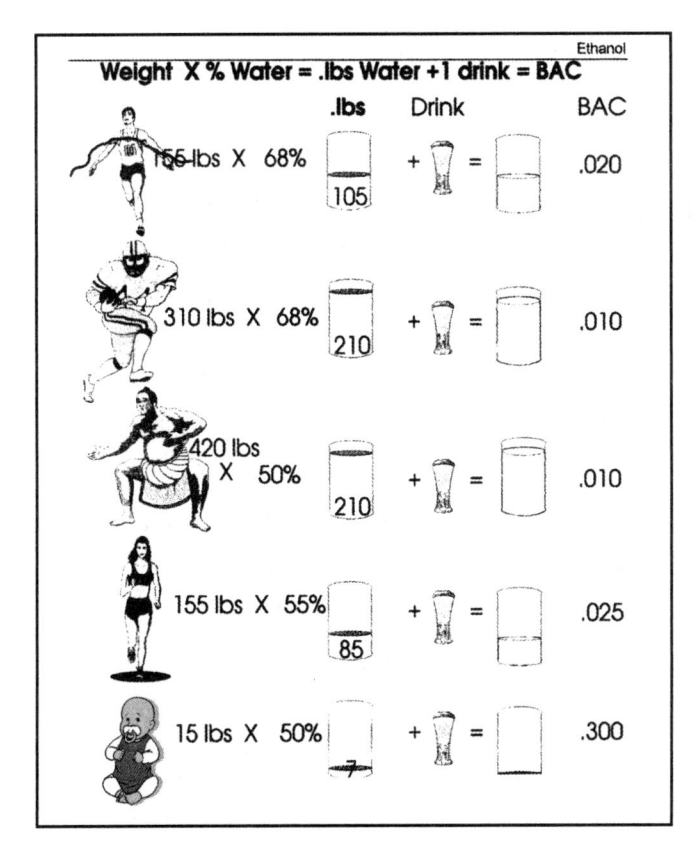

▲ Result of a drunk driver.

Alcohol Tolerance

Alcohol tolerance refers to the fact that the more one consumes the more it takes to attain the same level of intoxication. This is deceptive to the person consuming alcohol as a physical tolerance may develop. A person leaving a bar may see two doors but from prior states of intoxication knows which door is the real one. This in no way increases their ability to operate a motor vehicle although they think they might be more capable. Some people have suggested that they can drive better under the influence. This attests to the above false tolerance perceptions of the drunk.

Legal Limit

All U.S. States now recognize that a person driving with a Blood Alcohol Concentration (BAC) of 0.08% is deemed to have his or her mental and physical facilities impaired so as to be unable to operate a motor vehicle. Other countries and regions have the following BAC limits expressed as a percent (%):

Country	BAC	Country	BAC
Australia	0.05	Russia	0.00
Canada	0.08	Saudi Arabia	0.00
China	0.03	South America	0.05
Denmark	0.05	Sweden	0.02
France	0.05	Taiwan	0.05
Japan	0.03	Turkey	0.05
Jordan	0.00	United Kingdom	0.08

There has been a trend in recent years to lower the BAC legal driving limit. Thirty years ago, the limit in the U.S. was 0.15%, which is almost twice as high as it is now. The U.S. National Safety Council and others have been trying to lower the BAC level for years. Most of the countries listed above have a lower BAC than does the U.S. It is thought now by many that all drivers are impaired at the 0.05 BAC level. Sweden has been a leader in this regard and for many years has had their limit at 0.02.

The Cost of Drunk Drivers

It is impossible to place a value on human life. More than 18,000 individuals die each year in the U.S. as a result of drunk drivers. This translates into many millions of dollars in loss of property and human life.

Driving under the influence of multiple drug use

We have a way to measure BAC in humans and with alcohol we can determine when the alcohol level results in driver impairment. But what if the driver is under the influence of more than one drug? There was a recent case in Alaska where a driver was apprehended and toxicology results showed that he had alcohol, marihuana, cocaine and heroin in his system. With the use of multiple drugs, each drug in combination with another may be synergistic. That means that one dose of each is worse than two doses of a single drug. Case in point: Marilyn Monroe may have died an accidental death due to the synergism of alcohol and barbiturates.

Testing for alcohol in drivers

Prior to the invention of breath alcohol instruments in the 1950's the only means to test for alcohol was from a blood sample. As suspects would rather blow in a breath instrument than have blood drawn intravenously, the "Breathalyzer" invented by Dr. Robert Borkenstein was a huge advancement for the apprehension of drunk drivers.

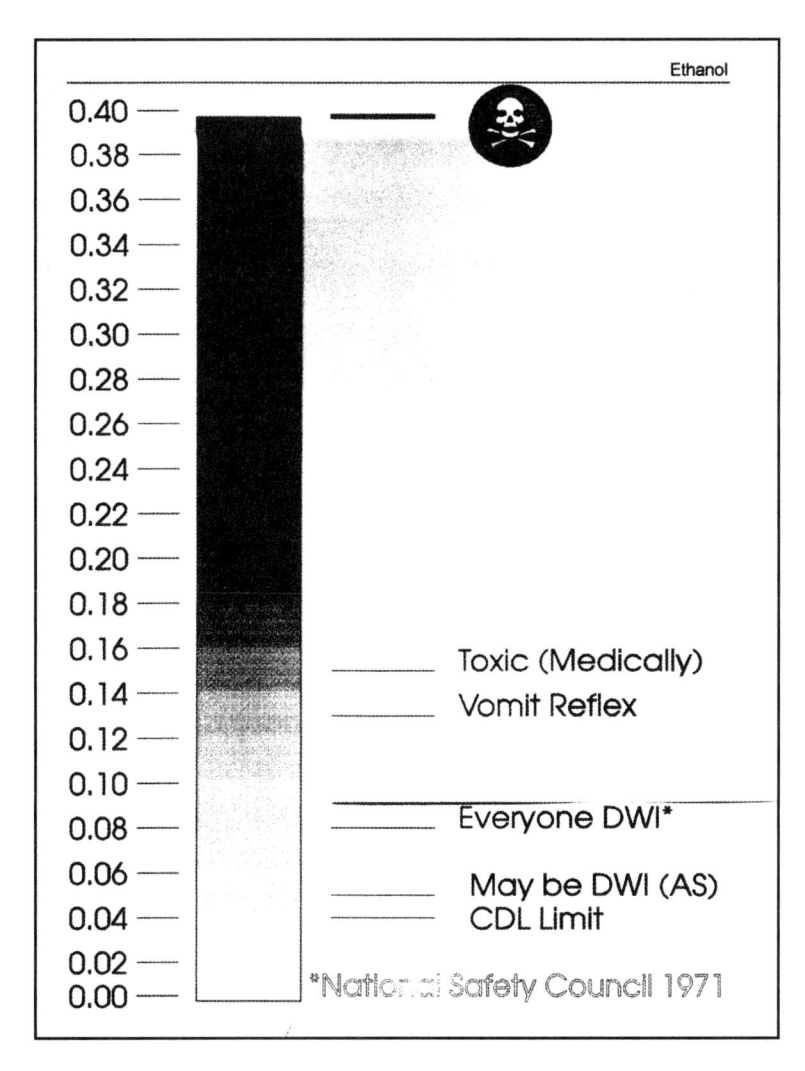

◀ Shows a range from 0.00 to 0.40, showing CDL limit at bottom and skull (indicating death) at top. From the National Safety Council 1971. Courtesy of the Alaska Breath Alcohol Manual.

▲ Road sign showing the BAC limit.

The advantages of breath testing are as follows:

- Suspect more willing to blow into an instrument
- Breath analyzers available for Law Enforcement to operate
- Instant results for Law Enforcement
- Provides additional evidence for court presentations that correlates with Law Enforcement observations of erratic driving
- Law Enforcement can observe how suspect acts and correlate that with the BAC level

References for further study on Alcohol

1. *Medicolegal Aspects of Alcohol, Fifth Edition*, available from Lawyers and Judges Publishing Co. www.lawyersandjudges.com
2. National Safety Council Committee on Alcohol and Other Drugs (CAOD) www.network.nsc.org/index.php/caod/index
3. International Association for Chemical Testing (I.A.C.T.) www.iactonline.org

The Authors want to thank Lawyers & Judges Publishing Co., the Texas Department of Public Safety Breath Alcohol Testing Program and the Alaska Department of Public Safety Breath Alcohol Testing Program for their input into this section.

Forensic Science Today

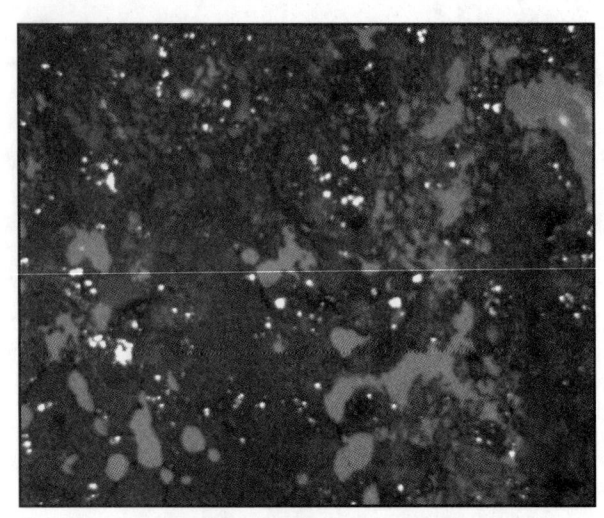

Histological specimen (a thin slice of body tissue) shows solid toxic compounds in patient's lung tissue specimen.

WHAT DO TOXICOLOGISTS DO?

Forensic toxicologists interact with other professionals in a variety of interesting cases. They work in medical examiner/autopsy laboratories, police laboratories, drug testing laboratories, or at the horse- or dog-racing track (to test racing animals for drugs). They also work in hospitals, universities, industry laboratories, and with agencies that monitor the use of drugs in sports.

Forensic toxicologists can choose from a variety of specializations within their field. The first specialty is "postmortem toxicology," which determines whether drugs or other chemicals contributed to a person's death. Postmortem toxicologists use their knowledge of analytical chemistry techniques such as gas and liquid chromatography, mass spectrophotometry, UV-Visible spectrophotometry, flame emission and atomic absorption spectrophotometric techniques, gas chromatograph-mass spectrometry, antigen-antibody immunoassay methods, and other more traditional methods of analysis. These chemistry techniques are explained later in this chapter.

Another type of forensic toxicologist is one who works together with law enforcement agencies to investigate crimes where drugs or alcohol were involved in commission of the crime. Here, the toxicologist analyzes the samples and gives an opinion in court as to how the drugs or chemicals affected the person at the time of the crime.

A third type of forensic toxicology involves conducting drug and urine tests of employees who work in jobs with hazardous work conditions, jobs impacting the safety of others, or certain government jobs. In addition to using traditional urine tests to screen for drugs, forensic toxicologists are also developing tests that use samples of an individual's hair, saliva, or even sweat. This is also true for "on the job" drug testing of horses and dogs with unscrupulous owners involved in the racing industry.

WHAT TOOLS AND TECHNIQUES DO FORENSIC TOXICOLOGISTS USE?

Forensic toxicologists have a variety of high-tech laboratories and instruments at their disposal to identify and analyze toxic chemicals found in evidence. Some of these are described below.

Chromatography is a method of laboratory analysis used to separate and quantify individual chemical compounds found in a substance. In the chromatography technique the substance being identified is exposed to two phases of materials. During the stationary phase, the molecules are separated due to their difference in attraction to the materials in the stationary phase (an adsorbent material) versus their solubility in the mobile phase (some liquid or gas.) The molecular mass of the material also affects the separation. A larger compound with the same attraction to the adsorbent and same solubility in the solvent will move slower, resulting in it showing up lower on a thin-layer chromatography test and exiting at a later timeframe on a gas chromatography test. By comparing the rates that substances move in these phases, substances with similar behavior can be identified. Chromatography results do not allow for the unique identification of substances.

Chapter 18
Forensic Toxicology

Thin-Layer Chromatography uses a glass or plastic plate coated with a thin layer of finely ground adsorbent. The adsorbent may be made up of a silica gel or alumina, mixed with a binder. The sample is deposited on the plate, and the locations of the separated components are noted after the solvent evaporates. The location of the components may be determined by spraying with a chemical that develops a unique color or by exposing the compounds to an ultraviolet light.

Gas Chromatography separates liquids and gases. During the stationary phase, the material is coated on fine clay or glass beads and packed into a steel or copper column. The column is then heated so that the liquid inside is heated above its boiling point. The gas then pushes the compounds through the column; and special detectors then measure the emerging fractions.

Mass Spectrometry is a method of identifying the chemical structure of unknown substances, and can be used in conjunction with the gas chromatograph. After separating the chemical components of the unknown substance in the gas chromatograph, each component enters a vacuum chamber. In the vacuum chamber, the component molecules collide with a beam of high-energy electrons. This causes the component molecules to lose electrons and thus become positively charged ions. Because the ions are highly unstable, they immediately break apart into smaller pieces, which then pass through a magnetic or electric field, where they are then separated according to their masses and identified. Each compound is identified by the unique set of pieces they will break into due to their differences in molecular structure. It is similar to taking a cup and a saucer and breaking each into pieces using a hammer. The pieces from the cup do not look like the pieces from the saucer, nor can one be put back together to make the other.

Spectrophotometry is a technology that characterizes different chemical substances by measuring how much radiation from the electromagnetic spectrum is absorbed by a particular chemical. The *spectrophotometer* creates a graph, also known as an absorption spectrum, showing the absorption of light as a function of wavelength or frequency. This is done by passing radiation from a light source first through a frequency selector and then through the chemical sample being tested. The radiation that exits the sample then passes through a detector that converts electromagnetic radiation into an electrical signal, and finally the signal is recorded by the instruments. Additionally, *Infrared Spectrophotometry* measures the absorption of infrared light by different chemicals, and *Ultraviolet Spectrophotometry* measures how different substances absorb ultraviolet (UV) light.

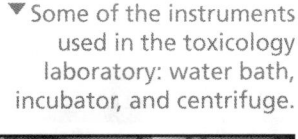

▼ Some of the instruments used in the toxicology laboratory: water bath, incubator, and centrifuge.

EDUCATION AND TRAINING TO BECOME A FORENSIC TOXICOLOGIST

If you want to be a forensic toxicologist, you need to get a bachelor's degree at a university in a physical science (such as biology, chemistry, or physics). It is also a very good idea to take courses in pharmacology. In order to enter the field at a more advanced level and higher pay scale, you can also pursue a Master's degree or Ph.D. in forensic toxicology. Other relevant courses include medicinal chemistry, pharmacology, pharmacokinetics, or clinical chemistry. Of course, you can always work in one of these fields first, and then go into forensic toxicology. Remember that your career will often take a variety of interesting twists and turns before you find your true niche. If you want professional certification, look into the American Board of Forensic Toxicology and the Forensic Toxicology Certification Board. Additionally, the Society of Forensic Toxicology maintains internet links to most major toxicologist certification boards and professional societies. The International Association of Forensic Toxicologists also provides excellent international reference materials and other details on career opportunities for forensic toxicologists.

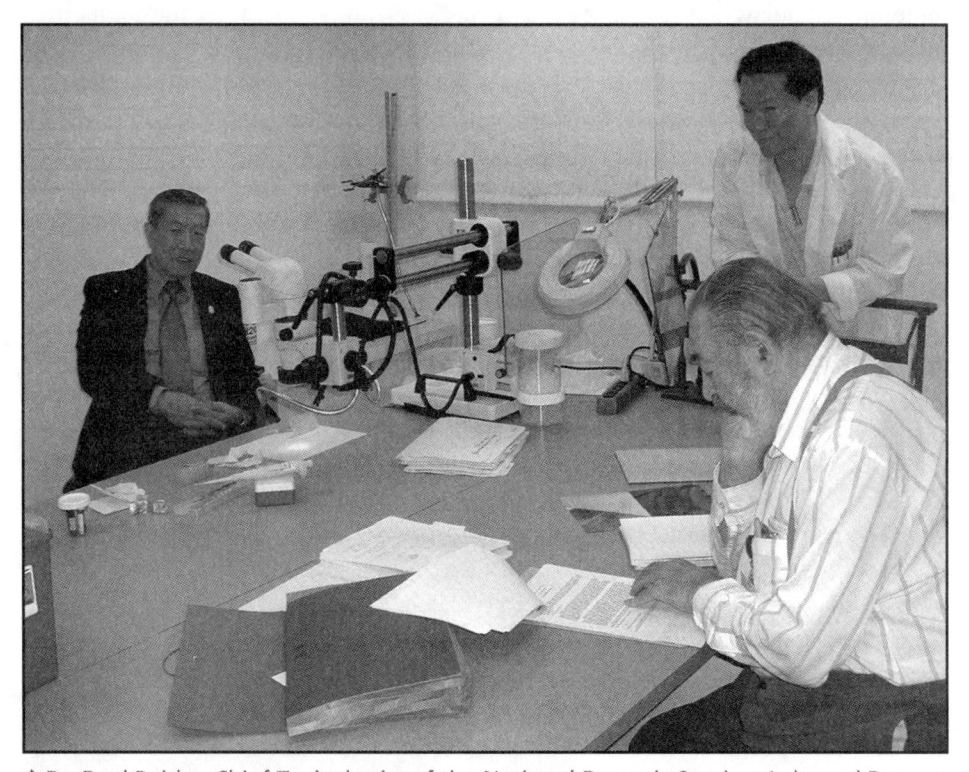

▲ Dr. Fred Reider, Chief Toxicologist of the National Forensic Services Lab, and Dr. Henry Lee examine evidence and study a toxicology report. (Courtesy of Dr. Michael Reider.)

Chapter 18
Forensic Toxicology

REAL CASE SCENARIO

Police officers in Juneau, Alaska apprehended a driver for erratic driving. The driver was so impaired that he was unable to get out of the car on his own. The toxicology report later revealed that he had high levels of marijuana, heroin, alcohol, and cocaine in his blood. Many foolish drivers operate motor vehicles after using a combination of several drugs. Drugs, including those available by prescription, exacerbate the effects of other drugs with which they may be combined, making driving or operation of machinery even more dangerous. Even normal dosages of some prescription drugs can impair a person's ability to drive; in these cases, it is the individual's legal responsibility not to get in a car when taking such medication. Impaired drivers are often apprehended with a low blood alcohol, but they learn the hard way that police officers don't stop with the breathalyzer: toxicology reports will always reveal a high concentration of another drug or combination of drugs.

TECHNICAL DEFINITIONS

Toxicology: The study of poisons, drugs, chemicals, and their effects.

Forensic Toxicologist: A scientist who identifies and analyzes poisons, drugs, and chemicals and their effects on the human or animal body in relation to legal issues.

Postmortem Toxicology: The study of how drugs or other chemicals contributed to a person's death.

Chromatography: A method of laboratory analysis to separate and quantify individual chemical compounds within a substance.

Thin-Layer Chromatography: A type of chromatography that deposits the sample and solvent on a plate of glass thinly coated with adsorbent.

Gas Chromatography: A type of chromatography that separates the components with carrier gases.

Mass Spectrometry: A method of identifying the chemical structure of unknown substances by aiming high-energy electrons at the substance to determine mass.

Blood Alcohol Concentration: The concentration of ethyl alcohol in the blood. BAC is used as a measure of the degree of intoxication in an individual.

Spectrophotometry: A technology that characterizes different chemical substances by measuring how much radiation from the electromagnetic spectrum is absorbed by a particular chemical.

Chapter 19

Forensic Anthropology

CHAPTER OBJECTIVES

✔ Explain the role of forensic anthropologists in solving crimes.
✔ Discuss the questions forensic anthropologists ask when examining human remains.
✔ Identify the steps needed to pursue a career in forensic anthropology.
✔ Define the following vocabulary words and phrases: *anthropology*, *forensic anthropology*, *skeletal remains*, and *postmortem interval*.

Anthropology is the study of humankind. Physical anthropology, a specialty within anthropology, is the study of the human skeleton and how it has developed and evolved throughout the history of the human race. Forensic anthropologists are physical anthropologists who specialize in recovering and examining human skeletal remains where legal questions are involved, especially when a crime has been committed.

Forensic anthropology first gained legitimacy in 1939 when a landmark paper entitled, "Guide to the Identification of Human Skeletal Material" was published by the FBI. Less than a decade later, a professor of gross anatomy named Mildred Trotter led the effort to repatriate (return home) the bodies of dead World War II soldiers in Hawaii. In 1948, Trotter organized a laboratory to examine bones of the dead soldiers and identify them based on x-rays, dental records, and health records provided by the army and families of the deceased soldiers. Professor Trotter also created new structural charts and mathematical tables that were used to estimate the height and structure of the skeletons to aid in their identification. However, it was not until 1979 that the first textbook was written on forensic anthropology by Dale Stewart, the former curator of anthropology at the Smithsonian Institution. Today, a wide variety of resources provide a fascinating insight into forensic anthropology. (For more information see the "Resources" section at the back of this book.)

▶ Bone fragments found at a murder scene.

Forensic Science Today

KEY TERMS

Anthropology
Forensic Anthropology
Skeletal Remains
Postmortem Interval

Today's forensic anthropologists can determine the age, sex, height, ancestry, and even left- or right-handedness and other behavioral information about the decedent (such as whether he smoked a pipe). Forensic anthropologists also work with pathologists and medical examiners to establish the cause of death. Or, they may help determine how much time has elapsed since death by examining the state of body decomposition, and by working with forensic entomologists (insect specialists) who examine the remains of insects found on the human body.

Where do forensic anthropologists work? Usually, they work in a laboratory at universities or major research institutions. In addition, state and local medical examiner offices work with forensic anthropologists as medical investigators; and state and federal law enforcement agencies hire physical anthropologists as laboratory personnel or even to serve as special agents. Additionally, forensic laboratories often contact physical anthropologists at local universities for assistance.

SECRETS HIDDEN IN BONES

Bones hold the secrets of the past, and can reveal many hidden mysteries in the stories they tell. It is the job of the forensic anthropologist to solve the mysteries contained in the bones. The examination of recovered bones can reveal whether the bones are human, the age and height of the person, their sex, injuries to the skeleton, the nature of wounds, and a history of accidents or injury. Bones can also give information about medical conditions and childbearing history.

Forensic anthropologists examine bones for clues by looking at marks left on the bone surface, and also by going in at the cellular level. Bones contain DNA, the most accurate way of determining an individual's identity. In addition, forensic anthropologists can determine whether skeletal remains were deposited relatively recently or whether they belong to a person who died many decades or even centuries ago.

Logic Tree: Anthropological Individualization

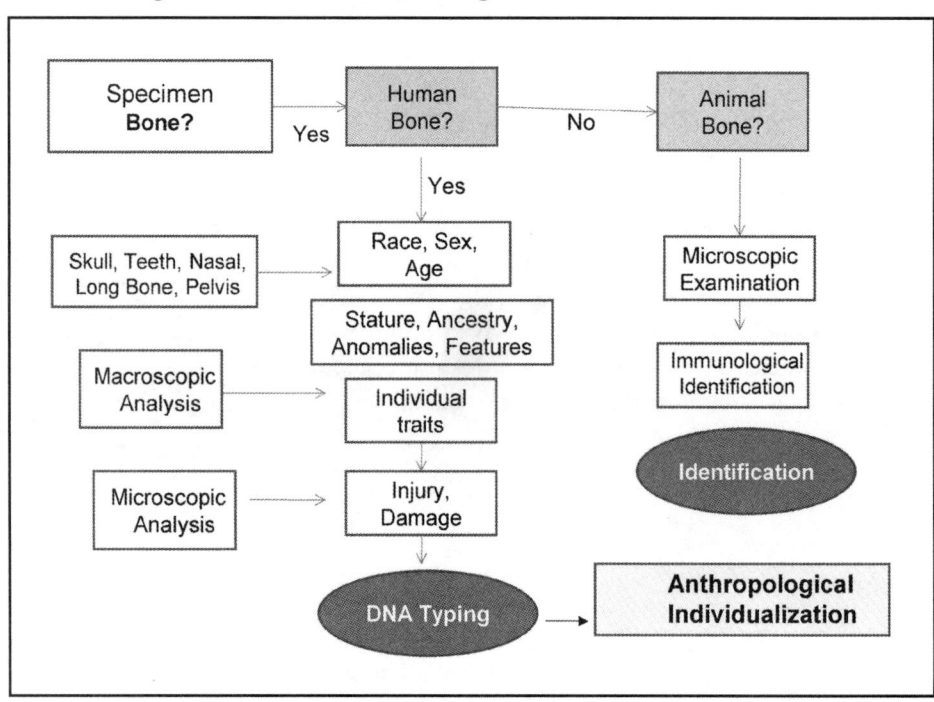

Chapter 19
Forensic Anthropology

HOW FORENSIC ANTHROPOLOGISTS ANALYZE HUMAN REMAINS

Forensic anthropologists ask a series of questions about skeletal remains in order to build an anthropological profile to aid in identification:

Are the remains human? Close to a quarter of cases seen by most forensic anthropologists are actually non-human, belonging to some other species of animal. Human bones can primarily be distinguished from non-human bones based on size and structure of the bones. Even if only a fragment of bone exists, examination under a microscope will quickly reveal whether the bone came from a human.

How many individuals are represented by these bones? Forensic anthropologists next determine the minimum number of individuals that could be present in skeletonized form. The simplest way is to look for duplication of bones—obviously three skulls mean the presence of three people. Even if there is no duplication of bones, forensic anthropologists look for differences in the age, size, structure, and preservation of the bones to see how many individuals are present. Finally, DNA analysis of the bones can definitively determine the minimum number of individuals found.

▲ Unknown bones were found in a wooded area.

What was the sex of the individual? It is easier for a forensic anthropologist to determine the sex of bones of an adult human rather than a child, since sex differences do not appear until puberty. Therefore, skeletal remains of children are often un-sexed. For adult skeletons, the best way to determine sex is by examining the pelvis. The pelvis of a female is generally wider than that of a male, in order to accommodate the passage of a baby during childbirth. In addition, the skulls of males and females are quite different: Generally, male skulls have a square chin (as contrasted with the rounder chin bone of the female), and the foreheads of males also protrude further than in females.

What was the age of the individual? Forensic anthropologists also seek to determine how old the individual was when he or she died. This is easiest done on skeletons of infants and children, since skeletal growth is quite standard, and sections of bone fuse at fairly regular rates. However, once puberty is attained, it becomes much more difficult to determine the age of the individual apart from knowing that he or she was an adult and not a child. Some clues can be found in the traits of the pubic symphysis, the sternal ends of right rib numbers three through five, cranial suture closure, and teeth.

Forensic Science Today

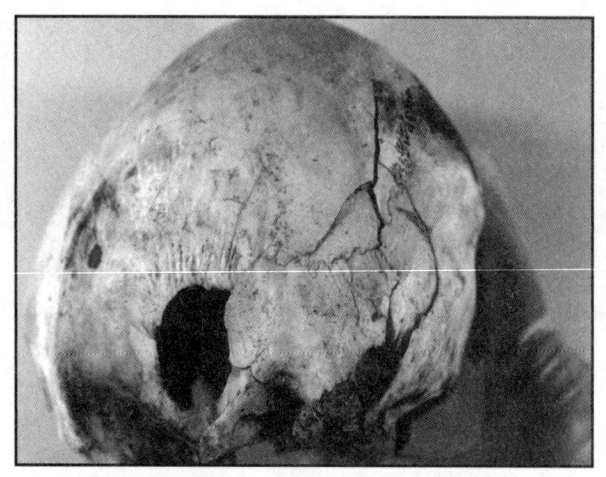

▲ A human skull showing evidence of a gun shot wound to the head.

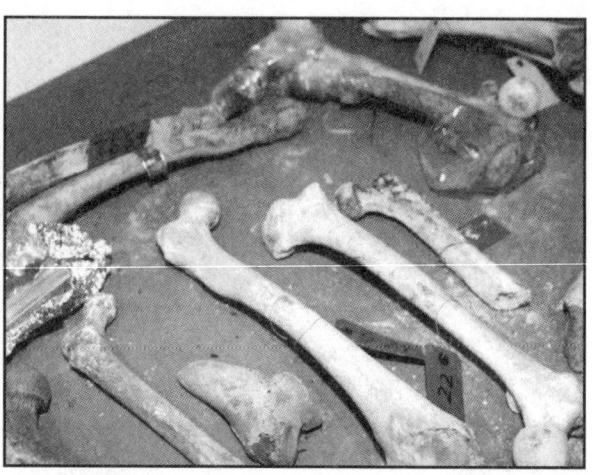

▲ Bone evidence submitted to a forensic laboratory for identification.

The Case of the Guilty Sausage Maker

Forensic anthropology was first used in the United States to solve a crime in 1897. When a Chicago murderer made his victim's body into sausage, there was no body left to trace the crime to him. However, upon a search of his sausage factory, detectives found traces of bone. Police called in George Dorsey, the curator of physical anthropology at the Field Museum of Natural History in Chicago, who specialized in the study of ancient human remains. Mr. Dorsey identified the bone fragments as belonging to a human, and his testimony in court secured the conviction. This was also the first time an anthropologist was called to testify in a United States court.

How long ago did the individual die? Forensic anthropologists can also be called upon to determine the postmortem interval (how much time has passed since the individual died). This is a rather challenging task for the forensic anthropologist since so many variables play into the state of human remains. For instance, in a hot humid environment, the human body can become skeletonized in a matter of weeks, while in the mountains or other cooler climates the process will take much longer. Other factors that come into play include the presence of animal scavengers, temperature, relative humidity, rainfall, the depth of burial, the weight and build of the decedent, clothing and other materials around the decedent (such as garbage bags), and insect activity (see also Chapter 20, "Forensic Entomology"). Finally, forensic anthropologists examine each case for the presence of remaining soft tissue such as skin or muscle, and look at whether any remaining soft tissue has been preserved through mummification, how much grease remains in the exposed bone, and whether the bones have been bleached by the sun.

Chapter 19
Forensic Anthropology

IDENTIFICATION OF HUMAN REMAINS

1. Are the remains human?
2. Are the remains recent?
3. Who did the remains belong to?
4. Estimation of age
5. Determination of sex, race, stature
6. Manner and cause of death
7. Means and facts related to death

▶ Unknown body parts were discovered by police. The identification of these body parts are essential for investigation.

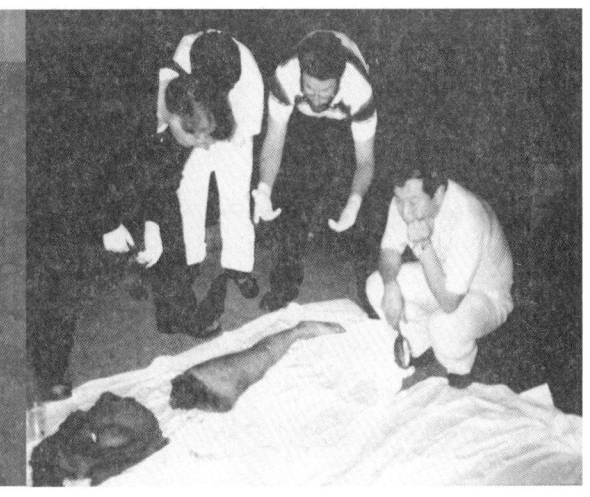

HOW CAN I BECOME A FORENSIC ANTHROPOLOGIST?

Forensic anthropologists usually major in anthropology during college as undergraduates. Courses in the physical sciences, especially biology, are extremely useful. Students interested in forensic anthropology as a career should also include courses in statistics, archaeological recovery methods, human anatomy, and skeletal biology. Then, students must go on to earn a Ph.D. in anthropology with an emphasis on the study of human anatomy and osteology. The broader the training in physical or biological anthropology, the better, so long as there is an emphasis in skeletal biology. Once you've begun practicing as a forensic anthropologist, you will need to gain practical experience in the field before the courts will accept you as an expert witness.

In the early 1970s, Forensic Anthropology became a separate section within the American Academy of Forensic Sciences, and it now has its own certification program. The American Board of Forensic Anthropology (ABFA) certifies experts in the field who must submit case reports for review. In addition to obtaining a Ph.D., forensic anthropologists must also pass a written and practical exam in order to be certified.

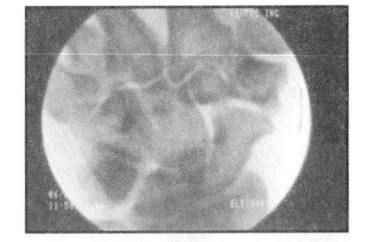

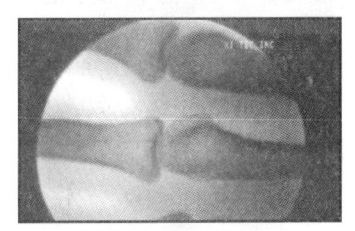

▲ X-rays of a victim's hand that were used for identification.

INTERESTED IN FORENSIC ANTHROPOLOGY?

A web site you should review is **www.archaeology.org**. Archaeology digs are very similar to crime scene investigations. If you are looking for travel and summer work, look up **www.findadig.com** for a summer experience.

SEARCH

Forensic Science Today

Mayor of the Body Farm

One of the world's top anthropologists is Dr. William Bass, who has helped solve hundreds of cases around the world using forensic anthropology. Dr. Bass was the creator of the "Body Farm," made famous by the novel of the same name by popular crime writer Patricia Cornwell. Officially known as the University of Tennessee Forensic Anthropology Center, Dr. Bass's Body Farm contains several acres of land dotted with rotting human corpses. Here, Dr. Bass and his colleagues study the effects of weather, insect activity, and shelter on decaying human flesh. Not only has Dr. Bass's landmark research brought forward new advances in forensic science, but his work on individual cases has also brought many killers to justice. If you would like to learn more about some of the exciting cases Dr. Bass has worked on, we recommend that you read his fascinating book called *Death's Acre—Inside the Legendary 'Body Farm.'* (For more information see the "Resources" section at the back of this book.)

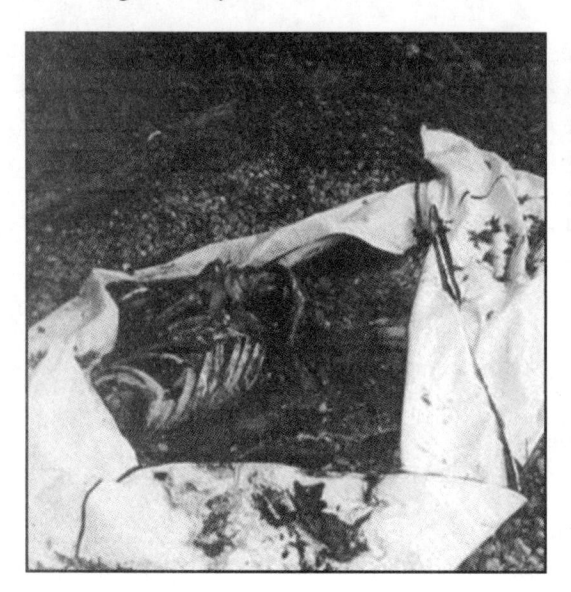

Three views of experiments with human remains at Dr. Bass's body farm.

TECHNICAL DEFINITIONS

Anthropology: The study of humankind.

Forensic Anthropology: A specialty within the field of anthropology that focuses on identification of human remains for legal purposes.

Skeletal Remains: The completely decomposed remains of an animal that have been reduced to bone through removal of tissues and fats by insects and other natural processes.

Postmortem Interval: The amount of time that has passed since a person has died.

Chapter 20

Forensic Entomology

CHAPTER OBJECTIVES

- ✔ Understand and discuss the role that forensic entomology plays in solving crimes.
- ✔ Explain what insects can tell investigators about a murder.
- ✔ Identify and describe the four types of carrion insects.
- ✔ Identify and describe the five stages of insect-aided body decomposition.
- ✔ Begin planning for a career in forensic entomology, if desired.
- ✔ Define the following vocabulary words and phrases: *entomology*, *entomologist*, *forensic entomologist*, *decomposition*, *carrion insects*, *necrophagous insects*, *predatory insects*, *parasitic insects*, and *omnivorous insects*.

TAKE NOTE!
Some of the material contained in this chapter is of a biologically sensitive nature, as it relates to human decomposition and insects that aid in the process. In other words, don't read this while eating breakfast!

Insects give us a wealth of information about a crime, particularly when a murder has occurred. Entomology is the study of the life cycles and distribution of insects,; and a forensic entomologist is a scientist who studies the life cycles of insects found on a decomposing body. By looking at what types of insects are found on a body, and where they are in their various stages of development (eggs, larvae, maggots, adults), forensic entomologists can usually pinpoint the time period of death. They can also tell us how long a body has been left at the crime scene after death occurred.

Forensic Science Today

KEY TERMS

Entomology
Entomologist
Forensic Entomologist
Decomposition
Carrion Insects
Necrophagous Insects
Predatory Insects
Parasitic Insects
Omnivorous Insects

WHAT CAN INSECTS TELL US ABOUT A MURDER?

In the world of forensic entomology, a human corpse found at a crime scene is a fascinating ecosystem that quickly changes form as it undergoes the various stages of natural decay. This "micro-habitat" provides food, a home, and a place to lay eggs for a large variety of insects. These insects that colonize the human body will then also become food for "predatory" insects (i.e., insects that hunt and eat other insects) that arrive later.

Because the life cycle and behavior of most common insects have been carefully studied by entomologists, the waves of insect infestation of a deceased human body follow certain predictable patterns and time intervals. The forensic entomologist is often called upon to identify the type of insect present on a body or at a crime scene, and to note where these insects are in their life cycle. This information is then applied to estimate the possible amount of time that the human remains have been undiscovered and unprotected after death.

Insects also provide clues about different places the body may have been located before being discovered (i.e., secondary crime scenes). Different insects live in different places. So, if a body dumped in the countryside has insects on it that only live in the city, this tells us that the body had at one time been in an urban area for a period of time after death before being moved to the country.

▲ A decomposed body showing maggot (larvae) activity was discovered by police. A forensic entomologist was called to assist in determining how long the victim's body was at this site and to reconstruct the crime through insect activities.

Chapter 20
Forensic Entomology

COLLECTION AND PRESERVATION OF INSECT SPECIMENS

Equipment
- hand net
- forceps and digging tool
- thermometer
- vials, jars, and plastic bags

Supporting Data Needed
- previous weather for the area
- on-site weather data (3 to 5 days)
- photos and video of crime scene
- time of collection
- medical examiner autopsy report, when available

Types of Carrion Insects

There are four main types of "carrion" insect species—that is, those insects that feed on dead organisms (including dead humans). These are:

Necrophagous species: These feed directly on the human corpse. Two types of insects are the most common: "blow flies" and "flesh flies." They have an incredible sense of smell, and can arrive at the scene of a dead human body as quickly as *10 minutes* after the deceased person has taken his or her last breath! Female flies will land on the body, feed on any body fluids seeping out, and lay their eggs in the body's cavities.

Predators and parasites: Next come the insects that are attracted to the necrophagous species that have already arrived to feed on the dead body. Predators are insects that hunt and eat other insects. Parasites are insects that rely upon other insects in a variety of ways, such as laying their eggs in other insects' egg sacs. The predator insects include different types of beetles that prey on both the eggs and maggots (larvae) of flies feeding on the dead body. The parasites that lay their eggs inside or on top of the maggots of the necrophagous insects include ants, bees, and wasps. What's interesting about these parasites is that they often are attracted to a certain kind of fly, so even after the original fly has left the body, forensic entomologists can determine what kind of fly was the first to arrive on the scene based on their parasites that have been left behind. These can then provide important clues as to the time of death and how much time has passed since death.

Omnivorous insects: Some types of wasps, beetles, and ants feed on *both* the human body *and* on other insects. Some of these insects are so good at catching others that are already on the body that they actually slow down the overall rate of decomposition!

Normal insects: The fourth type of insect found on and near decomposing human bodies is that which already inhabits the area where the body was deposited, and uses the corpse as an extension of its home. This includes hunting spiders that weave webs on body parts, and soil-dwelling organisms that feed on body fluids seeping into the soil underneath the body. These organisms may stay in the soil where the body was found for years after death, even after the body is long gone.

Forensic Science Today

Figure 1. Diagram illustrating the stages of development in a blow fly. (*Based on Maggot Gross Cycle, courtesy of William Lord, Oklahoma University Forensic Program.*)

Blow Fly Development

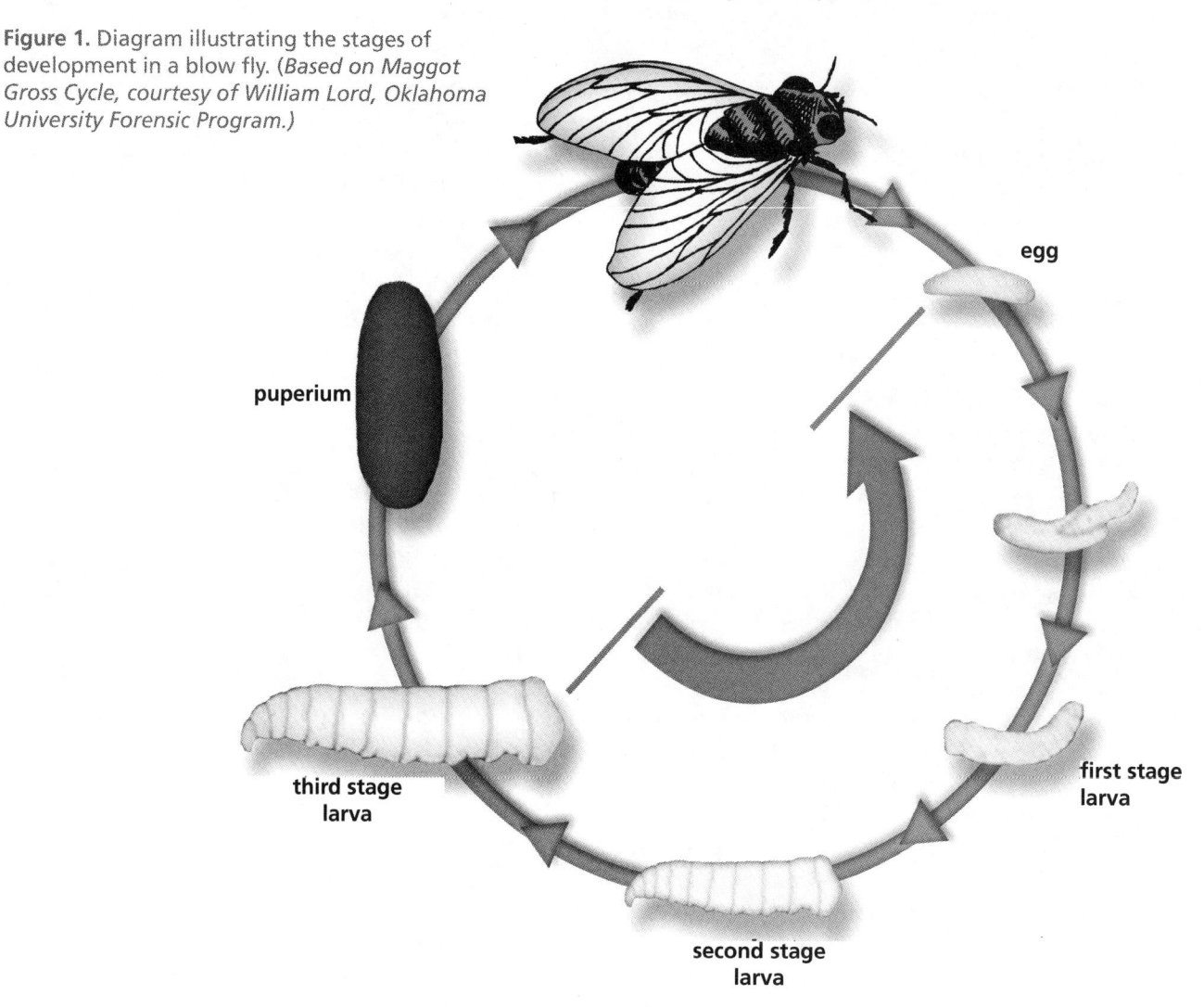

- egg
- first stage larva
- second stage larva
- third stage larva
- puperium

Stages of Decomposition

The rates of human decomposition depend upon the circumstances and location of the body. For instance, corpses in hot climates tend to decompose much more quickly than in cold climates. Decomposition is also affected by the dampness of the region. Thus, bodies in Texas decompose very differently than those in Maine or New Hampshire. Likewise, bodies in the New Mexico desert decompose in a different manner than in Hawaii, even though both states share hot climates—it is the humidity that makes the difference.

In general, there are five different stages of insect-aided body decomposition:

1. **The "Fresh" Stage.** The fresh stage begins at the moment of death, when the body is, so to speak, "fresh." During this stage, the body temperature falls to that of the air and soil around it. Insects such as blowflies land on the body, feed on body fluids that can be found, and lay eggs in body cavities (openings) and wounds. The eggs take twelve to eighteen hours to hatch. Once the eggs hatch, the maggots then feed on the body. Flesh flies will fly over the corpse and lay more eggs on the body. As described above, predators and parasites arrive next.

2. **The "Bloated" Stage.** Next comes the bloated stage, where bacteria inside the human body release gases during the process of "putrefaction." These gases first make the abdomen bloated, and later make the whole body blow up like a balloon. During this stage, the body becomes very hot due to the activities of bacteria, up to 127 degrees Fahrenheit. Meanwhile, blowflies and flesh flies are joined by houseflies, which all lay countless eggs that generally hatch at the same time. This results in masses of maggots crawling in and around the decomposing body, feeding upon body tissue like muscles and fat. As maggots get bigger and their numbers increase, even more predatory insects are attracted to the scene, contributing further to decomposition. Body fluids begin to seep from the body and into the soil below, which makes some insects that normally live in the soil leave very quickly. Other insects that like the body fluids will soon take their place.

3. **The "Decay" Stage.** The decay stage generally begins when the skin of the corpse begins to crack open as maggots feed on the skin, and gases escape the body. The release of gases makes the body lose its bloated appearance, and this is when the body begins to emit a very strong, foul, "dead body" odor. During the decay stage, the number and types of insects increase on the body. Now, more and more beetles arrive and quickly remove the flesh from the body. As the flesh is consumed by beetles, the flies begin to leave, and most of the maggots have hatched into flies and left as well.

4. **The "Post-Decay" Stage.** During the post-decay stage, there is not much left of the body besides bones, cartilage, skin, and hair. Different types of beetles come on the scene and replace those that have left. Additionally, many different types of "mites" (very small insects) begin to inhabit the soil under the body. These feed on by-products of the body's decomposition, as well as other insects feeding on the body.

5. **The "Skeletal" Stage.** When the body becomes fully "skeletonized," bones and hair are all that remain of the body. This is due to the work of all the different insects described above. At this point, there are no more insects left on the skeleton apart from any bugs that would normally inhabit the region.

HOW CAN I BECOME A FORENSIC ENTOMOLOGIST?

First, you have to like bugs! If you meet that criterion, you must obtain your undergraduate Bachelor of Science degree with a major in biology. It is also important to get a good background in the other physical sciences such as chemistry and physics. Then, to be considered an expert in the field to testify in court, you will need to earn a postgraduate degree such as a Ph.D. in entomology. After completing this degree, you will most likely need to work in the general field of entomology before being asked to consult on criminal cases. In the meantime, it is a good idea to keep exploring your interest in forensic entomology by taking continuing education courses in the field and becoming certified by the American Board of Forensic Entomology.

One of the leading forensic entomologists in the United States is M. Lee Goff, who helps solve crimes using insect evidence in his home state of Hawaii. Much of the information in this chapter is derived from his research. If you want to read about some of the exciting cases he has helped to solve and learn more about forensic entomology after finishing this chapter, we recommend Goff's book, *A Fly for the Prosecution—How Insect Evidence Help Solves Crimes*. (See "Resources" at the end of this book for more information.)

CASE PROFILE
How Maggots Solved a Murder*

*Adapted from Case Histories Of The Use Of Insects In Investigations by Wayne D. Lord, Federal Bureau of Investigation, Washington, DC, with the author's permission.

A partially clad body was found by passing motorists in a wooded area in November, approximately fifty feet from the highway. The victim was a young adult female. Her body was cold, there was no sign of rigor mortis, and only minor external decompositional changes were evident.

It was determined that the victim had died of multiple stab wounds to the chest and neck. During the autopsy, several large maggots were seen migrating away from the corpse. This behavior is characteristic of fully developed, post-feeding blow fly larvae. Several more maggots were also removed from the neck wounds and clothing.

The victim was later identified as a young woman who had been reported missing eighteen days prior to the discovery of her body. She was last seen alive during the early morning hours in late October in the apartment of the primary suspect. The suspect had recently been released from prison, with a history of sex offenses and violent crimes.

Strong circumstantial evidence pointed to the suspect. Witnesses recalled hearing a woman screaming and seeing a man and woman struggling in the vicinity of the suspect's apartment during the early morning hours eighteen days prior to discovery of the victim's body. Hair from the victim's head and pubic region was found on bed sheets in the suspect's apartment. One of the victim's shoes was later found in a secluded urban woodlot a short distance away. Moreover, samples of blood and hair taken from the suspect's car also were identified as belonging to the victim. Finally, an analysis of auto carpet fibers, removed from the victim's clothing, further demonstrated that she had been transported in the suspect's vehicle.

However, an accurate estimate of the time of death was critical in establishing the sequence of events surrounding the death in order to prove the suspect's guilt beyond a reasonable doubt to the jury. Several conflicting estimates of postmortem interval, ranging from two to eight days, were given by case investigators and medical examiners. This was based primarily on the physical appearance of the body and the extent to which autolytic self-digesting changes had occurred within various organs.

Forensic entomology experts were called in to determine the postmortem interval with more precision. The maggots observed in and around the body during the autopsy were identified. However, none of these maggots could be reared to the adult stage since none of them had been maintained alive. Soil samples also failed to yield any additional specimens. Therefore, investigators examined numerous photographs depicting the crime scene, the surrounding vegetation and terrain, and the body itself.

The forensic entomologists also obtained climatological data—including maximum and minimum temperature, cloud cover, rainfall, wind speed and direction, and relative humidity from a National Weather Service observatory located less than a quarter of a mile from where the victim's body was found. Reports describing the condition of the body when found and detailing autopsy procedures and results were also reviewed.

The forensic entomologists identified the largest fly larvae inhabiting the remains as being fully engorged third instar larvae and post-feeding larvae of the blow fly species, Calliphora vicina. One specimen of this species showed morphological signs of pupariation.

Entomologists calculated the number of days necessary for Calliphora vicina to develop from egg to prepupa, based on the temperature records for the relevant period. Because the developing flies were exposed to a cool

continued on next page...

(10 degrees Celsius) average daily temperature, and because not enough larvae were present within the corpse to elevate the body's (microenvironmental) temperature, the postmortem interval was estimated to be fifteen days.

The postmortem interval of fifteen days, together with other evidence acquired during the police investigation, gave probable cause to arrest and charge the suspect with first degree murder, kidnapping, and felony rape. The suspect was found guilty of all charges and sentenced to a lengthy prison term.

It was later learned that the perpetrator had murdered the victim during the early morning eighteen days prior to discovery of the victim's body. The perpetrator had hidden the body in a nearby woodlot for three days by covering it with tree branches, a mattress, and other debris. He had then transported the corpse to the location where it was found. Oviposition by adult Calliphora vicina may have been prevented during the first few days following the victim's death due to the materials the perpetrator used to cover her body. Therefore, the fifteen-day postmortem interval was valid, even though the victim had in actuality been murdered eighteen days prior to her discovery.

Further Information: Maggots

Maggots grow rapidly and pass through three growth steps (instars) before becoming fully grown. Each of these instars is punctuated by a moulting period. Typically, large numbers of larvae hatch together and move as a group around the corpse. By doing this, they disseminate bacteria and secrete enzymes that enable them to consume virtually all of the soft tissues of the corpse. This growth will take several days to several weeks depending upon species, environmental conditions, and the number of larvae present.

TECHNICAL DEFINITIONS

Entomology: The study of insects.

Entomologist: A scientist who studies insects.

Forensic Entomologist: A scientist who studies the life cycles of insects found on a decomposing body to determine how much time has passed since death (the postmortem interval).

Decomposition: The process by which human and other animal remains disintegrate from insect activity and other natural factors.

Carrion Insects: Any insect species that feeds on dead organisms.

Necrophagous Insects: Species of insects that feed directly on the corpse.

Predatory Insects: Species of insects that hunt and eat other insects.

Parasitic Insects: Insects that rely upon other insects as a place to lay their eggs and for other purposes.

Omnivorous Insects: Insects that feed on both the human corpse and other insects.

Chapter 21

Forensic Dentistry

CHAPTER OBJECTIVES

✔ Identify the two different types of forensic examinations that forensic odontologists are usually called in to perform.
✔ Explain how forensic odontologists help identify human remains.
✔ Explain how forensic odontologists analyze bite marks on victims, suspects, and evidence.
✔ Understand the role of new technologies in the field of forensic odontology.
✔ Identify the steps to pursuing a career in forensic odontology, if desired.
✔ Define the following vocabulary words: *antemortem*, *postmortem*, *odontology*, and *Lucis*.

Odontology is the technical term for forensic dentistry, and odontologists are forensic dentists. Therefore, odontology is the application of dental science to problems of human identification. When working a case, the forensic odontologist will examine a particular piece of dental evidence and look for all the unique and individual characteristics in that piece of evidence. The odontologist will then look for pre-existing dental records containing those same characteristics. If the unique characteristics of the evidence match those contained in dental records, then identification is possible.

WHAT DO FORENSIC ODONTOLOGISTS DO?

Forensic odontologists usually perform two different types of forensic examinations: (1) identifying unknown human remains, and (2) analyzing bite marks on crime victims, suspects, and evidence.

Identifying unknown human remains

Forensic odontologists compare dental x-rays of unidentified human remains with the antemortem (prior to death) dental records of known persons, in order to identify remains that are difficult or impossible to identify by more conventional means. This is usually called for when a skeleton is found, or in the case of plane crashes. Forensic odontologists can identify a person by using a complete jaw, a fragment of a jaw, or even just a few teeth.

▶ Dental remains found at a fire scene can be used for victim identification.

Forensic Science Today

KEY TERMS

Odontology
Forensic Dentistry
Skeletal Remains
Antemortem
Postmortem Interval
Lucis

Dental x-rays provide an excellent source of evidence for the purposes of identification. They show the relative positions of the teeth, and contain the shapes and sizes of the fillings and other material contained in the tooth. Dental x-rays also expose the roots of the teeth, unique anomalies, and pathology such as cysts. Thus, dental x-rays supply many different areas for comparison.

Forensic odontologists may also participate in autopsies. When examining human remains, forensic odontologists chart dental and cranial (skull) features, take x-rays, and write up reports of their findings. Then, law enforcement agencies can use this information to help identify the remains that have not yet been identified. Law enforcement personnel can also code the information contained in the forensic odontologist's report into a computer identification program containing both antemortem and postmortem dental records. The computer program can then compare the records entered in by law enforcement personnel to match the unknown remains with known samples, photographs, and records already contained in the database.

Sometimes forensic odontologists will also need to work at the scene of a crime or mass disaster (such as airplane crashes, terrorist attacks, fires, bombings, earthquakes, and floods). On other occasions, they are called in to identify remains of individuals in mass graves after political atrocities, such as massacred indigenous (native) people in Guatemala or genocide victims in Rwanda.

Analyzing bite marks on crime victims, suspects, and evidence

Forensic odontologists also locate, examine, and compare bitemarks. Criminals sometimes bite their victims during abuse and violent crimes including assault, rape, and/or homicide. Likewise, victims may sometimes bite their attackers in self-defense. Both perpetrators and victims can leave potential bite mark evidence at a crime scene, impressed in food remains or even a piece of chewing gum! (Note that chewing gum is also an excellent source of DNA evidence contained in saliva).

▼ Forensic odontologists use a dental model to test a bite mark on human skin.

Computer imaging technology makes comparison of bite marks a very interesting job. A computer can generate an outline of teeth, and then overlay it onto a digital photograph of a human bite mark on a victim. This is called a "compound overlay." Then, the image can be zoomed in for a close-up look at the bite edges of a suspect's teeth that are placed next to the injury pattern on the victim. If the bite marks match, then we have possible proof that the suspect committed the crime. At trial, the forensic odontologist will provide expert testimony in the courtroom during trial, and explain the techniques that were used to come to his or her conclusions.

Chapter 21
Forensic Dentistry

HOW CAN I BECOME A FORENSIC ODONTOLOGIST?

First, you need to become a dentist. In order to become a dentist, you must go to college to earn a Bachelor of Science degree. Then, you go to dental school to earn your doctorate in dentistry, which is known as either the DDS (Doctor of Dental Science) or DMD (Doctor of Dental Medicine). Next, you must gain the experience and additional education to specialize in forensic odontology. This is necessary in order to become qualified as an expert in the field so that you can testify in court. Remember, the final determination as to whether you are considered an expert is made by the court. The American Academy of Forensic Science (AAFS) has an affiliate odontology organization, the American Board of Forensic Odontology (www.abfo.org). Once a dentist takes all the required courses to specialize in forensic odontology, he or she can sit for the Board examination.

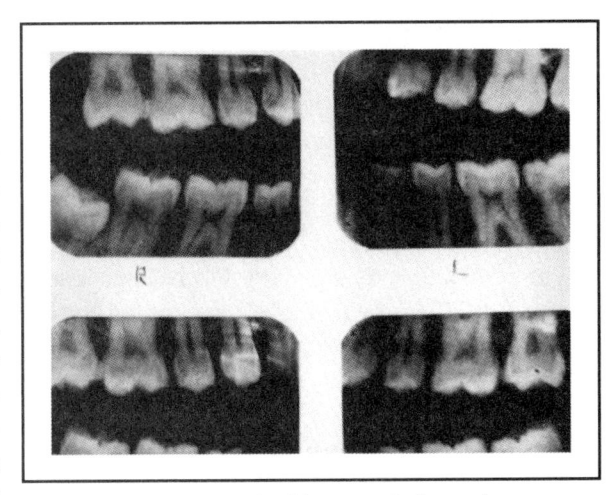

▲ Dental x-rays are valuable records for unknown body identification.

EXPERT PROFILE: DR. LOWELL LEVINE, FORENSIC ODONTOLOGIST

Dr. Lowell Levine, Co-director of the New York State Police Medicolegal Investigations Unit, is one of the country's top dental experts and has made remarkable contributions to the field of forensic odontology. He has aided the investigations into the assassinations of President John F. Kennedy and Martin Luther King, Jr.; he has traveled to Russia to examine the remains of assassinated Russian Tsar Nicholas II and his family; and he has traveled to Brazil to help identify the remains of Josef Mengele, the Nazi doctor responsible for sending 400,000 people to Auschwitz. Also, in South America, Dr. Levine has assisted investigations into the "disappeared" people of Argentina, to help identify the remains of civilians kidnapped and murdered for political reasons. Back home, Dr. Levine has worked on many murder cases involving bitemarks, including that of the notorious serial murderer Ted Bundy. He has also served as the President of the American Academy of Forensic Sciences, the American Board of Forensic Odontology, and the Forensic Sciences Foundation.

How did Dr. Levine achieve all this? After earning his Bachelor's degree in biology and chemistry, Dr. Levine earned his dental degree from New York University's College of Dentistry and went into private dental practice. He then taught behavioral science at New York University and dentistry at NYU's Dental College, and later became the coordinator of the Forensic Consultant Unit in the New York State Police Department (NYSPD). In 1986 he became a consultant in forensic odontology for the NYSPD and the Co-director of the Forensic Sciences Unit in Albany, New York.

CASE PROFILE
How a Murderer's Teeth Gave Him Away

The body of twenty-eight year-old Connecticut resident Carla Terry was found by a man who was walking his dog and noticed a hand sticking out of the snow on the side of the road. Paramedics were called to the scene but failed to revive Terry with CPR. Afterward, they cleaned off her body, wiping away critical evidence in the process. Although investigators attempted to trace tire tracks near the scene, the original tracks were lost after police cars and other vehicles had driven to the area several times.

Despite the compromised evidence, there was still enough to eventually convict Alfred Swinton, whose name was printed on a card found in the victim's pocket. Forensic odontologist Dr. Constantine "Gus" Karazulas of the Connecticut State Forensic Science Lab was called in to examine two bite marks that had been found on the victim's body. Dr. Karazulas took dental imprints of Swinton's teeth in an effort to match them to the bite marks. Unfortunately, the marks on the body were not clear enough and the case went cold.

However, nearly ten years later, Dr. Karazulas learned of a new photographic technology developed by digital image expert Barbara Williams. This technology was contained in a digital enhancement program called "Lucis," which lightens dark spots in photos and reveals what might be hidden by shadows. Williams worked with a photograph of the bite mark found on the victim's body ten years earlier and was able to lighten the bite marks. This allowed Dr. Karazulas to finally measure the bite marks, which indeed precisely matched the measurements of the dental impressions made by Swinton's teeth.

Next, Dr. Karazulas teamed up with a scientist skilled in Adobe Photoshop to create a nearly transparent image of Swinton's dental impressions. When this image was fit over the photograph of the bite marks on the victim's body, the match was perfect.

With this new evidence, the case went back to court and Dr. Karazulas testified on the stand as an expert witness. Not only did he demonstrate how the image of Swinton's teeth impression showed an identical match to the bite mark on the victim's body, but he also made a bite mark on his own skin and let the courtroom see that the mark took about ten minutes to disappear. This demonstrated that the bite marks were made by the murderer very close to the time of the victim's death, since they had no time to disappear. Once the blood stops flowing after death, a bite mark will stay on the body for a prolonged period of time.

This evidence was enough to prove Alfred Swindon guilty beyond a reasonable doubt. He was convicted in 1991.

Chapter 21
Forensic Dentistry

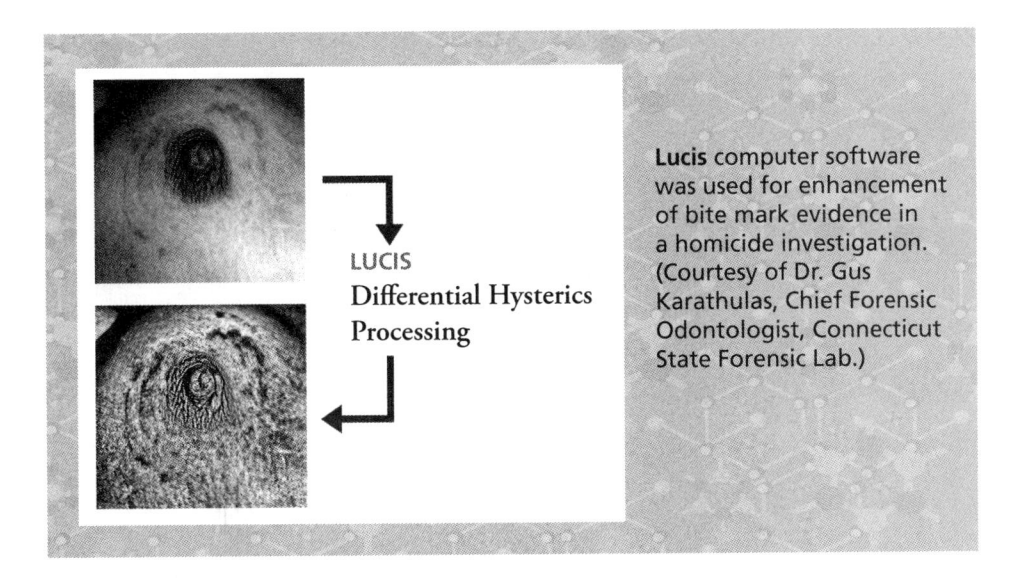

LUCIS
Differential Hysterics
Processing

Lucis computer software was used for enhancement of bite mark evidence in a homicide investigation. (Courtesy of Dr. Gus Karathulas, Chief Forensic Odontologist, Connecticut State Forensic Lab.)

TECHNICAL DEFINITIONS

Antemortem: Before death.

Postmortem: After death.

Odontology: Forensic dentistry.

Lucis: Computer software for enhancement of bitemark evidence.

Forensic Engineering

Chapter 22

CHAPTER OBJECTIVES

✔ Understand how forensic engineers use their engineering skills to reconstruct events.
✔ Give examples of the questions forensic engineers ask about an event.
✔ Describe different situations where forensic engineers are needed.
✔ Identify the steps needed to pursue a career in forensic engineering, if desired.
✔ Define the following vocabulary words and phrases: *engineer*, *forensic engineer*, and *accident reconstructionist*.

An *engineer* is a scientist who applies the principles of mathematics and science to design, construct, and solve electric, mechanical, civil and other problems. *Forensic engineers* use their engineering skills to reconstruct events such as building or bridge collapses, train wrecks, car accidents and airplane crashes. Forensic engineers are usually asked to assist in civil (non-criminal) matters, though sometimes they may also assist law enforcement in criminal cases, especially in cases involving large fires, airline crashes, or bombings where there may be a loss of life.

Forensic engineers are involved in different types of analyses that apply engineering principles to legal issues. Such analyses include reconstructing events from the physical evidence record in cases involving material failures, building and structure collapses, vehicular accidents, and other tragedies. Forensic engineers also conduct quality evaluations of construction or manufacturing; and they analyze maintenance procedures as well as environmental issues.

Forensic engineers can work on a huge variety of problems. Sometimes they go down to the molecular level; sometimes they look at an entire communications network or transportation system. Sometimes they examine a microscopic piece of metal; sometimes they examine whole skyscrapers.

Forensic Science Today

KEY TERMS

Engineering
Forensic Engineer
Accident Reconstructionist

Forensic engineers are called in to answer these types of questions:

- Did the driver hit the brakes before running over the victim with his car?
- Did the airplane crash as a result of an explosion?
- Why did the World Trade Center collapse when it was struck by airplanes?
- Was sabotage involved in the train derailment?

WHEN ARE FORENSIC ENGINEERS NEEDED?

There are many different circumstances in which forensic engineers are called to duty. One such situation is when an airplane crashes and it is necessary to find out why and how it happened. Plane crashes can be caused by weather conditions, structural failures, electrical problems, improper design or maintenance, or pilot error. After a plane crash, engineers working for the FAA (Federal Aviation Administration) are called to investigate the case.

Forensic engineers also examine the circumstances of automotive accidents. In many traffic accident cases, forensic engineers must study the evidence very closely to figure out what happened. They look at skid marks left by tires on the pavement to determine how fast the vehicle was traveling when it crashed and how far the vehicle went after hitting the brakes before coming to a complete stop. Engineers examine airbags and evidence surrounding their deployment to determine if the airbags inflated and deflated properly and were triggered under appropriate circumstances.

Additionally, forensic engineers consult on cases involving amusement park rides, off-road vehicles (boats, ATVs, snowmobiles), building and bridge collapses, building fires, construction accidents, product failures, and a huge variety of other issues. Forensic engineers work for a variety of organizations, ranging from large corporations to small firms to government agencies. Some even work for themselves by offering their services to organizations and individuals on a contract basis.

TYPES OF FORENSIC ENGINEERING INVESTIGATIONS

- Fire and arson
- Industrial accident
- Traffic accident reconstruction
- Transportation disaster
- Civil Engineering
- Structure and material failure
- Product and material liability issues
- Environmental system failure analysis
- Computer and electronic crime analysis

▲ Police investigate a motor vehicle accident.

Chapter 22
Forensic Engineering

◀ Material failure and metal fatigue are the cause of the bridge collapse.

FORENSIC ENGINEERING

- Application of engineering in the legal system
- Investigation of failures, catastrophes, and accidents
- Analysis of products, merchandise, machines, instruments, vehicles, and structures
- Review plans, designs, blueprints, reports, and diagrams related to claims and litigation
- Reconstruction of physical cause of incident, accident, and events

◀ September 11, 2001: Forensic engineering was vital to solve the puzzle of why the twin towers of the World Trade Center collapsed so easily after being struck by commercial airliners flown by international terrorists.

CASE PROFILE
TWA Flight 800

On July 17, 1996, TWA Flight 800 exploded shortly after departure from JFK Airport in New York, en route to Paris, France, killing all 230 people on board. While the plane was still climbing, an explosion occurred on the plane that caused it to crash into the Atlantic Ocean near the southern coast of Long Island, New York. The flight crew had not reported any problems to air traffic control before the explosion. Witnesses reported an explosion in the air followed by debris falling to the ocean. Some witnesses saw a fireball in the sky lasting five seconds; others reported a second explosion upon impact with the water.

What caused TWA Flight 800 to crash? Airplanes can crash due to mechanical failures such as engine malfunction; severe weather such as wind shear, turbulence, or lightning; human error in the cockpit and/or air traffic control tower; or any combination of the above. To discover what happened to TWA Flight 800, forensic engineers with the National Transportation Safety Board (NTSB) participated in one of the longest and most expensive accident investigations in U.S. history.

Airline accident reconstruction requires a multidisciplinary team approach. Investigators must identify the victims and call in forensic engineers to reconstruct what happened. All items recovered from the crash, no matter how seemingly insignificant, are marked with flags to indicate their position. The recovered material is brought to a large area such as an aircraft hanger so that the pieces of the plane can be put back together like a giant puzzle. Additionally, the "black box," which records pilot conversation and instrumental readouts during the last minutes and seconds of the flight, can provide vital clues as to what caused the crash.

Four years after the crash of TWA Flight 800, the NTSB released a report dealing with safety issues such as fuel tank ignition sources, fuel tank flammability, design and certification standards, and the maintenance and aging of aircraft systems. The report concluded that the crash was caused by an explosion in the fuel tank, resulting from ignition of the flammable mixture of fuel and air contained in the tank. While the NTSB could not determine the source of the ignition with certainty, it concluded that the most likely cause was a short circuit allowing excessive voltage to enter the fuel tank through electrical wiring.

Some witnesses from the ground, however, claimed to have seen a missile flying through the sky toward TWA Flight 800. A member of the flight crew on a Saudi Arabian Airlines flight reported seeing a bright greenish object streaking past the plane only seconds prior to the TWA explosion. While some theorize that a rocket-launcher fired from the ground may have hit the flight, others postulate that Flight 800 was struck by a U.S. naval vessel operating in the vicinity. The NTSB, on the other hand, concluded that there was no evidence of a missile or bomb detonation, and the FBI agreed that there had been no criminal act.

What do you think?

Chapter 22
Forensic Engineering

◀ Wreckage of an aviation accident involving a small private plane. (left)

▶ Crash site from an accident involving a large commercial airliner. (right)

HOW DO I BECOME A FORENSIC ENGINEER?

If you are interested in pursuing a career as a forensic engineer, you must earn a Bachelor's of Science degree in engineering or a related science. Not many universities offer courses in forensic engineering, so you must first become qualified as a general engineer and then take courses or workshops specializing in forensic engineering, or work with other forensic engineers. It is definitely recommended that you earn an advanced graduate degree after obtaining your Bachelor's degree. Good skills to build in addition to a science background include writing, speaking, and evidence handling techniques. It is also very useful to have a strong understanding of legal procedures and different types of evidence allowed in the courtroom.

TECHNICAL DEFINITIONS

Engineer: A scientist who applies the principles of mathematics and science to solve problems.

Forensic Engineer: A scientist who uses engineering skills to reconstruct accidents and other events.

Accident Reconstructionist: An engineer who reconstructs the sequence of events during accidents and crimes from an engineering perspective.

Chapter 23

Forensic Psychology

CHAPTER OBJECTIVES

- ✔ Understand the difference between a psychologist and psychiatrist, as well as the contributions each can make to forensic science.
- ✔ Describe several examples of the role a forensic psychologist or psychiatrist might play in criminal law.
- ✔ Describe several examples of the role a forensic psychologist or psychiatrist might play in the civil arena.
- ✔ Explain how forensic psychologists and psychiatrists might testify on the stand in trial.
- ✔ Understand the steps needed to pursue a career in forensic psychology or psychiatry, if desired.
- ✔ Define the following vocabulary words and phrases: *psychologist*, *psychiatrist*, *competency*, and *psychological profiling*.

A psychologist is a professional who has earned a bachelor's and master's or doctorate degree in psychology and has clinical experience in counseling. A psychiatrist is a medical doctor who specializes in brain chemistry and can prescribe medication for treatment of mental illness and other conditions. Therefore, forensic psychologists and psychiatrists specialize in studying human psychology, psychiatry, and behavior in order to examine the suspect to determine his or her psychological conditions, to profile criminals as well as victims, and to answer other legal questions related to human behavior.

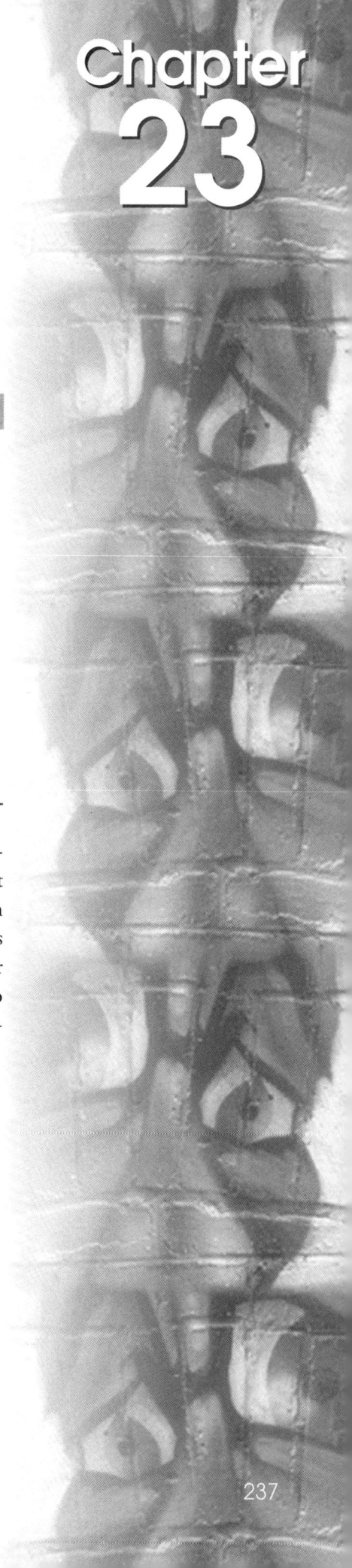

KEY TERMS

Psychologist
Psychiatrist
Competency
Psychological Profiling

WHAT DO FORENSIC PSYCHOLOGISTS AND PSYCHIATRISTS DO?

Forensic psychologists and psychiatrists deal with a wide variety of legal issues in both criminal and civil matters. In the criminal arena, forensic psychologists and psychiatrists may answer the following questions:

- Does this suspect fit the psychological profile of the person who committed this crime?
- Is the suspect competent to stand trial? (Under the U.S. Constitution, people must be mentally competent to stand trial; i.e., they must understand the nature of what they did, the nature of the proceedings against them, and the consequences of their actions).
- Is the convicted criminal competent to be executed? (Again, under the U.S. Constitution, people convicted of crimes cannot be executed if they are not mentally competent.)
- Was the suspect mentally competent when he confessed his crime to police? (In other words, is his or her confession valid and therefore admissible evidence in court?)
- Is the suspect competent to waive legal representation? (In other words, is he or she mentally fit to act as his or her own attorney during trial?)
- Was this person truly responsible for this crime? (Sometimes mental illness can decrease a person's responsibility for a crime by reason of "diminished capacity;" at other times a person can be found "not guilty for reason of insanity," or, in some states, "guilty but insane" leading to confinement in a mental institution rather than a prison sentence.)
- Will this convicted prisoner be a danger to society when he or she is discharged into the community on parole?
- What is the psychological profile of a criminal?

In the civil arena, forensic psychologists and psychiatrists may answer these types of questions:

- Was a deceased person mentally competent when making out a will giving millions of dollars to the cat instead of the children?
- Are the prospective adoptive parents of this child mentally fit to be parents?
- Should custody of the kids in a divorce go to one parent but not the other?
- Is this person mentally competent to refuse life-saving medical treatment?
- Is this person mentally disabled such that he or she is unable to work and must receive Social Security payments?

Chapter 23
Forensic Psychology

CASE PROFILE
PORTRAIT OF A SICK MIND

In a particularly disturbing murder case, a young woman was found dead on the roof of a building. Her body had been mutilated in a manner consistent with a sexual killing, including bitemarks. (See Chapter 21 for more information on bitemarks.) The perpetrator had defecated next to the body of his victim, and covered the feces with the victim's clothing. There was no evidence of a potential perpetrator left on or near the body, except a single hair consistent with Negroid origin. A forensic psychologist was called in to give a profile of the likely killer. Contrary to the hair type found, the type of person who would commit this crime, said the profiler, would most likely be white, unemployed, and relatively young. He probably would have some connection and familiarity to the building, which is why he would choose it to commit the murder in that location. When the building's residents were interviewed, suspicion fell on the landlord's son, who also happened to be white and unemployed. The case was solved when his teeth marks perfectly matched those found on the victim's body. While the bitemarks solved the case, the suspect would not have been found if it were not for the expertise of the psychologist who predicted the criminal profile of the perpetrator with great accuracy.

Forensic psychologists and psychiatrists are well versed in the law, and work regularly with police, lawyers, and judges. They are highly specialized, and may provide expert testimony during a courtroom trial. On the stand, they must identify the issue and the legal and psychological/psychiatric criteria that determine the answer, and they must explain the process from which they derived their professional opinion. They work in private practice, for the government, or in hospitals.

HOW CAN I BECOME A FORENSIC PSYCHOLOGIST OR PSYCHIATRIST?

Forensic Psychologist

In order to become a psychologist, you must major in psychology or behavioral science during college and go on to complete your master's degree. To become a forensic psychologist, you must have additional expertise derived by earning a doctorate in psychology, which generally takes four to six years. Then, you can do a post-doctoral fellowship (also commonly referred to as a "post-doc") in forensic psychology, and/or study forensic psychology independently and get training and experience in the field. To become certified as a specialist in forensic psychology, you will take an exam and apply to the American Board of Professional Practice in Psychology.

Forensic Psychiatrist

To become a psychiatrist, you must first go to college, studying the human and physical sciences. Then, you go to medical school and earn your Medical Doctor degree. After that, you spend an additional four years training as a resident in psychiatry. This is where you begin to specialize as a psychiatrist. You can also spend an additional year or two in post-residency training in the field of psychiatry and the law, and/or study independently and gain experience on the job. Post-residency specialization fellowships in forensic psychiatry are certified by the Accreditation Council on Fellowships in Forensic Psychiatry. Next, you take an examination in forensic psychiatry to be certified by the American Board of Psychiatry and Neurology.

TECHNICAL DEFINITIONS

Psychologist: A professional who studies human behavior and is licensed to provide therapeutic counseling services.

Psychiatrist: A medical doctor who specializes in mental disorders and may prescribe medication.

Competency: A legal term indicating that a person is mentally fit and has full legal capacity.

Psychological Profiling: An analysis based upon psychological and statistical factors of a criminal used to determine the general characteristics and behavior of the most likely suspect for a crime.

Chapter 24

Crime Scene Reconstruction

CHAPTER OBJECTIVES

- ✔ Understand the scientific and logical nature of crime scene reconstruction.
- ✔ Explain who is responsible for reconstructing a crime, and how.
- ✔ Understand the requirements for crime scene reconstruction.
- ✔ Identify and explain the five stages in reconstruction.
- ✔ Identify five different types of reconstruction.
- ✔ Define the following vocabulary words and phrases: *reconstruction*, *conjecture*, *hypothesis*, and *theory*.

From crime scene to court, solving a crime involves a series of distinct steps: recognizing what is and what is not physical evidence; documenting, preserving and collecting physical evidence; identifying the characteristics of the evidence; comparing evidence to known samples and other collected evidence; individualizing the evidence; and finally, reconstructing and interpreting the events. Many parts of this book have been focused on the first few steps regarding physical evidence. Now, as our introduction to forensic science draws to a close, we will learn about how forensic scientists and criminal investigators reconstruct crimes using a combination of logical analysis and forensic evidence to uncover the facts.

Crime scene reconstruction is the process of determining the sequence of events, criminal activities, and logical predictions about what occurred during and after the crime. As such, reconstruction is a scientific fact-finding process; and each step in the process should follow a logical analysis model. It involves the scientific analysis of a crime scene, the interpretation of crime scene pattern evidence, the laboratory examination of forensic evidence, and also a systematic study of related information and the formulation of a logical hypothesis.

Forensic Science Today

KEY TERMS

Reconstruction
Conjecture
Hypothesis
Theory

The importance of crime scene reconstruction is one major reason for maintaining the integrity of a crime scene. Reconstruction requires an intact crime scene that has not been mismanaged or mishandled by investigating officers, as does the verification of witness statements, the investigation of alibis, and the development of investigative leads. If the scene was altered, this can seriously affect the ability to reconstruct the crime.

It is important to understand that crime scene reconstruction is very different from 're-enactment," "re-creation," or "criminal profiling." Re-enactment generally refers to having the victim, suspect, witness, or other individual re-enact events, based on their knowledge and recollection of the crime. Re-creation, on the other hand, is to replace the necessary items or actions back at a crime scene through original scene documentation. And criminal profiling is an analysis based upon statistical and psychological factors of the criminal, in order to determine the general characteristics of the most likely suspect for a crime. While each of these types of analyses may be useful in certain aspects of criminal investigation and crime reconstruction, they rarely provide a direct solution of a crime.

SCIENCE + LOGIC = THE NATURE OF RECONSTRUCTION

Crime scene reconstruction is partly science and partly logic. The steps used in crime scene reconstruction and the stages of forensic examination are closely related. Both types of analysis follow basic scientific principles and theory formation. They involve consideration and incorporation of all investigative information and analysis, as well as examination of physical evidence into a reasonable explanation of the crime and related events. Logic, careful observation, and considerable experience, both in crime scene investigation and evidence examination, are necessary for proper interpretation, analysis, and, ultimately, reconstruction.

Logic Tree: Robbery/Burglary

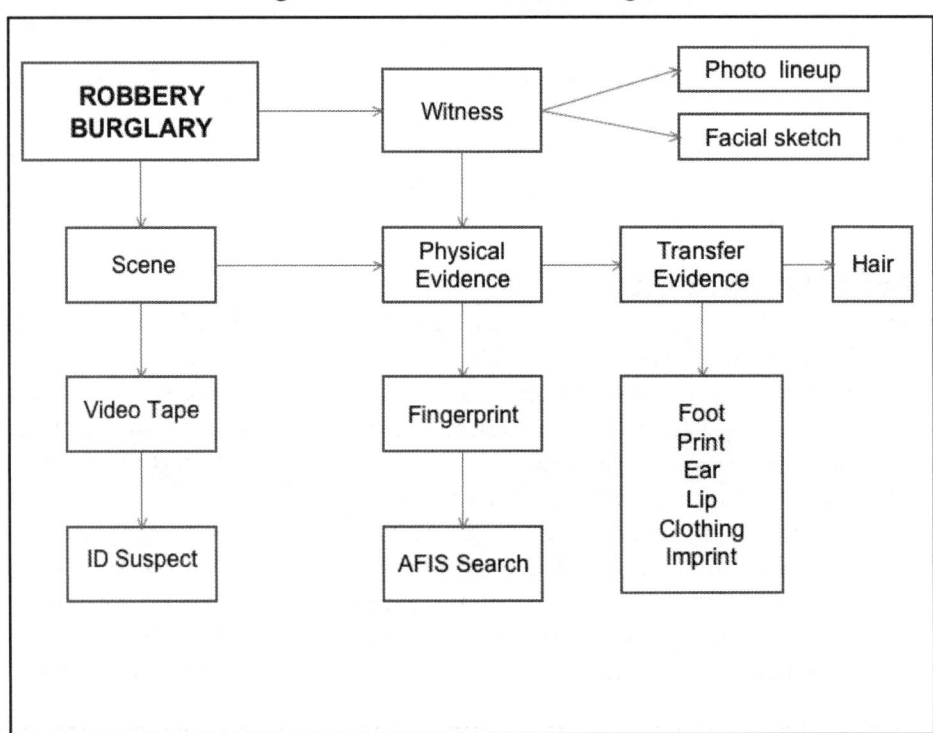

RECONSTRUCTION: A Hypothetical Case Study

The following is a mock case study of crime reconstruction. Notice how this case study demonstrates how evidence and logic come together in the different stages of crime scene reconstruction.

On a February morning, the body of a man was discovered by a campus security officer in the restroom of a local college. The victim had been shot three times by a gun. No weapon, spent cartridge cases, or projectiles were observed at the scene during the initial crime scene investigation. A detailed, close examination of the crime scene patterns revealed the following information:

Bloodstain Pattern Evidence

- One large pool of blood was found adjacent to the victim's head. This pool of blood was still in a liquid state at the time of the examination. This information indicated that the victim had been shot a short time prior to the discovery of his body.
- A trail of blood drops was noticed from one side of the bathroom stall leading to the body. This indicated the location where the victim first received his injuries and his subsequent movements.
- No bloodstains or damaged areas were found on the victim's clothing, which suggests that no struggle or major contact occurred after he was shot.

Firearms Evidence

- A close search revealed a spent bullet, which was found under the victim's body. This bullet was identified as a copper jacketed 9 mm type projectile that had most likely been discharged from a semi-automatic 9 mm weapon.
- Only one discharged cartridge case was recovered from under the victim's overcoat at the scene. This information suggested that the individual who committed the crime had knowledge of criminal investigation procedures and the importance of physical evidence.
- During a subsequent search of the security guard's vehicle, a 9 mm gun was found. This gun showed signs of having been discharged recently.
- A search of the security guard's boots revealed three spent bullet casings in his left boot, and one intact cartridge in his right boot. These cartridge cases were positively identified as having been discharged from the security guard's weapon. This also raised the question of how many total shots were discharged from the weapon.

Other Physical Evidence

- Bullet trajectories were reconstructed and indicated that the victim was first shot while in a standing position. Based on the location of the bullet holes in his clothing and the bullet hole in the bathroom wall, his initial position was reestablished. The gunshot residue (GSR) distributions and their patterns indicated that three shots were discharged at a close range.
- No GSR was detected on the hand swabs of the security guard. This information could suggest either that the security guard did not fire the gun, or alternatively, that he did fire the gun but then cleaned his hands after the shooting. A search of the garbage can was conducted, some paper towels were recovered, and laboratory analysis confirmed the second interpretation, that the security guard had indeed fired the gun and then cleaned his hands.
- Hair and fiber evidence further supported the physical contact of the security guard with the victim.
- Fingerprint evidence also confirmed the involvement of the security guard and his locations and activity at the scene.

Based on these and other analyses and direct crime scene observation, a comprehensive reconstruction report was issued by the local Police Forensic Science Laboratory. Subsequently, the suspected security guard pled guilty to this crime.

Forensic Science Today

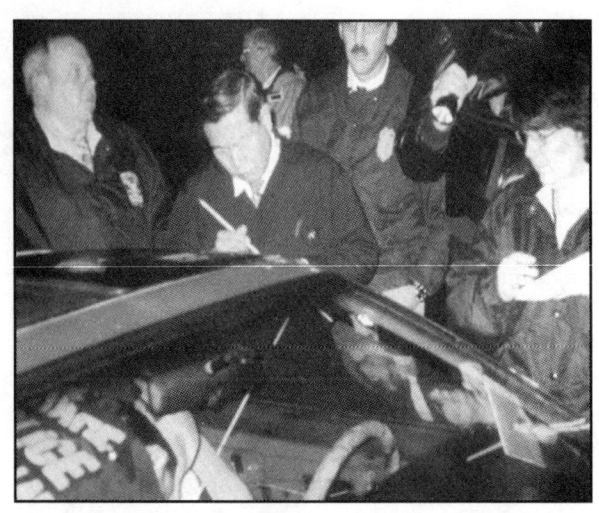

▲ Crime scene investigators reconstruct the bullet trajectories on a vehicle involved in a shooting. (Courtesy of Kenneth Zercie, Assistant Director of CSP Forensic Laboratory.)

WHO CONDUCTS THE RECONSTRUCTION?

Because reconstruction is the one of the last steps of investigating a crime, it is never conducted by responding police officers or crime scene technicians who first arrive on the scene. While police detectives sometimes find that gut instinct and "hunches" can lead to new investigative possibilities, they should never form theories while the criminal investigation is still ongoing. Rather, reconstruction is primarily the responsibility of forensic investigators who work in forensic laboratories or investigative units and have access to all available evidence and information. Reconstruction is also an important part of the court process as prosecutors use the information provided to them by forensic laboratories and detectives to build their case, and as forensic investigators testify on the stand to reconstruct the crime for the jury.

REQUIREMENTS FOR RECONSTRUCTION

When conducting a crime scene reconstruction, all available crime scene photographs, autopsy photographs, videotapes, measurements, notes, reports, and physical evidence related to the crime must be thoroughly examined. Complete and accurate documentation of a scene and the physical evidence will provide the best opportunity for later reconstruction. Whenever possible, the person responsible for reconstruction ideally should visit the crime scene as soon as the incident is reported, and make direct observation of the scene and patterns such as blood, glass, tire marks, and gunshot residue. Any reconstruction can only be as good as the information provided. This information may come from the crime scene, physical evidence, records, statements, witness accounts, and other databases.

Sometimes it is simply not possible to completely reconstruct the crime. In these situations, all available evidence and information may still leave gaps in the sequence of events. However, partial reconstructions—reconstructing certain facts or aspects of the event—can still be extremely valuable. Any information developed through the reconstruction process itself can often lead to the successful solution of a case.

Finally, crime scene reconstruction requires a good deal of teamwork. Investigators, forensic laboratory personnel, and other forensic specialists we learned about in earlier chapters of this book must cooperate and work together to document every important aspect of a crime scene, carry out the analysis of the physical evidence, and conduct a thorough and unbiased investigation of a case. Such cooperation will ensure that all the information necessary for reconstruction is available.

STAGES IN RECONSTRUCTION

Crime scene reconstruction is a scientific fact-gathering and logical evaluation process. These are the five separate stages commonly used in the process of reconstruction:

1. **Collecting Data.** All information obtained at the crime scene or from the victim, including the condition of the evidence, obvious patterns and impressions, and the condition of the victim, are reviewed, documented, and organized.

Chapter 24
Crime Scene Reconstruction

2. **Making a Conjecture.** A possible explanation of the events involved in a crime may be arrived at before making a detailed analysis of the evidence. However, this must not become the only explanation or "theory" being considered at this stage. It is important to keep an open mind, since there may be several possible explanations for what has been observed.

3. **Forming a Hypothesis.** Further accumulation of data is based on the examination of the physical evidence and the continuing crime scene investigation. This process leads to the formulation of an educated guess as to the probable course of events—a *hypothesis*.

4. **Testing the Hypothesis.** Once a hypothesis is formulated, further testing must be done to confirm or disprove the overall interpretation or specific aspects of the reconstruction. This stage includes comparisons of samples collected at the scene with known standards. At this time, additional instrumental, biological, chemical, and microscopic analyses may also be conducted. Verification of collected information through logical analysis is carried out at this stage.

5. **Forming a Theory.** Finally, only after a hypothesis has been thoroughly tested and verified by analysis can it be considered a plausible theory. Additional information may be acquired during the investigation about the condition of the victim or suspect, the activities of the individuals involved, accuracy of witness accounts, and other information about the circumstances surrounding the events. All of the investigative information, analyses and interpretation of physical evidence, and experimental results must be considered in testing and when attempting to verify the hypothesis.

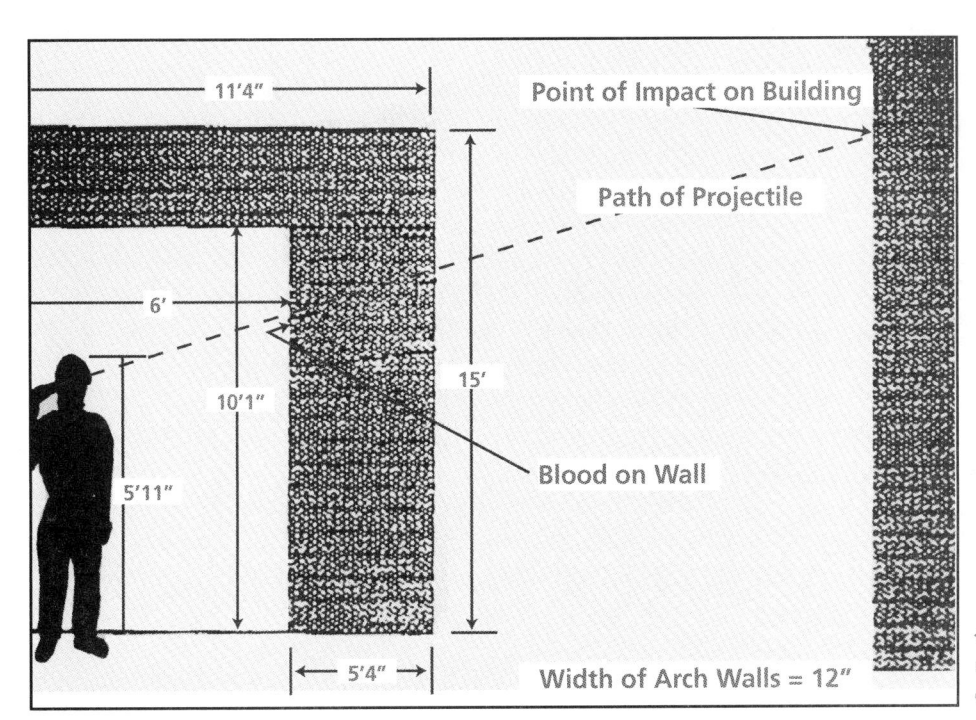

◀ A diagram shows the reconstructed bullet trajectory of a police shooting case.

CASE PROFILE
The Sherman Murder Case

The Victim
Ellen Sherman, Age 38.
Found strangled in her bedroom at dusk on Sunday, August 4, 1985. She was 5 ½ months pregnant at the time of her death. The man who discovered Ellen's body opened the bedroom door and found the room extremely cold because the air conditioner had been turned on full-force.

The Victim's Husband
Edward Robert Sherman, Age 42.
On Friday evening, Edward had left to go to Maine. When his wife's body was discovered he was sailing with shipmates in the Buzzards Bay area of Massachusetts.

Crime Scene
Ellen Sherman was found on her bed, face up and nude. Two deep ligature marks were plainly observed around her neck. A third thinner abrasion mark on her neck seemed pressed into her flesh. To the side of her bed were torn, blue, bikini-style underpants that were stretched out of shape. To the side of the bed was a bra and white slip. The underpants showed no traces of blood or semen upon initial inspections.

When police arrived at the murder scene, the bedroom was at a normal summer temperature. The first witness and the paramedics had not kept the room closed.

The Sherman's Business
They jointly ran a successful business called Ad Graphics, which hired artists and other employees. Ellen did most of the work at the company and owned 51% of the firm. Edward taught as a marketing instructor at Mohegan Community College.

The Sherman's Marriage
The Shermans were married in 1969 in New York. They had an unusual and troubled marriage. Edward Sherman became involved in a number of extramarital affairs that his wife allowed. At her husband's urging, Ellen had brief affairs with three other men.

In 1977 Edward became romantically involved with a married woman at the college where he taught. The woman divorced her husband in 1978, about the same time Edward convinced his wife Ellen to have an abortion. In 1982 his mistress also had an abortion.

For a while both women tolerated Edward's behavior. In 1984 his mistress became pregnant and gave birth in November. News of the baby upset Ellen and she won a commitment from her husband to have another child. She became pregnant in early 1985.

The acrimony between the two women grew and each pressured Edward to break off the relationship with the other woman. On Father's Day, June 16, Ellen and Edward had a family fight which lasted all day. Over the summer, Sherman's relationship with both women was on and off.

Ellen confided in friends that she was considering divorce. She told her friend Len Fredriksen that she would be able to come away with the business, the couple's house, and child support payments.

Evidence
The autopsy found food in Ellen's stomach. It had not yet moved into the small intestines. The food, linguine and a red seafood sauce, matched leftovers found in the refrigerator.

continued on next page...

Chapter 24
Crime Scene Reconstruction

Ellen Sherman was writing a novel which portrayed a husband who has a mistress, kills his wife, turns the air-conditioning on full force, and leaves the corpse behind in their locked house.

Sherman was oddly dressed for his sailing trip, wearing long pants and two long-sleeved shirts that were buttoned at the wrists. The other four sailing buddies wore shorts and summer shirts.

Witnesses

Len Fredriksen found Ellen's body. Edward Sherman called June (Ellen's friend) from the sailboat. Edward said he was concerned because he had been unable to reach his wife. He asked June to check on her. June asked Len to go to the house. Len Fredriksen was an employee of Ellen's and had also had an affair with her.

Jamie Albright was the 8-year-old daughter of one of Edward's sailing buddies. On Friday evening the group stopped at the Albright home before heading to Maine. Sherman asked to use the phone. Jamie listened on the extension as Sherman had what appeared to be a fake conversation. He said "We're leaving now …good-by honey…I love you too." On the other end Jamie heard nothing but a dial tone.

Stanley Mueller picked up Sherman to go sailing. Originally the group was scheduled to leave at 10:00 P.M. At Sherman's urging the time had been moved up to 7:30.

Following the appropriate stages in reconstruction described above, reconstruct what happened in the murder of Ellen Sherman. *Hint: see the end of this chapter to find out what really happened.

TYPES OF RECONSTRUCTION

There are five general types of reconstruction. These classifications are based on the nature of the crime, the questions that need to be answered, the types of events that have taken place, or the degree of involvement of the reconstructionist. The five types of reconstruction are as follows:

1. **Reconstruction of a Specific Type of Crime**
   - Homicide reconstruction
   - Arson scene reconstruction
   - Rape case reconstruction
   - White-collar crime reconstruction
   - Accident reconstruction
   - Other specific crime scene reconstructions

2. **Reconstruction of Specific Events**
   - Sequence determination
   - Directional determination
   - Position determination
   - Relational determination
   - Conditional determination
   - Identity determination

Forensic Science Today

3. **Degree of Reconstruction**
   - Total case reconstruction
   - Partial case reconstruction
   - Limited event reconstruction
   - Specific pattern reconstruction

4. **Specific Type of Physical Evidence Reconstruction**
   - Serological evidence reconstruction
   - Blood spatter pattern reconstruction
   - Firearms evidence reconstruction
   - Handwriting or document reconstruction
   - Gunshot residue pattern reconstruction
   - Injury pattern determination and reconstruction
   - Fingerprint position reconstruction
   - Footprints or shoeprints reconstruction
   - Tire mark reconstruction
   - Toolmark reconstruction

5. **Special Determinations**
   - Criminal profiling
   - MO, motive, and psychological determinations of perpetrators
   - Organized versus disorganized crime scene determination
   - Primary versus secondary crime scene determination
   - Scene profiling
   - Physical characteristics determination
   - Genetic profiling

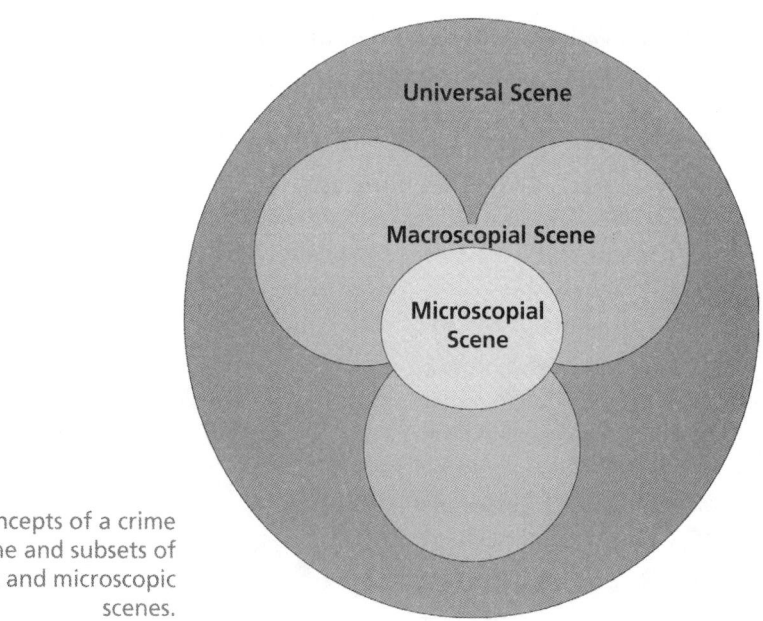

▶ The concepts of a crime scene and subsets of macroscopic and microscopic scenes.

CASE PROFILE (PART II)
How the Sherman Murder Case was Reconstructed

(1). Initial determination of the time of death was wrong. Investigators found that the air conditioner in the room was set on high to delay the body decomposition.

(2). Manner of the death is homicide. The cause of death is ligature strangulation.

(3). The ligature pattern on the victim's neck was identified by matching to the waistband of her panties.

(4). Evidence on Edward Sherman's computer hard disk revealed information about his plot to murder his wife.

TECHNICAL DEFINITIONS

Reconstruction: The scientific, objective process of determining the sequence of events, criminal activities, and logical predictions about what occurred during and after the crime.

Conjecture: A possible explanation of events in a crime scene that is not based on any scientific examination of evidence. One of the first stages of crime scene reconstruction; one of many explanations.

Hypothesis: An educated guess as to the probable course of events in a crime scene, based on the accumulation of data, examination of evidence, and continuing investigation. An intermediate stage of crime scene reconstruction.

Theory: A hypothesis that has been tested and proven to be true. A final stage in crime reconstruction.

Chapter 25

Jurisprudence: Law and Forensics

CHAPTER OBJECTIVES

- ✔ Understand and differentiate between the roles played by prosecutors and defense attorneys.
- ✔ Identify the steps needed to pursue a career as an attorney, if desired.
- ✔ Define the following vocabulary words and phrases: *jurisprudence*, *attorney*, *prosecutor*, *public defender*, and *Juris Doctor degree*.

Working as a criminal lawyer—either for the prosecution or the defense—is a good opportunity to apply the knowledge of forensic science in your career. Jurisprudence is the philosophy of law; and it also refers to the law as it is applied to a particular area. In the area of criminal law, a solid understanding of forensic science and a high ethical standard are essential, no matter which side you may be working for.

WHAT DO LAWYERS DO?

Lawyers (also known as attorneys) represent opposing sides in a legal dispute by applying their knowledge of the law to the facts and circumstances at hand. In the criminal field, lawyers may represent the state as prosecutors or the defendant as private or court-appointed defense attorneys. Attorneys also represent victims of crimes in civil cases for monetary damages (where the judge awards an amount of money that the defendant must pay to the victim to compensate for the crime or wrongful action committed). Finally, lawyers may also act as judges in criminal or civil trials.

Before a criminal case goes to trial, lawyers must not only study all the points of law applicable to the case, but they must also know their case inside and out, backwards and forward. All good trial lawyers know that preparation is the key to courtroom success, which means becoming very familiar with all the facts of the case. Lawyers must also become very familiar with forensic science so that they can utilize the evidence to the best extent in court.

Forensic Science Today

KEY TERMS

Jurisprudence
Attorney
Prosecutor
Public Defender
Juris Doctor Degree

Attorneys often consult a forensic expert to testify at trial. The expert will get on the stand and explain his or her background to the jury, and will then explain how the physical evidence was analyzed to come to a conclusion. The trial attorney will guide the expert's testimony to show how it proves either the defendant's guilt or innocence, depending on which side the attorney represents.

HOW DO I BECOME A LAWYER?

To become a lawyer, you first need a Bachelor's degree from an accredited university. While in college, it is also a good idea to strengthen the skills you will need in law school, especially your writing and analytical skills. Next, you must take the LSAT (Law School Aptitude Test) and achieve a high enough score to get in to the accredited law school of your choosing. The higher your LSAT score and undergraduate GPA, the more likely it is you will be accepted into your top choice of law school. Once a student is accepted into a law school, he or she must attend classes for three years before earning a Juris Doctor (J.D.) degree. After earning a J.D., law graduates must apply to the State Bar of the state in which they choose to live and practice law. This involves a lengthy application process, and passing the state Bar Exam. It generally takes three months to study for the Bar Exam, which is offered twice a year. Once you have passed the bar and completed your application, you can be admitted to practice law in your state.

TYPES OF EVIDENCE COMMONLY USED IN THE COURT OF LAW

Evidence—Anything that tends to prove or disprove an alleged fact.

- **Direct Evidence:** Establishes the element of a crime through an eyewitness or record of the criminal act.

- **Indirect (Circumstantial) Evidence:** Based on one fact being inferred from another fact.

- **Physical Evidence:** Actual tangible "things" or "patterns" resulting from an act.

Chapter 25
Jurisprudence: Law and Forensics

Prosecution or Defense?

Criminal lawyers work either for the prosecution or the defense. Which side you choose is up to you; it is also not uncommon for prosecutors or defense attorneys to switch sides mid-career. While in law school, you can do an internship or work as a law clerk at the local prosecutors or public defenders office. Many law schools also offer clinical experience in these environments to second and third year law students who can argue cases in court under the supervision of a practicing attorney. You can also choose to get work experience at a private criminal defense firm. All of these options offer an excellent opportunity to learn the ropes. Once you pass the Bar Exam and begin practicing law, it is also a good idea to take Continuing Legal Education courses in evidence and forensic science, when offered. Whether you choose to work for the prosecution or defense, integrity and a strong work ethic are traits of key importance.

Police testify in court only concerning the investigative facts. Forensic scientists testify in court only concerning the scientific findings.

TECHNICAL DEFINITIONS

Jurisprudence: The philosophy of law.

Attorney: Lawyer.

Prosecutor: An attorney who works for the government to bring criminals to justice.

Public Defender: An attorney who works for the government to represent and defend those accused of crimes whom cannot afford to hire a private attorney.

Juris Doctor Degree: The doctorate degree that is earned upon graduation from law school.

Glossary and Resources

Glossary

Accelerant—Any substance that aids in the development or spread of a fire, such as gasoline.

Accident Reconstructionist—An engineer who reconstructs the sequence of events during accidents and crimes from an engineering perspective.

Age Progression—The use of photographs of missing victims or suspects to show what a person who has not been seen in many years might look like today.

Algor Mortis – The change of temperature that occurs after death due to the lack of energy being produced and expended.

Angle of Impact—The angle at which a drop of blood approaches and hits the target surface.

Antemortem—Before death.

Anthropology—The study of humankind. *See also Forensic Anthropology.*

Arson—A fire that has been set intentionally.

Attorney—Lawyer.

Automated Fingerprint Identification System (AFIS)—A computerized database of fingerprints from people around the world.

Autopsy—A postmortem surgical procedure conducted by a medical examiner to determine the cause, manner, and mode of death by examining the internal organs of a corpse.

Bit—Comes from the phrase binary digit and refers to the ones and zeros that make up the foundation for storage of information in a computer.

Blood Alcohol Concentration—The concentration of ethyl alcohol in the blood. BAC is used as a measure of the degree of intoxication in an individual.

Blood Spatter—A bloodstain caused by medium- or high-velocity force impact of blood upon the target surface.

Blood Type—Also known as Blood Group, the system of classifying human blood according to the presence or absence of genetically determined antigens upon which immunological compatibility is based. This system has largely been replaced by DNA testing in most forensic laboratories.

Boiling Point—The temperature at which a liquid is transformed into a gas (at standard atmospheric pressure).

Bullet—A projectile that is pushed forward through the barrel of a gun upon detonation.

Byte – A collection of 8 bits of information. Some examples are: a) 0 0 0 0 0 0 0 0, b) 1 1 1 1 1 1 1 1, c) 0 1 0 1 0 1 0 1, and d) 1 0 0 1 0 1 1 0.

C

Caliber—The size of the interior diameter of the barrel of a handgun or rifle.

Carrion Insects—Any species of insect that feeds on dead organisms.

Cartridge Case—The metal container that holds the cartridge in a gun.

Cartridge—The ammunition package contained in a gun consisting of the bullet, gunpowder, and primer.

Chain of Custody—The correct procedure for ensuring that evidence is properly tracked, labeled, and identified at all times so that it may be admissible as evidence in a court of law.

Charring—Patterns left behind by a fire burning walls and other objects.

Chemist—A scientist who specializes in chemistry. Qualified forensic chemists often work as technical advisors to criminal investigation teams, especially in the case of clandestine laboratories.

Chromatography—A method of laboratory analysis to identify and quantify individual chemical compounds within a substance. *See also Gas Chromatography, Pyrolysis Gas Chromatography, Thin-Layer Chromatography.*

Clandestine Drug Lab—An illicit, secret laboratory where drugs are illegally made.

Classification of Evidence—Different ways of classifying and placing evidence in an organized fashion.

Color—The perception by the eye of various wavelengths of refracted light components.

CODIS (Combined DNA Index System)—The official law enforcement database in which DNA profiles of convicted felons as well as DNA found at unsolved crime scenes are kept for comparison.

Combustible—A material that easily catches fire.

Comparison—The process of comparing an item of evidence with known standards or alibi standards to determine where the evidence came from. *See also Known Standard.*

Competency—A legal term indicating that a person is mentally fit and has full legal capacity.

Composite—Drawings of individuals or evidence that are made up of a combination of parts that have been individually described.

Compression Toolmark—A toolmark that is caused by one object pressing into a surface. *See also Sliding Toolmark.*

Conditional Evidence—Evidence that demonstrates conditions of a crime scene, body, or physical stages of evidence.

Conjecture—A possible explanation of events in a crime scene that is not based on any scientific examination of evidence. One of the first stages of crime scene reconstruction; one of many explanations. *See also Hypothesis, Theory.*

Contact (Point-Blank) Range—When a gun is discharged directly against the target surface.

Controlled Substance—A drug available by prescription only, illegal if possessed in unlawful quantities or by unauthorized persons.

Glossary

Coroner—A county official, not always a medical doctor, who investigates and declares the official cause of death with the assistance of a forensic pathologist. *See also Medical Examiner, Forensic Pathologist.*

Cortex—The primary portion of the hair surrounding the medulla, covered by the cuticle.

Crime Scene Investigator—A professional trained in the field of forensic crime scene investigation who goes to the scene of a crime and documents, collects, and preserves evidence for forensic testing in a laboratory.

Crime Scene Survey—The initial "walk-through" conducted by investigators upon arriving at a crime scene.

Criminalistics—A catch-all term encompassing many different types of forensic specialties used to solve crime.

Cuticle—A layer of scales covering the hair shaft.

Decomposition—The process by which human and other animal remains disintegrate as a result of insect activity and other natural factors.

Density—The amount of mass per unit volume of a substance.

Digital Evidence—A broad term meant to include all electronic devices that contain binary information that may be useful in an investigation. Examples include: a) computers, b) cell phones, c) gaming stations, d) digital cameras, e) GPS Systems, f) MP3 Players, etc.

Directionality—The direction in which a drop of blood had been traveling prior to hitting the target surface and creating a bloodstain.

Deoxyribonucleic Acid (nuclear DNA)—A unique genetic identifier found in nucleated cells of the human body, composed of two strands of inherited material forming a double helix.

Document Examiner—A forensic professional who specializes in examining document evidence.

Drug-Facilitated Rape (DFR)—Sexual violence committed with the aid of drugs given to the victim by the perpetrator. DFR drugs create an amnesia-like state in the victim. *See also Rohypnol, Gamma Hydroxybutyrate (GHB).*

Electrostatic Lifting—A method used to lift dust residue imprints.

Engineer—A scientist who applies the principles of mathematics and science to solve problems. *See also Forensic Engineer.*

Entomologist—A scientist who studies insects. *See also Forensic Entomologist.*

Entomology—The study of insects.

Evidence—Anything that can be used to prove the facts of a matter in a court of law. Evidence can be something like a piece of clothing, the testimony of an eye-witness, tire tracks in the dirt, or even microscopic human cells.

Expert Witness—Someone who testifies in court as a qualified expert in a particular subject area.

Facial Reconstruction—The use of the human skull and knowledge of cranial anatomy to make a sculpture or draw a sketch of what the victim may have looked like while alive.

False Positive—A false field test result indicating that blood is present at a scene when in actuality no blood is present.

File Signature – A common series of bytes at the beginning of files of the same type. There is often, but not necessarily, a common series of bytes at the end of files of the same type.

File System—A set of rules that govern how data is to be saved, tracked, retrieved, and deleted from a digital storage device.

Forensic Anthropology—A specialty within the field of anthropology that focuses on identification of human remains for legal purposes. *See also Anthropology.*

Forensic Engineer—A scientist who uses engineering skills to reconstruct accidents and other events. *See also Engineer.*

Forensic Entomologist—A scientist who studies the life cycles of insects found on a decomposing body to determine how much time has passed since death. *See also Entomologist, Entomology.*

Forensic Pathologist—A medical doctor who studies disease and injury to determine the cause and manner of death. A forensic pathologist sometimes also performs autopsies. *See also Coroner, Medical Examiner.*

Forensic Science—The direct application of scientific knowledge and techniques to matters of law. Forensic scientists use logic and follow careful scientific steps in order to solve issues in criminal and civil litigation.

Forensic Toxicologist—A scientist who identifies and analyzes poisons, drugs, and chemicals and their effects on the human or animal body in relation to legal issues. *See also Toxicology, Postmortem Toxicology.*

Forgery—A faked or altered document.

G

Gamma Hydroxybutyrate (GHB)—A common Drug-Facilitated Rape (DFR) drug used by sexual predators to facilitate rape. *See also Drug-Facilitated Rape (DFR), Rohypnol.*

Gas Chromatography—A type of chromatography that separates liquids and gases. *See also Chromatography, Pyrolysis Gas Chromatography, Thin-Layer Chromatography.*

Gauge—The size of the interior diameter of the barrel of a shotgun.

Graphical User Interface (GUI)—An interface that uses icons, pictures and symbols to allow people to interact with a computer. Older and more primitive computers and electronic devices use a text interface. GUI lets you "drop and drag," and "point and click." Text interface requires learning operating system and programming languages like MS DOS.

Grooves—Indentations in the interior of a gun barrel caused by rifling. *See also Lands, Striation.*

Gunshot Residue (GSR)—The residue left behind on the fingers, hands, and clothing of someone who has shot a gun. If a gun is fired on contact or near contact, then GSR may also be found on the victim's body or clothing.

H

Handwriting Analysis—A specialty within forensic science in which handwriting experts compare samples of writing to determine the document's true author.

Hash Value—A fixed length hexidecimal value that is calculated by putting a stream of data through a mathematical algorithm. The length of the Hash Value is dependant upon which Hash algorithm that is used.

Head Space—Space left at the top of a container used for storing and transporting chemical evidence, to allow for evaporation without causing explosion under pressure.

Hexidecimal—A numeral system that uses 16 (instead of 10) as a base. It uses letters in addition to numbers to represent these base values. Counting to 15 in hexadecimal looks like this: 0, 1, 2, 3, 4, 5, 6, 7, 8, 9, A, B, C, D, E, F. This number system is especially useful in computer science because four bits of information represents 16 combinations of 1's and 0's. Each combination can be assigned a unique value from the zero to F range of choices.

Hypothesis—An educated guess as to the probable course of events in a crime scene, based on the accumulation of data, examination of evidence, and continuing investigation. An intermediate stage of crime scene reconstruction. *See also Conjecture, Theory.*

Identification—The process of identifying the class characteristics of an item of evidence by taking measurements and analyzing its physical, chemical, and biological properties.

Ignition Device—A device used to start a fire or explosion.

Image Modification—Methods of digital enhancement of existing photographs to make them more useful to investigators.

Impression—A three-dimensional indentation or pattern that is produced when an object such as a shoe, tire, or tool presses into a softer substance like snow, mud, or sand.

Imprint—A two-dimensional pattern found on a hard surface, left behind by residue from a shoe, tire, or other object leaving behind a pattern.

Index of Refraction—The ratio of the velocity of light in a vacuum to its velocity in the substance.

Individualization—The process of showing that an item of evidence is unequivocally derived from a unique source.

Inorganic—Not derived from living matter. *See also Organic.*

Juris Doctor Degree—The doctorate degree that is earned upon graduation from law school.

Jurisprudence—The philosophy of law.

Known Hair Standard—Samples of hair collected from a person's head to compare with hairs found at a crime scene.

Known Standard—Something from a known source to which evidence from an unknown source is compared. *See also Comparison.*

Known Writing Sample—A true and genuine sample of the writing of a person (or printing of a computer printer) that is compared to the document that has been seized as evidence to determine authenticity. *See also Handwriting Analysis.*

Lands—The portions of a gun barrel interior that have not been cut by rifling, which appear raised relative to grooves. *See also Grooves.*

Latent Fingerprints—Fingerprints that are not visible to the naked eye. *See also Patent Fingerprints, Plastic Fingerprints.*

Leucomalachite Green—A chemical reagent that is used to field test at crime scenes for the presence of blood, which immediately turns a greenish-blue color if blood is present.

Linkage Theory—A theory of crime scene investigation based on the idea that if investigators can establish positive links between the physical evidence and/or victim and/or crime scene and/or suspect, then the case can be resolved.

Livor Mortis—The discoloration of the skin after death caused by gravity-induced pooling of red blood cells in the small blood vessels. (Also known as lividity.)

Locard's Theory of Exchange—The principle that transfer evidence is created whenever two surfaces come into contact with one another. *See also Transfer Theory.*

Luminol—A chemical reagent used to field test at crime scenes for the presence of blood, which must be applied to suspected bloodstains in total darkness, and makes bloodstains luminesce if blood is present.

M

Malware—Any software designed to infiltrate a computer system and make changes without the owner's knowledge and consent. These programs are malicious software. Over time these two words were combined into Malware. Sometimes "computer virus" is used as a synonym for Malware. In fact, a virus is a specific type of malware. Other types include worms, trojan horses, rootkits, and spyware.

Mass Spectrometry—A method of identifying the chemical structure of unknown substances by aiming high-energy electrons at the substance to determine mass.

Medical Examiner—A forensic pathologist who is a local government official responsible for performing autopsies to determine the cause and manner of death. *See also Coroner, Forensic Pathologist.*

Medulla—The central core of the hair.

Melting Point—The temperature at which a solid is transformed into a liquid (at standard atmospheric pressure).

Mineralogical—Having to do with the mineral properties of a substance.

Mitochondrial DNA—The DNA found in the mitochondria of a cell, inherited from the mother only.

N

Necrophagus Insects—Species of insects that feed directly on a corpse.

O

Odontology—Dentistry.

Omnivorous Insects—Insects that feed on both the human corpse and other insects.

Operating System—A collection of computer programs that control memory, handle system requests, control the devices used to input and output data, and provide a Graphical User Interface (GUI).

Organic—Derived from living matter. *See also Inorganic.*

P

Parasitic Insects—Insects that rely upon other insects as a place to lay their eggs and for other purposes.

Patent Fingerprints—Fingerprints that are clearly visible. *See also Latent Fingerprints, Plastic Fingerprints.*

Glossary

Pathology—The study of disease and pathogens. *See also Forensic Pathology.*

Pattern Evidence—Evidence resulting when an object or a person come into contact with a surface, leaving behind a pattern, or resulting from a physical change in shape or pattern.

pH—A measure of the acidity or basicity of a substance. On a scale of 0 to 14, 7.0 is neutral; less than 7.0 is acidic; and greater than 7.0 is basic.

Phenolphthalin—A chemical reagent that is used to field test at crime scenes for the presence of blood, which turns pinkish-red within 15 seconds if blood is present.

Physical Evidence—Anything that may be recognized, documented, collected, and analyzed in order to determine the facts of what happened during a crime.

Plastic Fingerprints—Fingerprints left when a person presses a finger into a soft substance. *See also Latent Fingerprints, Patent Fingerprints.*

Point of Convergence—The point at which all lines coming from individual bloodstains come together, used to measure the point of origin of the blood at a crime scene. *See also Point of Origin.*

Point of Fire Origin—The point at which a fire is first ignited.

Point of Origin (Blood)—The place where a source of blood released blood droplets into the air prior to the blood hitting the target surface. *See also Point of Convergence.*

Polymerase Chain Reaction (PCR)—The method used to amplify (copy) small quantities of DNA samples so that they may be further analyzed.

Postmortem Interval—The amount of time that has passed since a person has died.

Postmortem Toxicology—The study of how drugs or other chemicals contributed to a person's death. *See also Toxicology, Forensic Toxicology.*

Postmortem—After death.

Predatory Insects—Species of insects that hunt and eat other insects.

Primary Crime Scene—The original place where the crime occurred. *See also Secondary Crime Scene.*

Primary Transfer—The transfer of evidence directly from one surface to another. *See also Secondary Transfer.*

Primer—A component of gunshot residue, the material that is detonated in the powder in the casing, resulting in the explosion that propels the bullet forward.

Profiling—An analysis based upon psychological and statistical factors of a criminal used to determine the general characteristics of the most likely suspect for a crime.

Propellant—Also known as gunpowder, the substance that propels the bullet forward upon detonation of the primer. *See also Primer.*

Prosecutor—An attorney who works for the government to bring criminals to justice.

Psychiatrist—A medical doctor who specializes in mental disorders and may prescribe medication.

Psychologist—A professional who studies human behavior and is licensed to provide therapeutic counseling services.

Public Defender—An attorney who works for the government to represent and defend those accused of crimes that cannot afford to hire a private attorney.

Forensic Science Today

Pyrolysis Gas Chromatography—A form of chromatography that analyzes solids whereby the sample is heated to a temperature at which the material decomposes into its gaseous components. *See also Chromatography, Gas Chromatography, Thin-Layer Chromatography.*

Q

Questioned Document—Any document whose origin or authenticity is in question. In other words, is this document real? Does it really come from where the suspect says it came from?

R

Recognition—The process of recognizing the potential value of items of evidence related to a crime.

Reconstruction—The scientific, objective process of determining the facts of a crime and sequence of events, criminal activities, and logical predictions about what occurred during and after the crime, based on a thorough analysis of all physical evidence, witness statements, and other facts.

Refractive Index—The ratio of the speed that light travels through a vacuum to the speed it travels through a substance.

Restriction Fragment Length Polymorphism (RFLP)—The variations in lengths of DNA fragments that are created by adding a special "restriction" enzyme, which cuts the DNA. *See also DNA.*

Ridge Impressions—The pattern left behind by an individual's fingerprint; another way of saying "fingerprint."

Rifling—Cut grooves going down the interior surface of the length of the gun barrel, which cause the bullet to spiral as it moves through the air.

Rigor Mortis—The temporary stiffening of the body after death after the heart stops pumping oxygen to the muscle cells.

Rohypnol—A common Drug-Facilitated Rape (DFR) drug used by sexual predators to facilitate rape. *See also Drug-Facilitated Rape (DFR), Gamma Hydroxybutyrate (GHB).*

Root—The bulb of the hair found in the hair follicle, the only part of the hair containing nuclear DNA.

S

Secondary Crime Scene—An additional crime scene other than the primary crime scene where evidence may also be found. *See also Primary Crime Scene.*

Secondary Transfer—The transfer of evidence from one surface to another through an intermediate object or person. *See also Primary Transfer.*

Serology—The study of fluids remaining after coagulation, particularly in blood.

Short Tandem Repeat (STR) Markers—A specific portion of variation DNA that is compared, together with other specific portions, with other DNA samples. *See also DNA.*

Sieve—A tool used by forensic scientists to strain out and separate the various components of soil.

Skeletal Remains—The completely decomposed remains of an animal that have been reduced to bone through removal of tissues and fats by insects and other natural processes.

Sliding Toolmark—A toolmark that is caused by one object sliding against a surface. *See also Compression Toolmark.*

Soil—A mixture of both organic and inorganic matter. It may contain a wide variety of substances derived from minerals, plant matter, glass, paint chips, plastic, and other trace materials.

Glossary

Solubility—The ability of one substance to dissolve into a solvent.

Spectrophotometry—A technology that characterizes different chemical substances by measuring how much radiation from the electromagnetic spectrum is absorbed by a particular chemical.

Spectroscopy—The study of spectra in order to ascertain the chemical composition of a substance and the physical properties of atoms, ions, and molecules.

Striation—Scratch marks made on a bullet as it is fired from a rifled gun barrel, caused by the shape and markings of the interior of the gun barrel.

Synthetic Drug—A drug that is made in a laboratory, and does not occur in nature.

Tandem Repeat Sequence—The repeating sequences of amino acids comprising DNA. *See also DNA.*

Target Surface—The surface upon which drops of blood land.

Terminal Velocity—The constant speed reached by a falling object when the downward gravitational force is equal to the frictional air resistance of the substance through which it is falling.

Tetramethylbenzidine—A chemical reagent that is used to field test at crime scenes for the presence of blood, which turns an intense blue color if blood is present.

Theory—A hypothesis that has been tested and proven to be true. A final stage in crime reconstruction after conjecture and hypothesis. *See also Conjecture, Hypothesis.*

Thin-Layer Chromatography—A type of chromatography that deposits the sample and solvent on a plate of glass thinly coated with adsorbent. *See also Chromatography, Gas Chromatography, Pyrolysis Gas Chromatography.*

Toolmark—Any cut, scrape, impression, gouge, scratch, indentation, or other marking left on an object or surface by another object being forced into or moved across it.

Toxicology—The study of poisons, drugs, chemicals, and their effects. *See also Forensic Toxicology, Postmortem Toxicology.*

Trailer—Material designed to spread a fire from one point to another.

Trajectory—The path a bullet takes in the air after being fired from a weapon.

Transfer Theory—Pattern evidence and trace evidence can be transferred between the victim and suspect, with other objects, and the crime scene.

Transfer/Trace Evidence—Evidence that is produced by physical contact between two objects or two surfaces, when material is transferred from one object to another.

Transient Evidence—Any type of physical evidence that is temporary in nature.

Unallocated Space—Area on a digital storage device that is currently not assigned to a computer file.

Vapor Point—The temperature at which a particular liquid chemical is transformed into a gas.

Vapor Pressure—The pressure exerted when a solid or a liquid is in equilibrium with its own vapor.

Vapor—A substance that, though present in the gaseous phase, generally exists as a solid or a liquid at room temperature.

Variable Number Tandem Repeat (VNTR)—The number of tandem repeat sequences in an individual's DNA. *See also DNA, Tandem Repeat Sequence.*

Resources

BOOKS

American Academy of Forensic Sciences. *So You Want to Be a Forensic Scientist!* Colorado Springs, Colorado: The Forensic Sciences Foundation, Inc. 1996.

Baden, Michael and Marion Roach. *Dead Reckoning: The New Science of Catching Killers.* New York, New York: Simon & Schuster. 2002.

Bass, Bill and Jon Jefferson. *Death's Acre: Inside the Legendary 'Body Farm.'* New York, New York: Penguin Group, Inc. (USA). 2003.

Doswell, Paul. *Investigating Murder Mysteries.* Great Britain: Harcourt Education, Ltd. 2004.

Evans, Colin. *Murder 2: The 2nd Casebook of Forensic Detection.* Hoboken, New Jersey: John Wiley & Sons, Inc. 2004.

Ferllini, Roxana. *Silent Witness—How Forensic Anthropology is Used to Solve the World's Toughest Crimes.* Buffalo, New York: Firefly Books (U.S.) Inc. 2002.

Genge, N.E. *The Forensic Casebook.* New York, New York: Ballantine Books. 2002.

Goff, M. Lee. *A Fly for the Prosecution—How Insect Evidence Helps Solve Crimes.* Cambridge, Massachusetts: Harvard University Press. 2001.

Heath, David. *Crime Lab Technician.* Mankato, Minnesota: Capstone Press. 1999.

Lee, Henry. *Physical Evidence in Forensic Science.* Tucson, Arizona: Lawyers & Judges Publishing Company, Inc. 2000.

Lee, Henry, and Jerry Labriola. *Famous Crimes Revisited—From Sacco-Vanzetti to O.J. Simpson.* Southington, Connecticut: Strong Books. 2001.

Lee, Henry, and Thomas W. O'Neill. *Cracking Cases—The Science of Solving Crimes.* Amherst, New York: Prometheus Books. 2002.

Lee, Henry, and Thomas W. O'Neill. *Cracking More Cases: The Forensic Science of Solving Crimes.* Amherst, New York: Prometheus Books. 2004.

Lee, Henry, Timothy Palmbach, and Marilyn T. Miller. *Henry Lee's Crime Scene Handbook*. San Diego, California: Academic Press. 2001.

Lee, Henry, and Frank Tirnady. *Blood Evidence—How DNA Is Revolutionizing the Way We Solve Crimes*. Cambridge, Massachusetts: Perseus Publishing. 2003.

Mann, Robert, and Miryam Williamson. *Forensic Detective: How I Cracked the World's Toughest Cases*. New York, New York: Ballantine Books. 2006.

Owen, David. *Hidden Evidence: Over 40 True Crimes and How Forensic Science Helped Solve Them*. London: Quintet Publishing. 2000.

Platt, Richard. *Crime Scene: The Ultimate Guide to Forensic Science*. London: Dorling Kindersley Ltd. 2003.

Rainis, Kenneth G. *Crime-Solving Science Projects—Forensic Science Experiments*. Berkeley Heights, New Jersey: Enslow Publishers, Inc. 2000.

Taylor, Karen T. *Forensic Art and Illustration*. Boca Raton, Florida: CRC Press, LLC. 2000.

Thomas, Peggy. *Forensic Anthropology—The Growing Science of Talking Bones*. New York, New York: Facts On File, Inc. 2003.

White, Peter (Ed). *Crime Scene to Court—The Essentials of Forensic Science*. Cambridge, United Kingdom: The Royal Society of Chemistry Information Services. 1998.

ADDITIONAL MATERIALS

CD-ROM *What Every Law Enforcement Officer Should Know About DNA Evidence: Best Practices for Identification, Preservation, and Collection of DNA Evidence at the Crime Scene—Beginning and Advanced Level Modules*. Sponsored by the National Commission on the Future of DNA Evidence and National Institute of Justice. Distributed by U.S. Department of Justice, Office of Justice Programs, National Institute of Justice, Washington, D.C. 20531.

CD-ROM *InterFIRE*. Available from American Re-Insurance Company. This is an interactive CD and website arson investigation program produced as a government and private industry partnership. For further information contact Rod Ammon, President, Stonehouse Media, Inc. by email at **rod@stonehousemedia.com**.

VIDEO *Motive, Means, and Opportunity, A Guide to Fire Investigation*. Produced by Joseph Toscano. For ordering information contact Jennifer Bradley, Corporate Marketing Dept., American Re-Insurance Company, 555 College Road East, Princeton, New Jersey, or email to **jbradley@amre.com**.

WEBSITES

All websites listed below are accurate as of the time of publication but are subject to change over time. If a link fails to take you to the correct website, we recommend a search of the topic on your web browser.

General Forensic Websites

Learn more about forensic science and famous cases, and take a virtual tour of a forensic laboratory. www.drhenrylee.com

A comprehensive forensic entomology website. www.forensicentomology.com

A comprehensive forensic science website. www.forensicdna.com

Carpenter's forensic science resources. www.tncrimlaw.com/forensic

Crime statistics. www.ojp.usdoj.gov/bjs/

TruTV. true crime website. www.crimelibrary.com

Forensic resource site edited by Andre A. Moenssens, Douglas Stripp Missouri Professor of Law Emeritus at the UMKC School of Law. www.forensic-evidence.com

Kruglick's forensic resource and criminal law search site. www.bioforensics.com/kruglaw/

Reddy's forensic home page. www.forensicpage.com

Swiss website on criminal profiling. www.criminalprofiling.ch/introduction.html

Zeno's forensic site. www.forensic.to/

Forensic Education

George Washington University Department of Forensic Sciences. Provides an extensive list of web links to issues and specialties in forensic science. www.gwu.edu/~forensic/resources/index.cfm

Council on Forensic Science Education. www.criminology.fsu.edu/COFSE/default.html

Extensive online list of colleges and universities across the United States that offer courses and degree programs in forensic science. www.aafs.org/default.asp?section_id=resources&page_id=colleges_and_universities

John Jay College of Criminal Justice. www.jjay.cuny.edu

National Center for Forensic Sciences: A program by the National Institute of Justice hosted by the University of Central Florida. www.ncfs.org/home.html

University of New Haven. www.newhaven.edu/5765/

International Forensic Organizations

A global police communication system with criminal databases, analytical services, and support for police around the world. www.interpol.int

Australian and New Zealand Forensic Science Society. www.anzfss.org.au/

British Forensic Science Society. www.forensic-science-society.org.uk

Canadian Society of Forensic Science. www.csfs.ca

International Association of Bloodstain Pattern Analysts. www.iabpa.org

International Association for Chemical Testing. www.iactonline.org

International Association of Forensic Genetics. www.isfg.org

International Association of Forensic Nurses. www.forensicnurse.org

International Association of Forensic Science. www.iafs2008.com

International Association of Forensic Toxicologists. www.tiaft.org

International Association for Identification. www.theiai.org

International Association for Microanalysis. www.iamaweb.com

National Institute of Forensic Sciences (Australia). www.nifs.com.au/home.html

Royal Society of Medicine (United Kingdom). www.roysocmed.ac.uk

National Forensic Organizations and Resources

American Academy of Forensic Psychology and American Board of Forensic Psychology. www.abfp.com

American Academy of Forensic Sciences. www.aafs.org

American Board of Criminalistics. www.criminalistics.com

American Board of Forensic Anthropology. www.theabfa.org/index.html

American Board of Forensic Odontology. www.abfo.org

American Board of Forensic Toxicology. www.abft.org

American Board of Pathology. www.abpath.org

American Society of Crime Laboratory Directors (ASCLD), a nonprofit organization dedicated to providing management leadership to laboratories. www.ascld.org

American Society of Crime Laboratory Directors Laboratory Accreditation Board (ASCLD-LAB), the body that offers accreditation to laboratories worldwide. There are currently 357 Laboratories accredited nationally and internationally. Seventy or more are accredited by ISO standards. www.ascld-lab.org

American Society of Forensic Odontology. www.asfo.org

American Society of Questioned Document Examiners. www.asqde.org

Association of Firearm and Tool Mark Examiners. www.afte.org

Association of Forensic DNA Analysts and Administrators. www.afdaa.org

C.A. Pound Human Identification Laboratory. www.anthro.ufl.edu/capoundlab.shtml

Laboratory Services of the Federal Bureau of Investigation (FBI). www.fbi.gov/hq/lab/labhome.htm

National Association of Document Examiners. www.documentexaminers.org

National Association of Medical Examiners. www.thename.org

National Institute of Justice: The research, development, and evaluation agency of the U.S. Department of Justice. www.ojp.usdoj.gov/nij

Society of Forensic Toxicologists. www.soft-tox.org

Website of Karen T. Taylor, forensic artist. www.karenttaylor.com

Young Forensic Scientists Forum. www.aafs.org/yfsf/index.htm

Regional and State Forensic Organizations

Alaska State Crime Lab. www.dps.state.ak.us/CrimeLab/

California Association of Criminalists. www.cacnews.org

California Association of Toxicologists. www.cal-tox.org

Connecticut Forensic Science Laboratory. www.ct.gov/dps/cwp/view.asp?a=2155&Q=296214&dsftns=43911

Mid-Atlantic Association of Forensic Scientists. www.maafs.org

Midwestern Association of Forensic Scientists. www.mafs.net

New Jersey Association of Forensic Scientists. www.njafs.org

Northeastern Association of Forensic Scientists. www.neafs.org

Northwest Association of Forensic Scientists. www.nwafs.org

Southeastern Association of Forensic Document Examiners. www.safde.org

Southeastern Association of Toxicologists. www.sat-tox.org

Southern Association of Forensic Scientists. www.southernforensic.org

Southwestern Association of Forensic Scientists. www.swafs.us

DNA Websites

Information about PCR analysis, including animation of the process. http://allserv.rug.ac.be/~avierstr/principles/pcr.html

Innocence Project, working for post-conviction DNA testing for wrongfully convicted prisoners throughout the United States and around the world. www.innocenceproject.org

STR DNA internet database. www.cstl.nist.gov/div831/strbase

Crime Prevention

A network of over 2,000 police chiefs, sheriffs, prosecutors and crime survivors dedicated to preventing crime and violence. www.fightcrime.org

Brady Campaign to Prevent Gun Violence. www.bradycenter.org

Men who want to end sexual violence. www.menagainstsexualviolence.org

Mending the Hoop: a Native American program that provides training and technical assistance to American Indians and Alaskan Native relations to eliminate violence in the lives of women and their children. www.msh-ta.org

Mobilizes youth to prevent men's violence against women. www.mencanstoprape.org

National Crime Prevention Council. www.ncpc.org

Self-defense tips and tricks for women. www.safetyforwomen.com

Taking action against substance abuse and gun violence. www.jointogether.org

Restorative Justice

Information about justice that focuses on the healing of the wounds of victims and their communities. www.restorativejustice.org

Restore Program, a model restorative justice program in Tucson, Arizona. www.restoreprogram.publichealth.arizona.edu/default.htm

Crime Victims and Survivors—General

National Center for Victims of Crime, provides assistance and information to all crime victims. www.ncvc.org

Helping survivors of violent crime find support, justice, and healing. www.witnessjustice.org

A database of grassroots organizations throughout the United States that provide services to crime victims. www.crime-victims.org

Tips and advice for survivors of crime provided by the Seattle Police Department. www.cityofseattle.net/police/prevention/crimesurvivor/default.htm

Child Abuse Treatment and Prevention

Family Violence Prevention Fund. www.endabuse.org

Resources for the treatment and prevention of child abuse. www.childhelpusa.org

Sexual Violence: Rape, Incest, and Sexual Abuse

Rape, Abuse, and Incest National Network: Provides an online directory of counseling services across the country for survivors of sexual violence. Also staffs a 24-hour hotline to let victims know where the nearest rape crisis center is located (1-800-656-HOPE). www.rainn.org

International message board for survivors of sexual violence, female and male, around the world. www.pandys.org

Overcoming the sexual victimization of boys and men. www.malesurvivor.org

Survivors of Incest Anonymous. www.siawso.org

Legal Resources

Consumer's guide to legal help provided by the American Bar Association. www.findlegalhelp.org

Helping people find solutions to legal issues. www.public.findlaw.com

Internet directory of lawyers. www.lawyers.com

Network of attorneys and associated professionals dedicated to facilitating civil court actions brought by crime victims. www.victimbar.org

Provides legal products and resources such as books, slide calculators, software, and reference works. www.lawyersandjudges.com

About the Authors

Dr. Henry C. Lee is one of the world's foremost forensic scientists. Dr. Lee's work has made him a landmark in modern-day criminal investigations. He has been a prominent player in many of the most challenging cases of the last 45 years. Dr. Lee has worked with law enforcement agencies in helping to solve more than 6000 cases. In recent years, his travels have taken him to England, Bosnia, China, Brunei, Bermuda, Middle East, South America and other locations around the world.

Dr. Lee's work figured prominently in the Kennedy assassination, Wood chipper murders, Vincent Foster's death investigation and O. J. Simpson trial, and in convictions of the "Woodchopper" murderer as well as hundreds of other murder cases. Dr. Lee has assisted local and state police in their investigations of other famous crimes, such as the murder of Jon Benet Ramsey in Boulder, Colorado, the 1993 suicide of White House Counsel Vincent Foster, the murder of Chandra Levy, the kidnapping of Elizabeth Smart and the reinvestigation of the Kennedy assassination.

Dr. Lee is currently the Chief Emeritus for the Scientific Services and was the Commissioner of Public Safety for the State of Connecticut from 1998 to 2000 and served as Chief Criminalist for the State of Connecticut from 1979 to 2000. Dr. Lee was the driving force in establishing a modern State Police Forensic Science Laboratory in Connecticut.

In 1975, Dr. Lee joined the University of New Haven, where he created the school's Forensic Sciences program. He has also taught as a professor at more than a dozen universities, law schools, and medical schools. Dr. Lee lectures throughout the country and world to police, Universities and civic organizations. He has authored hundreds of articles in professional journals and has co-authored more than thirty books, covering such area as DNA, Fingerprints, Trace Evidence, Crime Scene Investigation and Crime Scene Reconstruction. His recent books, *Famous Crimes Revisited*, *Cracking Cases*, *Blood Evidence*, and *Dr. Henry Lee's Forensic Files* have been well received by the public.

Dr. Lee has been the recipient of medals and awards, including the 1996 Medal of Justice from the Justice Foundation, and the 1998 Lifetime Achievement Award from the Science and Engineer Association. He has also been the recipient of the Distinguished Criminalist Award from the American Academy of Forensic Sciences; the J. Donero Award from the International Association of Identification, Alice Island Medal in 2005, Medal of Justice from Justice Foundation, Police Medal from Taiwan Government, Gusi Award from Philippines in 2008 and in 1992 was elected a distinguished Fellow of the AAFS and Life Active Distinguished Member of IAI in 2008.

Dr. Lee was born in China and grew up in Taiwan. Dr. Lee first worked for the Taipei Police Department, attaining the rank of Captain. With his wife, Margaret, Dr. Lee came to the United States in 1965, and he earned his B.S. in Forensic Science from John Jay College in 1972. Dr. Lee continued his studies in biochemistry at NYU where he earned his Masters Degree in 1974 and Ph.D. in 1975. He has also received special training from the FBI Academy, ATF, RCMP, and other organizations. He is a recipient of sixteen honorary Doctorate Degrees from Universities in recognition of his contributions to Law and Science. Dr. and Mrs. Lee have been married for forty-seven years and have two grown children, a daughter, Sherry, and a son, Stanley.

George Taft grew up in Aquebogue, New York. He has been in the Forensic Science field for over forty years in the Texas and Alaska Crime Laboratories. George was most recently Director of the State of Alaska Crime Laboratory. He has lectured internationally, is a member of the American Academy of Forensic Sciences, the International Association for Identification and a Diplomat in the American Board of Criminalistics. George has served on the Board of the American Society of Crime Laboratory Directors and is currently serving on the American Society of Crime Laboratory Directors Laboratory Accreditation Board (ASCLD-LAB).

He has spoken to numerous secondary school groups in regard to forensic science. George has a degree in Chemistry from Alfred University and a degree in Anthropology from the University of Alaska Anchorage.

Kimberly A. Taylor is a freelance writer who graduated cum laude from Wellesley College and earned her Juris Doctor degree from the University of Arizona College of Law. After practicing law for several years she now devotes herself full-time to raising a family and writing. Kimberly has also written several chapters on legal ethics in the *High Court Case Summaries* law study guide series, and co-edited the book *Toxic Mold Litigation*.

Jeanette Hencken is in her sixteenth year as a chemistry and forensic science teacher at Webster Groves High School in St. Louis County, MO. She graduated from the University of Missouri-St. Louis with a Bachelor's degree in chemistry and is a certified chemistry and physics teacher. She has received the St. Louis Area Chemistry Teacher of the Year award in 2003. Jeanette is also a member of the Editorial Advisory Board and contributing author of "The Forensic Teacher."

Jeanette has worked on Steering Committees for eight Forensic Science Conferences in St. Louis, Missouri, Anchorage, Alaska and Rapelje, Montana. The conferences were organized to educate teachers and law enforcement personnel about forensic science and methods used to teach forensic science. She has also given presentations and workshops at the National Science Teachers Association, Science Teachers of Missouri, and Forensic Science Education conferences. She is also the owner of the consulting business, Crime Science Curriculum.

Index

A

accelerant, 33, 36, 155-157, 159-163, 166, 257

accident reconstruction, 232, 234, 247

age progression, 7, 183-184, 189, 257

analysis of physical evidence, 6, 10, 29, 36, 40, 43, 244, 264

arson and fire evidence, 36, 155, 160-161

attorney, 2-3, 6, 8, 22, 30, 38-39, 42, 93-96, 117, 147, 205, 238-239, 251-253, 257, 263, 267, 273

Automated Fingerprint Identification System (AFIS), 45-46, 51, 257

autopsy, 2, 4, 7, 20, 26, 106, 145, 147-148, 187, 191-197, 200-201, 206, 219, 222, 226, 244, 246, 257, 260, 262

B

Belynda Kay Tillery, 188

bite marks, 7, 88, 193, 225-226, 228-229

blood evidence, 3, 77-78, 80-82, 268, 275

blood spatter, 5, 34, 59, 93, 101-102, 104, 106, 109, 148, 248, 257

bloodstain patterns, 77, 89, 101-102, 104, 106-108, 243, 270

Body Farm, 216, 267

boiling point, 160-161, 163-164, 167, 207, 257

bullet and cartridge databases, 146

bullet trajectories, 139, 142, 148-149, 153, 243-245

burn patterns, 155-156

C

C.S.I. (TV show), 1, 10, 186

careers, 1, 5, 9-11, 178, 180, 183, 189, 191, 199, 208, 211, 215, 217, 225, 231, 235, 237, 251

carrion insects, 192, 217-219, 223, 258

chain of custody, 13-14, 27, 29-30, 36-37, 68-69, 89, 97, 258

chemical evidence, 63, 70, 162-166, 194, 261

chromatography, 64, 71, 152, 155-156, 161-164, 166-167, 199-200, 206-207, 209, 258, 260, 264-265

clandestine drug labs, 67-68, 70, 75, 169, 258

class evidence, 40

classification of evidence, 6, 29-30, 36, 43, 258

Combined DNA Index System (CODIS), 85-86, 95, 98-99, 258

comparison, 29-30, 41, 43, 54, 58, 63-64, 70, 85, 90, 92, 99, 111, 126-128, 137, 141-142, 146, 203, 226, 258, 261

composite imagery, 7, 183

conditional evidence, 29-30, 34, 36, 43, 193, 258

conjecture, 241-242, 245, 249, 258, 261, 265

coroner, 17, 191-192, 197, 259-260, 262

crime scene documentation, 13, 23

crime scene investigation, 6-7, 10, 13-14, 16-17, 25, 27, 29, 41, 43, 193, 242-243, 245, 259, 262, 275

    search patterns, 20, 22

crime scene investigators, 2, 7, 14, 16-20, 22, 24-26, 29, 34, 37, 40-41, 43, 48, 54, 56, 59, 70, 78-79, 93, 101-104, 107, 112, 140, 144, 157, 200, 244, 259

crime scene reconstruction, 3, 7, 41, 149, 241-244, 249, 258, 261, 275

crime scene survey, 13, 27, 259

criminalistics, 1-2, 4-5, 7, 259, 270, 276

D

decomposition, 34, 187, 194-195, 212, 217-221, 223, 249, 259

DNA, 3, 5, 7, 10, 16, 30-32, 39-41, 51, 61, 63, 65, 77-78, 81-82, 84-99, 101, 112, 144, 174, 194, 212-213, 226, 257-259, 262-265, 268, 271-272, 275

    evidence, 30, 32, 78, 85, 88-90, 92, 95-97, 226, 268

        collection and preservation of, 85, 89, 268

        analysis of, 7, 51, 82, 85-86, 88-92, 96-97, 213

document examination, 7, 111

documents, 3-8, 14, 18, 22, 34, 43, 56, 70, 78, 111-120, 122, 129, 140, 144, 157, 159, 173, 177, 244, 248, 259-261, 264, 271-272

    types that can be used as evidence, 13, 111-112

Dr. Constantine "Gus" Karazulas, 228

Dr. Lowell Levine, 227

Dr. William Bass, 216

drug evidence, 67-70, 200

drugs, 6-7, 9, 16, 26, 35, 40, 67-75, 113, 143, 163, 166-167, 169, 173, 193-195, 199-200, 204-206, 209, 258-260, 263-265

Drug-Facilitated Rape (DFR), 67-68, 73-75, 200, 259-260, 264

Druggist's fold, 63, 164-165

E

electrostatic lifting, 123-124, 128, 259

expert witness, 1-2, 8-9, 11, 215, 228, 259

F

facial reconstruction, 7, 183-185, 188-189, 259

fiber evidence, 53, 60, 62-64, 243

fingerprints, 4, 7, 15, 32, 34, 40, 45-51, 86, 114, 117, 123-124, 127, 146, 195, 243, 248, 257, 264

firearms, 6-7, 139, 142-146, 150, 194, 243, 248, 271

forensic anthropology, 7, 189, 211-212, 214-216, 257, 260, 267-268, 270

forensic art, 7, 183-184, 186, 189, 268

forensic chemistry, 7

forensic engineering, 7, 231-233, 235

forensic entomology, 7, 214, 217-218, 221-222, 269

forensic odontology, 189, 225, 227, 270-271

forensic pathology, 4, 191, 197, 263

forensic serology, 7, 77

forensic toxicology, 7, 161, 166, 199-200, 206, 208, 263, 265, 270

forgery, 111-112, 122, 260

G

Gamma Hydroxybutyrate (GHB), 67-68, 74-75, 259-260, 264

gas chromatography, 64, 71, 152, 155-156, 161-163, 166-167, 199-200, 206-207, 209, 258, 260, 264-265

gunshot residue (GSR), 16, 139-140, 142, 149-153, 163, 166, 194, 243-244, 248, 260, 263

H

hair and fibers evidence, 36, 40, 53, 59-64, 90, 157, 243

handwriting analysis, 111-112, 122, 260-261

hit and run, 33, 57, 165

hypothesis, 14, 17, 152, 241-242, 245, 249, 258, 261, 265

I

identification, 2, 29-30, 36-38, 40-41, 43, 45-46, 51, 54, 71, 78, 81, 85, 87, 90, 130, 143-144, 146, 155, 164, 166, 184, 195, 206, 211, 213-216, 225-227, 257, 260-261, 268, 270-271, 275-276

image modification, 183-184, 189, 261

imprints, 34, 56-57, 107, 123-129, 157, 193, 228, 259, 261

impressions, 4, 7, 45, 47-48, 51, 57, 63, 123-135, 137, 140, 145-146, 153, 157, 193-194, 228, 244, 261, 264-265

individualization, 29-30, 41, 43, 78, 166, 212, 261

Index

Innocence Project, The, 88, 96, 272

J

JonBenet Ramsey, 196
jurisprudence, 8, 11, 251-253, 261

K

Karen T. Taylor, 183-189, 271
known standards, 6, 29-30, 41, 43, 53, 63, 70, 90, 97, 123-124, 126-128, 139, 245, 258, 261

L

latent fingerprints, 3, 7, 39, 45-51, 93, 124, 262-263
law school, 252-253, 261
Leucomalachite Green, 77-79, 84, 262
Linkage Theory, 13-14, 18-19, 22, 27, 262
lividity, 34, 193, 197, 262
Locard's Theory of Exchange, 13-14, 19, 27, 262
luminol, 77-79, 84, 125, 262

M

M. Lee Goff, 221
mass spectrometry, 71, 152, 163-164, 166-167, 199-200, 207, 209, 262
medical examiner, 2, 5, 7, 14, 17, 38, 51, 146, 187, 191-195, 197, 200, 206, 212, 219, 257, 259-260, 262
medulla, 53-54, 61-63, 65, 259, 262
melting point, 64, 163-164, 167, 262
mitochondrial DNA, 85-87, 99, 262

N

necrophagus insects, 262
nuclear DNA, 63, 65, 85-87, 99, 259, 264

O

O.J. Simpson, 23, 30, 267, 275
omnivorous insects, 217-219, 223, 262

P

parasitic insects, 217-218, 223, 262

patent fingerprints, 45-46, 51, 262-263
pathology, 4, 7, 191-192, 197, 226, 263, 270
pattern evidence, 17, 27, 29-30, 34, 43, 101, 124, 193-194, 241, 243, 263, 265
Phenolphthalin, 78, 263
physical evidence, 1-4, 6-7, 10, 13-18, 21-22, 24-27, 29-36, 40-41, 43, 45, 60, 63, 89, 94, 149, 151, 155, 157, 191, 193, 231, 241-245, 248, 252, 262-265, 267
plastic fingerprints, 45-46, 48, 51, 262-263
point of convergence, 101-102, 104-105, 109, 263
point of origin, 101-102, 104, 109, 155-157, 159, 263
   fire, 156-157, 159, 263
Polymerase Chain Reaction (PCR), 85-87, 91-92, 97, 99, 263, 272
postmortem interval, 194, 211-212, 214, 216, 222-223, 226, 263
postmortem toxicology, 199-200, 206, 209, 260, 263, 265
predatory drugs, 73
predatory insects, 217-218, 221, 223, 263
primary crime scene, 13, 20, 27, 56, 263-264
primary transfer, 13-14, 19, 27, 263-264
prosecutor, 3, 6, 8, 13, 17, 39, 51, 94, 96, 244, 251-253, 263, 272
psychiatrist, 8, 237-240, 263
psychological profiling, 237-238, 240
psychologist, 8, 237-240, 263
public defender, 251-253, 263
Pyrolysis Gas Chromatography, 64, 155-156, 162, 258, 260, 264-265

Q

questioned documents, 7, 111-113, 118, 122, 264, 271

R

R.A.I.N.N. (Rape Abuse & Incest National Network), 74, 273
rape, 15, 36, 67-68, 73-75, 77, 89, 95, 194, 200, 223, 226, 247, 259-260, 264, 273
recognition, 25, 29-30, 40, 43, 108, 264, 276

reconstruction, 3, 7, 17, 29-30, 34, 41, 43, 105, 108, 148-149, 183-185, 187-189, 232-234, 241-245, 247-249, 258-259, 261, 264-265, 275

Restriction Fragment Length Polymorphism (RFLP), 85-87, 89, 91-92, 99, 264

ridge impressions, 45, 47-48, 51, 264

rigor mortis, 34, 191-195, 197, 222, 264

Rohypnol, 67-68, 73, 75, 259-260, 264

S

searching for physical evidence, 13-14, 21, 24, 26, 243

secondary crime scene, 13-14, 16, 20, 27, 56, 248, 263-264

secondary transfer, 13-14, 19, 27, 263-264

serology, see forensic serology

sexual violence, 73, 75, 259, 272-273

Short Tandem Repeat (STR) marker, 85-87, 91-92, 94-95, 99, 264, 272

skeletal remains, 188, 211-213, 216, 226, 264

soil dirt and dust evidence, 48, 53

specialties in forensic science, 5, 7, 122, 260, 269

spectrophotometry, 161, 199-200, 206-207, 209, 265

spectroscopy, 58, 64, 71, 161, 163-164, 166-167, 265

T

Tandem repeat sequence (TRS), 85-86, 99, 265

target surface, 101-104, 109, 153, 257-259, 263, 265

terminal velocity, 101-104, 109, 265

Tetramethylbenzidine, 77-79, 84, 265

theory, 9, 13-14, 17-19, 22, 27, 241-242, 245, 249, 258, 261-262, 265

thin-layer chromatography, 163, 166-167, 199-200, 206-207, 209, 258, 260, 264-265

toolmarks, 31, 139-142, 146, 152-153, 157, 194, 248, 258, 264-265

trajectory, 24, 139-140, 143, 148, 153, 245, 265

transfer theory, 13-14, 18-19, 27, 262, 265

transfer and trace evidence, 1, 4, 7, 19-20, 27, 29-30, 35-36, 39, 43, 53, 55-57, 59-60, 64, 126, 141, 144, 164, 194, 262, 265, 275

transient evidence, 29-30, 34, 43, 265

TWA Flight, 234

V

vapor point, 163, 165, 167, 265

vapor pressure, 163, 165, 167, 265

Variable Number Tandem Repeat (VNTR), 85-86, 91-92, 99, 265

Vicki Lynn Hoskinson, 152

Virgilio Paz Romero, 189

W

Woodchipper murder case, 42, 111